Polymer Thin Films: From Fundamentals to Applications

Polymer Thin Films: From Fundamentals to Applications

Editor

Mohor Mihelčič

Basel • Beijing • Wuhan • Barcelona • Belgrade • Novi Sad • Cluj • Manchester

Editor
Mohor Mihelčič
University of Ljubljana
Ljubljana, Slovenia

Editorial Office
MDPI
St. Alban-Anlage 66
4052 Basel, Switzerland

This is a reprint of articles from the Special Issue published online in the open access journal *Coatings* (ISSN 2079-6412) (available at: https://www.mdpi.com/journal/coatings/special_issues/polymer_thin_film).

For citation purposes, cite each article independently as indicated on the article page online and as indicated below:

Lastname, A.A.; Lastname, B.B. Article Title. *Journal Name* **Year**, *Volume Number*, Page Range.

ISBN 978-3-0365-9268-8 (Hbk)
ISBN 978-3-0365-9269-5 (PDF)
doi.org/10.3390/books978-3-0365-9269-5

Cover image courtesy of Urška Gradišar Centa

© 2023 by the authors. Articles in this book are Open Access and distributed under the Creative Commons Attribution (CC BY) license. The book as a whole is distributed by MDPI under the terms and conditions of the Creative Commons Attribution-NonCommercial-NoDerivs (CC BY-NC-ND) license.

Contents

About the Editor . . . . **vii**

Mohor Mihelčič, Marta Klanjšek Gunde and Lidija Slemenik Perše
Rheological Behavior of Spectrally Selective Coatings for Polymeric Solar Absorbers
Reprinted from: *Coatings* 2022, 12, 388, doi:10.3390/coatings12030388 . . . . **1**

**Urška Gradišar Centa, Mohor Mihelčič, Vid Bobnar, Maja Remškar
and Lidija Slemenik Perše**
The Effect of PVP on Thermal, Mechanical, and Dielectric Properties in PVDF-HFP/PVP Thin
Film
Reprinted from: *Coatings* 2022, 12, 1241, doi:10.3390/coatings12091241 . . . . **17**

**Urša Opara Krašovec, Tjaša Vidmar, Marta Klanjšek Gunde, Romana Cerc Korošec
and Lidija Slemenik Perše**
In-Depth Rheological Characterization of Tungsten Sol-Gel Inks for Inkjet Printing
Reprinted from: *Coatings* 2022, 12, 112, doi:10.3390/coatings12020112 . . . . **27**

**Jingying Zhang, Weihua Qin, Wenrui Chen, Zenghui Feng, Dongheng Wu, Lanxuan Liu
and Yang Wang**
Integration of Antifouling and Anti-Cavitation Coatings on Propellers: A Review
Reprinted from: *Coatings* 2023, 13, 1619, doi:10.3390/coatings13091619 . . . . **43**

**Svetlana N. Khonina, Grigory S. Voronkov, Elizaveta P. Grakhova, Nikolay L. Kazanskiy,
Ruslan V. Kutluyarov and Muhammad A. Butt**
Polymer Waveguide-Based Optical Sensors—Interest in Bio, Gas, Temperature, and Mechanical
Sensing Applications
Reprinted from: *Coatings* 2023, 13, 549, doi:10.3390/coatings13030549 . . . . **65**

**Maria J. Romeu, Luciana C. Gomes, Francisca Sousa-Cardoso, João Morais,
Vítor Vasconcelos, Kathryn A. Whitehead, et al.**
How do Graphene Composite Surfaces Affect the Development and Structure of Marine
Cyanobacterial Biofilms?
Reprinted from: *Coatings* 2022, 12, 1775, doi:10.3390/coatings12111775 . . . . **95**

Martin Jaskevic, Jan Novotny, Filip Mamon, Jakub Mares and Angelos Markopoulos
Thickness, Adhesion and Microscopic Analysis of the Surface Structure of Single-Layer and
Multi-Layer Metakaolin-Based Geopolymer Coatings
Reprinted from: *Coatings* 2023, 13, 1731, doi:10.3390/coatings13101731 . . . . **113**

John Walker, Andrew B. Schofield and Vasileios Koutsos
Nanostructures and Thin Films of Poly(Ethylene Glycol)-Based Surfactants and Polystyrene
Nanocolloid Particles on Mica: An Atomic Force Microscopy Study
Reprinted from: *Coatings* 2023, 13, 1187, doi:10.3390/coatings13071187 . . . . **133**

Iulia Babutan, Otto Todor-Boer, Leonard Ionut Atanase, Adriana Vulpoi and Ioan Botiz
Crystallization of Poly(ethylene oxide)-Based Triblock Copolymers in Films Swollen-Rich in
Solvent Vapors
Reprinted from: *Coatings* 2023, 13, 918, doi:10.3390/coatings13050918 . . . . **145**

**Michael J. Grzenda, Maria Atzampou, Alfusainey Samateh, Andrei Jitianu,
Jeffrey D. Zahn and Jonathan P. Singer**
Microscale Templating of Materials across Electrospray Deposition Regimes
Reprinted from: *Coatings* 2023, 13, 599, doi:10.3390/coatings13030599 . . . . **161**

Shouhua Su, Juan Wang, Chao Li, Jinfeng Yuan, Zhicheng Pan and Mingwang Pan
Short-Branched Fluorinated Polyurethane Coating Exhibiting Good Comprehensive Performance and Potential UV Degradation in Leather Waterproofing Modification
Reprinted from: *Coatings* **2021**, *11*, 395, doi:10.3390/coatings11040395 . . . . **177**

About the Editor

Mohor Mihelčič

Mohor Mihelčič is a research associate at the Faculty of Mechanical Engineering, University of Ljubljana, Slovenia. He received his Ph.D. in Nanoscience and Nanotechnology at Jožef Stefan International Postgraduate School, Slovenia, in 2015. Mohor has co-authored 34 scientific publications and 47 conferences, as well as 2 patents. His research focuses on the preparation and characterization of multifunctional coatings and polymers such as spectrally selective thickness sensitive (TSSS) and thickness insensitive (TISS) coatings and hydrophobic, photocatalytic, thermo-, and electrochromic coatings.

Article

Rheological Behavior of Spectrally Selective Coatings for Polymeric Solar Absorbers

Mohor Mihelčič [1], Marta Klanjšek Gunde [2] and Lidija Slemenik Perše [1,*]

1. Laboratory for Experimental Mechanics, Faculty of Mechanical Engineering, Aškerčeva 6, SI-1000 Ljubljana, Slovenia; mohor.mihelcic@fs.uni-lj.si
2. Department for Materials Chemistry, National Institute of Chemistry, Hajdrihova 19, SI-1000 Ljubljana, Slovenia; marta.k.gunde@ki.si
* Correspondence: lidija.slemenik.perse@fs.uni-lj.si

Abstract: Since the world's energy demands are growing rapidly, there is a constant need for new energy systems. One of the cleanest, most abundant, and renewable natural resources available is solar energy; therefore, the development of surfaces with high absorption of solar radiation is increasing. To achieve the best efficiency, such surfaces are coated with spectrally selective coatings, which are strongly influenced by the pigments and resin binders. Spectrally selective paints have a very specific formulation, and since the applied dry coatings should exhibit high spectral selectivity, i.e., high solar absorptivity and low thermal emissivity, the rheological properties of liquid paints are of great importance. In the present work, we studied the effect of the rheological properties of liquid thickness-insensitive spectrally selective (TISS) paints on the spectral selectivity and adhesion of dry coatings on a polymeric substrate. The results showed that the functional and adhesion properties of dry coating on polymeric substrates is strongly dependent on the rheological properties of the binder and catalyst used for the preparation of the liquid paints. It was shown that the paints with good spectral selective properties (thermal emissivity $e_T < 0.36$ and solar absorptivity $a_S > 0.92$) and good adhesion (5B) can be prepared for polymer substrates.

Keywords: spectrally selective paints; polymer solar absorbers; rheological characterization

1. Introduction

In many countries around the world, hot water for domestic purposes is obtained electrically. However, the electrical energy capacity is limited due to increased usage. The increasing prices for electrical and conventional energy such as gas and oil, together with the unpredictability of prices and fossil resources, has led to increased development of other energy systems, among which the cleanest, most abundant, and renewable natural resource available is solar energy. At present, the general trend of solar thermal systems is towards simple solar systems with high quality standards and a long lifetime. To achieve this, the materials used should exhibit the best characteristics. Due to their advantages in terms of low cost, variability of properties over a wide range, processing, and specific weight, polymeric materials are very promising candidates for solar collector systems [1,2]. Most polymeric materials are used for unglazed solar collectors, and only a few polymer-based glazed water collectors have been developed and introduced into the market [3]. The main problems for polymeric materials are long-term stability, solar–thermal conversion efficiency, and high stagnation temperatures. One of the drawbacks of the polymer absorbers is also their safety. Recently, special emphasis has been dedicated to fire safety management [4] and the development of multifunctional spectrally selective coatings providing flame retardancy, which are eco-friendly [5]. However, the operating temperatures of collectors for domestic hot water applications and "solar combi-systems" (space heating in combination with domestic hot water preparation) in private residences range from

20 °C to 80 °C; thus, by an appropriate collector system design, including, e.g., overheating protection glazing, plastic materials could be excellent candidate materials [3].

Thickness-insensitive spectrally selective (TISS) paints are usually formulated by inserting metallic flakes and absorbing pigments into binder system [6]. The absorbing pigments (mixed inorganic oxides) are responsible for providing solar absorbance, while metallic flakes act as reflectors of the infrared radiation. The metal flakes change the optical properties throughout the solar and thermal region, and therefore influence the spectral selectivity of the coating. The benefit of such coatings is that they can be applied on substrates other than metal, such as, for example, polymeric substrates, while still providing good spectral selectivity. However, spectral selectivity depends strongly on the microstructure of all pigments in the dry coating, especially on the position and orientation of metal flakes [3,7–9]. To achieve the best performance of solar thermal absorbers, i.e., the highest absorptivity, the applied coating is usually black. However, in addition to high solar absorbance, black solar coatings are strong emitters of thermal (infra-red) radiation. At high temperatures, such absorbers produce substantial heat losses from the front cover of the solar collector. For the reduction of the collector heat losses, the optical properties of the coatings should be selective. This means that the coatings should exhibit high absorptance for solar radiation, but low emittance for thermal radiation. Good selective surfaces are expected to achieve average absorptance higher than 0.95, and average thermal emittance around 0.1 or lower [10]. These values are achievable with metal substrates, but are very difficult to achieve when using the polymeric surface of the collector. Since the adhesion of the coating on the polymeric substrate is difficult to achieve, at least some form of pre-treatment of the polymeric surface is usually necessary to ensure satisfactory adhesion of the coating.

As with many other paints, TISS liquid paints can also be applied on various substrates using several application techniques (spraying, coil coating, brushing, etc.). Regardless of the selected technique, the paints are subjected to complex flow conditions during use. Among the many processing conditions, the rheological behavior of liquid paint is the most important characteristic governing the behavior of the coating during application. The importance of the rheological properties of liquid paints for the optical and functional properties of dry coating has already been confirmed [5,10–14]. It has been shown that although two liquid solar paints exhibited similar shear thinning flow behavior with the same value of apparent shear viscosity at a specific shear rate, the homogeneity and optical properties of the dry coatings were tremendously diverse [11,12]. The results showed that rather than flow characteristics with a certain viscosity value, the whole rheological characterization, i.e., the viscoelastic properties, are of great importance [11]. To achieve the best final performance of the applied coating, the rheological properties of the liquid paint under the complex conditions encountered during the application must be considered. One of the most used parameters for the rheological characterization of liquid paint is viscosity, which could be independent of the shear conditions (Newtonian fluids). Most paints used in the various industrial processes exhibit more complex rheological properties, such as a strong shear rate dependence of viscosity (i.e., shear thinning, shear thickening), and very often also time dependence (thixotropy). During the storage, preparation and application of the liquid paint, various processes (mixing, spraying, leveling, etc.) occur in which the paint is subjected to a wide range of shear rate values [14]. After preparation, the paint is stored in the vessel, where no shear is applied. During mixing and pumping, the paint is exposed to shear rates between 1 and 100 s^{-1} or even higher, i.e., in the nip region of roll coaters (up to 10^5 s^{-1}) or spraying (up to 10^6 s^{-1}). The final dry coating is formed after the liquid paint has been applied on the substrate. The process is often referred to as leveling, which is performed in the range of shear rates between 10^{-2} and 1 s^{-1}. The process of leveling is especially important for achieving the best functionality of dry coating, i.e., spectral selectivity. Oh et al. [13] showed that loosely paced structure in wet coating state and more porous dried coating structure could be achieved when the motion of the particles in the structure formation of the coating components is restricted during the drying of the liquid paint on the substrate. Leveling could also be tailored by increasing the duration of the process, which is achieved for example when decreasing the amount of thickener or with the use of a lower amount of crosslinker or a weaker catalyst. The process of leveling tendencies of a liquid paint can be predicted by rheological characterization using a three-step test. The first step of low shear or low deformation is followed by immediate application of high shear or high deformation comparable to application, which is again followed by immediate application of low-shear motion and the measurement of the resultant viscosity/viscoelastic behavior of the paint over a period of several minutes. Due to the complex nature and non-Newtonian behavior of TISS paints and the extreme changes of shear rates during different steps from storage in the vessel through to application and final leveling on the substrate, a detailed rheological characterization over a wide range of shear rates is essential. It has already been reported in the literature that the rheological properties strongly influence the microstructure of the coating formed after the application. Shen et al. [12] showed that rheological characterization confirmed that the addition of co-binder increases the interactions between coating components, leading to better optical performance of the applied coatings.

Spectrally selective paints for solar absorbers are complex systems, and have attracted interest since the introduction of the concept of spectral selectivity [7–9,14]. However, to the best of our knowledge, there are very few investigations in the literature focused on the rheological properties of these paints during the preparation, application, and formation of dry coatings on a substrate. In the present work, we studied the rheological properties and spectral selectivity of TISS paints with various binder systems for the application on the polymeric substrates. In addition to this, the adhesion properties of the applied coatings on polymeric substrates were determined. Special emphasis was placed on determining the influence of rheological properties of the liquid paint on the final spectral selectivity and adhesion characteristics of dry coating. Among various binders, one was selected for further investigation of the influence of two different catalysts on the rheological properties, spectral selectivity, and adhesion, which were determined at various concentrations of catalyst added to the liquid paints.

2. Materials and Methods

All thickness-insensitive spectrally selective paints were prepared for application on high-density polyethylene (HDPE) polymer substrates with a spray gun. Liquid paints were prepared using a standard procedure in which large Al flakes (>50 µm) served as reflectors, enabling low emittance, while much smaller (<1 µm) black Mn-Fe pigment (SH-444) with a spinel structure provided high solar absorbance. Three fluoropolymer resin binders (Lumiflon 200, Lumiflon 9716 and Lumiflon 9721, Asahi Glass Co., Ltd., Tokyo, Japan) were used to prepare TISS paints following the same procedure. The pigment was mixed with a specific binder in a high-speed dissolver (Dispermat CNF2 (VMA-GETZMAN GMBH, D)). After the dispersion was prepared, it was milled in a ball mill by using glass balls with diameter 3 mm at 4000 rpm for 2 h. Two different commercially available polyisocyanate catalysts based on hexamethylene diisocyanate (HDI): Desmodur N75 and Desmodur N3300, both supplied by Covestro AG, Leverkusen, Germany, were used as catalysts for the final composition of the TISS paint. Desmodur N75, based on aliphatic HDI biuret, was chosen, as it exhibits much lower viscosity ~225 mPa·s (25 °C) than aliphatic HDI Desmodur N3300, for which, according to the producer, the viscosity is ~3000 mPa·s (25 °C). The reaction between isocyanate catalyst and lumiflon binder is explained elsewhere [15].

Prior to the application, liquid paints were characterized with a rotational controlled rate rheometer (Physica MCR302, Anton Paar, Graz, Austria), equipped with a cone and plate sensor system (CP 50/2°). Standard rotational flow tests were performed with a triangular method by changing the shear rate from 0–500–0 s^{-1}. Oscillatory stress sweep tests at a constant frequency of oscillation (1 Hz) were used to determine the linear viscoelastic range (LVR). Frequency tests were performed at constant small deformation in LVR by decreasing the frequency from 20 to 0.01 Hz. In addition to standard rotational and oscillatory tests, time tests were performed to simulate the three steps to which the paint is subjected: storage in the container, deposition on the substrate, and the formation of the dry coating. In the 1st and the 3rd step, the paint is subjected to conditions with no shear; therefore, these two steps were performed at constant small deformation in LVR, while in the 2nd step—during the application—the paint is subjected to high deformation; therefore, high stress was applied in this step under oscillatory conditions. All rheological measurements were performed at a constant temperature $T = 23\ °C$, which was also the temperature of the application of the paints on the substrate.

The spectrally selective properties of the paint coatings were determined from the infrared (IR) reflectance spectra. Reflectance in the visible (VIS) and near infrared (NIR) ranges was measured on a Perkin Elmer Lambda 950 UV/VIS/NIR spectrometer (PerkinElmer Inc., Waltham, MA, USA) with an integrating sphere (module 150 mm), while the reflectivity spectra in the middle IR spectral range were obtained on a Bruker IFS66/S spectrometer (Bruker Corporation, Billerica, MA, USA) equipped with an integrating sphere (OPTOSOL), using a gold plate as a standard for diffuse reflectance. Solar absorptance (a_S) and thermal emittance (e_T) values were determined from the reflectance spectra using a standard procedure [16]. The solar absorptance a_S is theoretically defined as a weighted fraction between absorbed radiation and incoming solar radiation. It was calculated according to [16]:

$$a_S = \frac{\int_{0.3}^{2.5} S(\lambda)(1 - R(\lambda))d\lambda}{\int_{0.3}^{2.5} S(\lambda)d\lambda} \quad (1)$$

where λ is wavelength, $R(\lambda)$ reflectance, and $S(\lambda)$ direct normal solar irradiance. It is defined according to ISO standard 9845-1 (1992), where an air mass is 1.5. Thermal emittance e_T is a weighted fraction between emitted radiation and the Planck black body distribution $r(\lambda,T)$ [16]:

$$e_T = \frac{\int_{2.5}^{15} r(\lambda, T)(1 - R(\lambda))d\lambda}{\int_{2.5}^{15} r(\lambda, T)d\lambda} \quad (2)$$

The values of e_T were calculated at 350 K, which is close to operating temperature of solar collectors. Emissivity and absorptivity values were determined by a set of samples, e.g., for a certain operating condition, 5 samples were sprayed and spectral selectivity was determined for each of the samples. The single value was then determined as an average of these measurements. The accuracy of the spectral selectivity was within ±2%.

After the preparation, liquid paints were deposited on polymeric substrates by a laboratory spraying gun.

3. Results and Discussion

3.1. Rheological Characterization

3.1.1. Binders

Before the preparation of TISS paints, rheological characterization of all three fluoropolymer binders was performed. The results show that in the shear rate range examined, the binders L9716 and L9721 exhibited Newtonian flow behavior with viscosity 4.5 Pa·s for L9716 and 3.9 Pa·s for L9721, respectively. In contrast to that, the binder L200 exhibited almost one order of magnitude higher consistency with non-Newtonian shear thinning flow behavior.

Three different fluoropolymer binders L200, L9716 and L9721 were milled with the same amount of black pigment (SH-444). After the milling, the so-prepared pastes were rheologically characterized under destructive and non-destructive conditions. The results of the destructive shear conditions showed that, in contrast to the flow behavior of the binders without black pigment (Figure 1), all three pastes exhibited shear thinning behavior (Figure 2). The addition of the pigment led to particle–particle interactions and interactions between the binder and the black pigment. It is clear that in all of the pastes, some pigment particles agglomerated, resulting in a 1st Newtonian plateau with constant high zero-shear viscosity. As shear rate increased, the particles started to de-agglomerate, and the viscosity strongly decreased with increasing shear stress. At this point the pigment particles were deagglomerated and well distributed inside the binder. Similarly to the binders, the flow behavior of the paste L200-444 significantly deviated from the other two. This paste exhibited the highest consistency; moreover, the presence of yield stress could be clearly observed at around 100 Pa. During the measurements by the triangular method, all pastes exhibited a hysteresis loop, indicating the time-dependent behavior of the pastes. Time dependency was the most pronounced for the paste with the binder L9716, while the paste L200-444 exhibited the lowest hysteresis, indicating that the structure had the lowest time dependency.

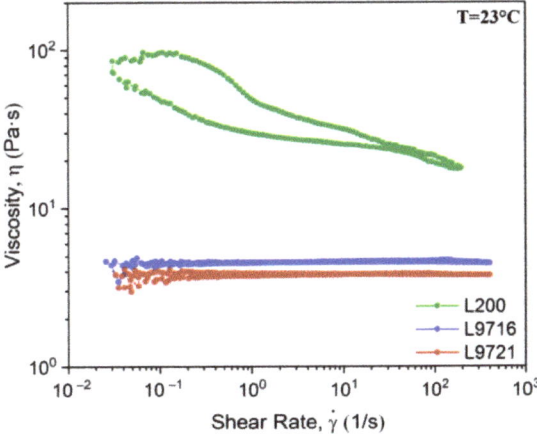

Figure 1. Flow characteristics of the binders used for the preparation of TISS paints.

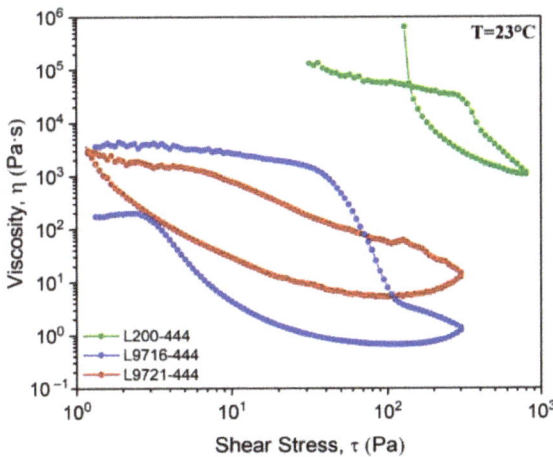

Figure 2. Flow characteristics of the pastes prepared by the binders with black pigment.

In addition to the flow characteristics, the viscoelastic properties of the pastes were determined under non-destructive oscillatory conditions in the linear viscoelastic range, which was previously determined with stress sweep tests at a constant frequency of 1 Hz. Similarly to what was determined with the flow tests, the frequency tests (Figure 3) showed that the highest consistency, i.e., the values of dynamic moduli (elastic G' and viscous G'' contribution to viscoelastic behavior) was observed for the paste with the L200 binder. At high frequencies, i.e., short times, the particles of black pigment were trapped inside the chains of the binder, and there was no time for them to sediment. Under these conditions, the prepared pastes exhibited a relatively stable structure. However, at lower frequencies, i.e., longer times, the chains of the polymer binder started to orientate and align, which enabled the particles to move between the chains, and they started to settle. Under such conditions, the pastes lost their stability, and the viscous contribution started to increase, while the elastic modulus remained constant. The increase in viscous contribution was the most clearly observed for the pastes with L200 binder. Despite this paste having the highest consistency, it could be concluded that the interactions of the particles with this binder at longer times were the weakest.

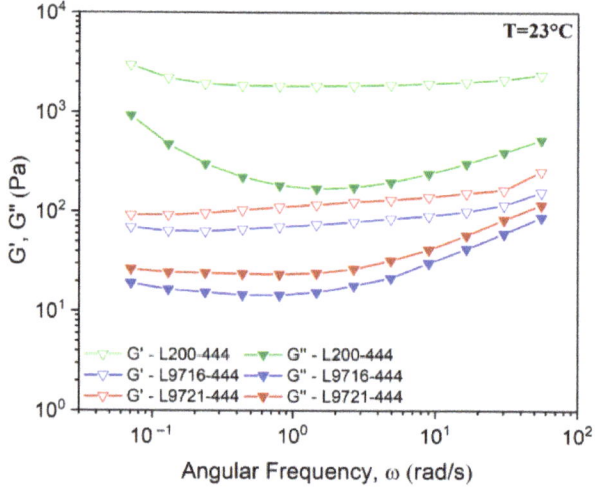

Figure 3. Oscillatory frequency tests in linear viscoelastic range for the binders with black pigment.

During its lifetime and application, the liquid paint is subjected to various conditions, from rest during storage to high shear during application, and again low shear during leveling after application on the substrate. To simulate such conditions, non-destructive conditions of linear viscoelastic response were applied in the first and third steps of the experiment, while destructive conditions of high shear were applied in the second step. The results (Figure 4) showed that during the first step (rest during the storage), the binders L200 and L9716 exhibited the structure of a strong gel with extreme prevalence of the elastic modulus G' over the viscous one G''. For the binder L200, the prevalence of the elastic modulus can be observed during all three steps of the experiment. Such behavior is not preferable for, e.g., spraying applications, since during the application at high shear the paint does not flow but exhibits the properties close to solid or gel. Consequently, the homogeneous application of this paint on the substrate is unlikely. To use this binder for spraying application the addition of rheological additives for increasing viscous character at high shear would be necessary. On the contrary, the binder L9716 exhibited a stable structure at rest and liquid-like characteristics at high shear therefore we do not expect any problems during the application. However, due to the sudden recovery of the dynamic moduli to initial values at the last step of the experiment some inhomogeneities during the process of leveling on the substrate could be expected. The binder L9721 exhibited the structure of a weak gel at rest and good flow characteristics during the application; however, a higher decrease in the viscosity in this step would be preferable. Similar has already been published for TISS coatings on metal substrates [11].

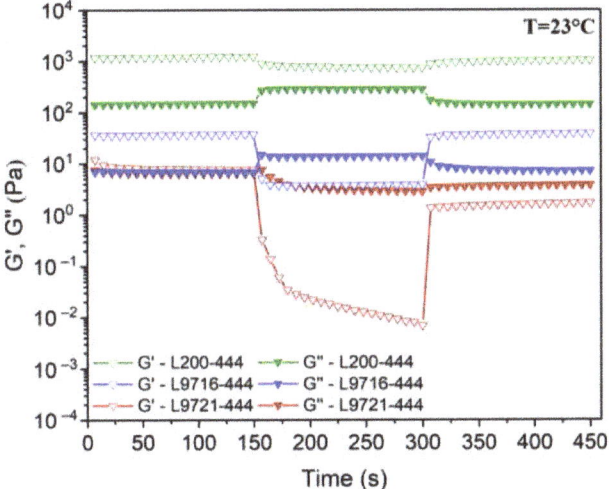

Figure 4. Three-step time test for various binders with black pigment: 1st step: constant low deformation in linear viscoelastic range ($\gamma = 1\%$, $\omega = 1$ Hz); 2nd step: constant high deformation in the range of destructive shear conditions ($\gamma = 100\%$, $\omega = 1$ Hz); 3rd step: constant low deformation ($\gamma = 1\%$, $\omega = 1$ Hz), the same as in 1st step.

3.1.2. TISS Liquid Paints

Thickness-insensitive spectrally selective (TISS) coatings were prepared by using all three of the above-mentioned binders milled with the same black pigment. All TISS paints were two-component systems, prepared by following the same procedure with the identical addition of the same additives and catalyst Desmodur N75. Rheological characterization was performed for all liquid paints prior to the application on the HDPE substrates using the spraying technique. The viscosity curves under destructive shear conditions (Figure 5) showed that all three TISS liquid paints exhibited shear thinning flow behavior. For all the paints, the first Newtonian plateau could be observed in the range of low shear stresses. The highest viscosity in this range was observed for the paint with L9716 binder. As the shear stress increased, the particles inside the paint started to align in the direction of flow, resulting in decreasing viscosity. The decrease in viscosity was less pronounced for the paint with L200 binder. Consequently, the viscosity of this paint in the range of high shear stresses was the highest. As shear stress started to decrease in the second step of the triangular flow experiment, the viscosity increased, with different values from those during increasing shear stress. A noticeable hysteresis can be observed for all three paints, indicating the time-dependent behavior of all paints. Due to the higher viscosity of the binder L200, the TISS paint prepared with this binder exhibited the highest viscosity, which was more pronounced especially at higher shear rates.

The results of the oscillatory tests, performed under non-destructive conditions of low deformation (in the range of linear viscoelastic response at $\gamma = 1\%$, Figure 6), showed that at longer times, none of the liquid paints exhibited a tendency towards sedimentation; however, for the paint with the L9721 binder (TISS-L9721-444) a small increase of dynamic moduli could be observed at lower frequencies, which was attributed to the drying of the paint due to the long duration of the experiment.

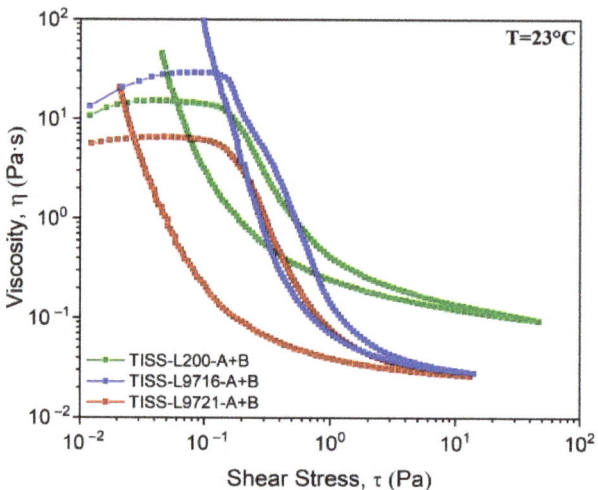

Figure 5. Viscosity curves of TISS paints, prepared with binders L200, L9716 and L9721. All liquid paints were measured after the addition of catalyst Desmodur N75.

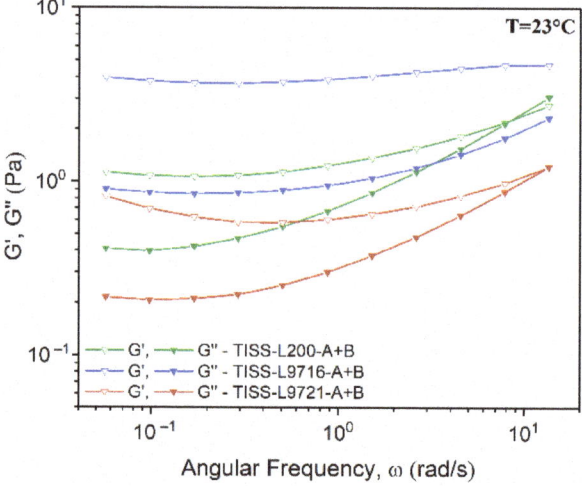

Figure 6. Oscillatory tests—frequency dependence of dynamic moduli for TISS liquid paints, prepared with various binders (L200, L9716 in L9721) and black pigment (Black-444). All liquid paints were measured after the addition of catalyst Desmodur N75.

Time tests, explained above, showed that all liquid paints exhibited a stable structure at the state of rest and appropriate flow characteristics under high shear during the application process (Figure 7). During the formation of the dry coating on the substrate (step 3), some inhomogeneities could be expected using the paints TISS-L9716-A+B and TISS-L9721-A+B, since the recovery to the initial values is almost immediate after decreasing the stress to the initial values. On the other hand, during the leveling of the paint TISS-L200-A+B, homogeneous final dry coating on the substrate could be expected.

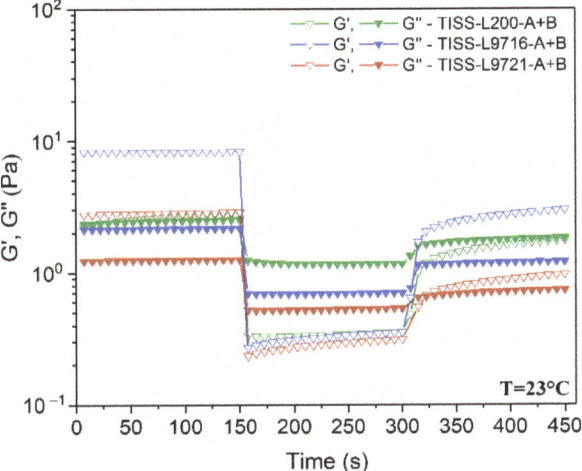

Figure 7. Three-step time test for TISS liquid paints, prepared with various binders (L200, L9716 in L9721) and black pigment: 1st step: constant low deformation in linear viscoelastic range ($\gamma = 1\%$, $\omega = 1$ Hz); 2nd step: constant high deformation in the range of destructive conditions ($\gamma = 100\%$, $\omega = 1$ Hz); 3rd step: constant low deformation ($\gamma = 1\%$, $\omega = 1$ Hz), the same as in 1st step. All liquid paints were measured after the addition of catalyst Desmodur N75.

3.1.3. The Influence of the Type and Concentration of Catalysts

Since TISS paint coatings prepared with fluoropolymer binder L200 turned out to be very promising for application on polymeric substrates, further optimization of the paint was performed with this binder. The influence of the type and concentration of catalyst was determined with two commercially available catalysts: Desmodur N75 and Desmodur N3300. The catalysts were added in three different proportions: 1:1.1; 1:1.3 and 1:1.5. The rheological tests were performed with the aim of defining the best catalyst and the appropriate concentration of its addition for obtaining the paint with the best rheological characteristics for application on the polymeric substrate.

The N75 and N3300 catalysts are aliphatic polyisocyanates. N75 is an HDI dimer, while N3300 is an HDI trimer. It has been reported in the literature [17] that both catalysts enable excellent appearance, UV resistance, and long pot life open time, and both meet current VOC regulations for key end-use markets. Although both catalysts exhibit good performance, the chemical resistance and mechanical properties of dry coating are better when using N3300, while N75 enables better adhesion, impact and flexibility and slower drying of the coating. During curing, the isocyanate groups from the catalyst react with hydroxyl groups from the Lumiflon binder. However, the mixing ratio of the components should be precisely determined to achieve the correct stoichiometry of the co-reactants. If the NCO/OH ratio is too low, some OH groups remain unreacted, leading to increased flexibility, better adhesion to substrates, and reduced solvent and chemical resistance [17]. On the other hand, if the NCO/OH ratio is too high, some NCO groups remain unreacted, resulting in a longer time for drying and surface hardening. However, the formed coating is harder, and solvent and chemical resistance is increased, while the flexibility is decreased and the adhesion to the substrate is reduced.

The results of flow tests (Figure 8) revealed that the type of the catalyst influenced the dependence of the viscosity on shear rate. In the same range of shear stresses, the decrease in viscosity with increasing shear stress was steeper when the catalyst Desmodur N3300 was used. With the use of this catalyst, the viscosity at lower shear was higher (compared to the TISS paints with the catalyst Desmodur N75), indicating the greater stability of the structure of this paint at rest. However, the values of the viscosity at high shear were more comparable, indicating that similar characteristics during the application of the paints could be expected. Whereas the type of the catalyst influenced the flow behavior of the TISS paints, the effect of the concentration was almost negligible. Small increases in the viscosity could be observed when higher concentrations of catalyst were added; however, the shape of the curves remained similar.

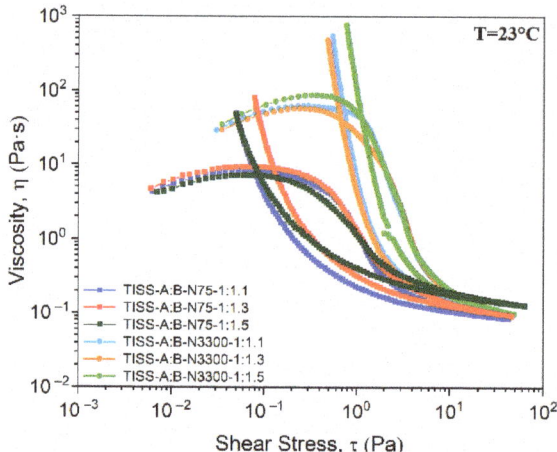

Figure 8. Flow curves of TISS paint coatings, prepared with the binder L200 and various concentrations of two catalysts Desmodur N75 and Desmodur N3300, respectively.

Similar results were also obtained with oscillatory frequency tests (Figure 9). Significant influence was observed using different types of catalyst, whereas the effect of the concentration was less important. The prevalence of elastic modulus over the viscous one and the fact that there is no dependency of either modulus on frequency of oscillation demonstrate that the structure of the paints with the catalyst Desmodur N3300 is relatively stable and no or only a small tendency towards sedimentation could be expected over longer durations. On the other hand, the paints with the catalyst Desmodur N75 exhibited a more liquid-like structure with lower values of dynamic moduli; hence, some tendency towards sedimentation could already be expected on a short time scale.

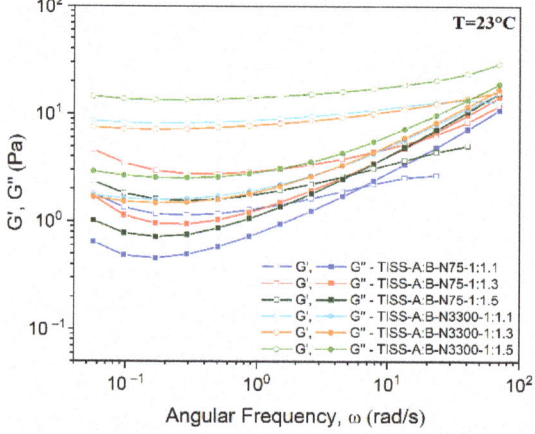

Figure 9. Oscillatory tests: frequency dependency of dynamic moduli for TISS paints, prepared with fluoropolymer binder L200 and various concentrations of two catalysts Desmodur N75 and Desmodur N3300, respectively.

The results of three-step time tests for the prepared TISS paints (Figure 10) showed that the paints with N3300 catalyst exhibited high consistency and stable structure at rest. During application, the rheological properties of these paints were suitable for successful application, since the viscous contribution prevailed over the elastic one. However, after the application, the recovery of these paints was fast, with the domination of the elastic modulus indicating fast, solid-like behavior and possible inhomogeneities in the dry coating on the substrate. The rheological behavior of the paints with N75 catalyst was similar, with different relaxation behavior in the third step of the experiment. After the high shear applied in the second step, the recovery of the structure in the third step was slower, with elastic modulus slightly prevailing over the viscous one. This indicates that there was enough time for the structure of these to form a homogeneous dry coating. Moreover, the structure of the paint was viscoelastic, with almost equal contribution of the elastic and viscous parts, enabling the formation of a uniform and smooth dry coating. According to these results, it could be expected that the coating with N75 catalyst would enable the formation of a better and more homogeneous dry coating compared to the catalyst N3300.

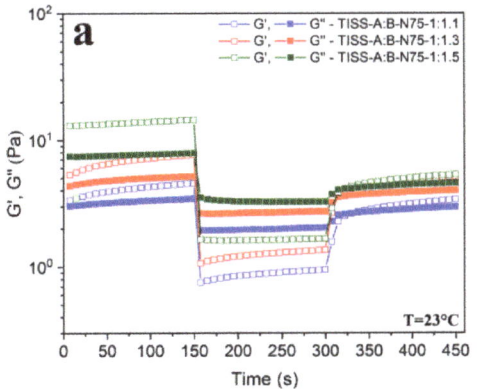
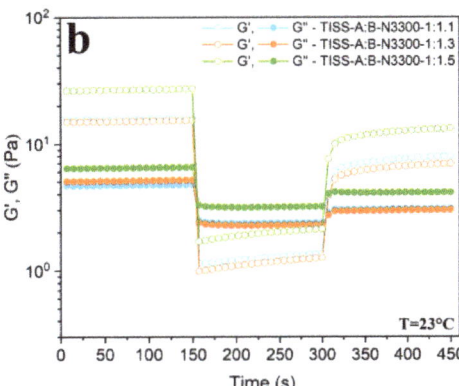

Figure 10. Three-step time test for TISS paints, prepared with fluoropolymer binder L200 and various concentrations of two catalysts: (**a**) Desmodur N75 and (**b**) Desmodur N3300, respectively: 1st step: constant low deformation in linear viscoelastic range; 2nd step: constant high deformation in the range of destructive conditions; 3rd step: constant low deformation, similar as in 1st step.

Moreover, due to the high flexibility of polymeric substrates, the coatings applied to such substrates should exhibit a viscoelastic structure with elongation comparable to that of a plastic substrate. If the structure of the applied coating is too stiff, with a high prevalence of elastic modulus, any dynamic deformation of the substrate will lead to premature adhesion failure of the applied coating. The results of the rheological three-step tests of TISS paints prepared with two catalysts at various concentrations (Figure 10) showed that all the paints prepared with N75 catalyst exhibited similar values of G' and G'' in the third step of the experiment, indicating the flexible viscoelastic structure of the coating after application. On the other hand, the elastic modulus of the paints prepared with N3300 catalysts strongly prevailed over the viscous one in the third step of the experiment, indicating that a more elastic and brittle structure of the applied coating could be expected.

3.2. Adhesion and Spectral Selectivity

In addition to the rheological characterization of liquid paints, dry coatings on the polymeric substrate were characterized by the determination of the functionality of the coating, i.e., spectral selectivity with solar absorptivity and thermal emissivity values. Moreover, the quality of the applied coatings, i.e., the adhesion, was examined using cross-cut tests. These tests were performed by cutting the surface of the coating with a special several-bladed knife. After the cutting, a sticky tape was glued on the surface and peeled-off. If there were no marks on the tape, the adhesion of the paint on the substrate was good, while marks left on the sticky tape indicated a loss of cohesion. The results were evaluated according to the DIN EN ISO 2409 standard as follows:

5B (ISO Class 0): Edges of cut are completely smooth; none of the squares of the lattice are detached.

4B (ISO Class 1): Detachment of small flakes at the intersection of the cuts; max 5% of the cross-cut area is affected.

3B (ISO Class 2): Flaked along the edges and/or intersection of the cuts; affected cross-cut area: 5%–15%.

2B (ISO Class 3): Squares are partly/wholly damaged; affected cross-cut area: 15%–35%.

1B (ISO Class 4): Squares are partly/wholly detached; affected cross-cut area: 35%–65%.

For all the coatings studied in the present research, the cross-cut areas of the coatings and the black marks left on the sticky tape after its removal were scanned and the average values are presented in Tables 1 and 2. Figure 11 presents photographs of the cross-cut tests for two coatings with different catalysts. From Figure 11 it can be clearly seen that the adhesion of the coatings on the polymeric substrate was acceptable for both catalysts used.

Table 1. Comparison of thermal emittance (e_T), solar absorbance (a_S) and adhesion to polymeric substrate of TISS coatings, prepared with the binder L200 and various concentrations of catalyst Desmodur N3300.

Sample	e_T	a_S	$a_S - e_T$	Adhesion
A:B-N3300-1:1.1	0.364	0.914	0.550	3B
A:B-N3300-1:1.3	0.427	0.923	0.496	3B
A:B-N3300-1:1.5	0.455	0.924	0.469	3B

Table 2. Comparison of thermal emittance (e_T), solar absorbance (a_S) and adhesion to polymeric substrate of TISS coatings, prepared with the binder L200 and various concentrations of catalyst Desmodur N75.

Sample	e_T	a_S	$a_S - e_T$	Adhesion
A:B-N75-1:1.1	0.353	0.912	0.559	5B
A:B-N75-1:1.3	0.395	0.906	0.511	3B
A:B-N75-1:1.5	0.403	0.911	0.508	3B

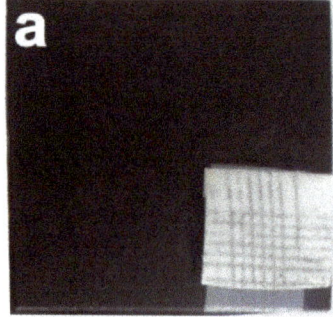
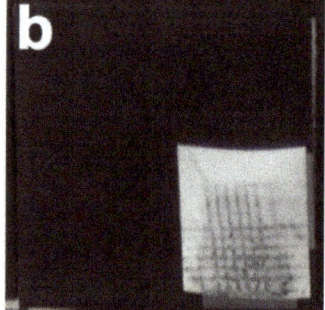

Figure 11. Photograph of cross-cut tests for two TISS coatings onto the polymeric substrate, prepared with the binder L200 and different catalysts: (**a**) Desmodur N75 and (**b**) Desmodur N3300.

As can be seen from the results, the solar absorptivity was above 0.91 for all of the coatings prepared using N3300 catalyst. However, these coatings also exhibited very high thermal emissivity ($e_T > 0.35$) and relatively weak adhesion (3B). Very good adhesion of dry coatings was achieved by the application of the paints with the lowest amount of

N75 catalyst. Moreover, all the coatings, prepared by using this catalyst, also exhibited good spectral selectivity—the difference between a_s and e_T was higher than 0.5 for all samples.

3.3. Morphology of the Applied TISS Paints

The morphological structure of the applied TISS paint on the substrate is presented in Figure 12. It can be seen from the figure that proper distribution of the metal flakes and black pigment in the binder is crucial for high solar absorptivity (smaller black pigments) and low thermal emissivity (larger Al flakes). Figure 12a presents the structure of the applied sample A:B-N3300-1:1.5. It can be seen from the Figure that Al flakes were completely covered with the binder with black pigment, which was homogeneously distributed inside the binder. Due to the complete coverage of the flakes, thermal emissivity of the coating was high (the highest of all samples prepared); however due to good distribution of black pigment inside the binder, solar absorptivity was high (higher than 0.92). However, appropriate distribution of all components in liquid TISS paint led to high spectral selectivity. This can be observed in Figure 12b, which presents the morphology of the sample A:B-N75-1:1.1. This coating exhibited the highest solar absorptivity and the lowest thermal emissivity of all samples prepared (Tables 1 and 2).

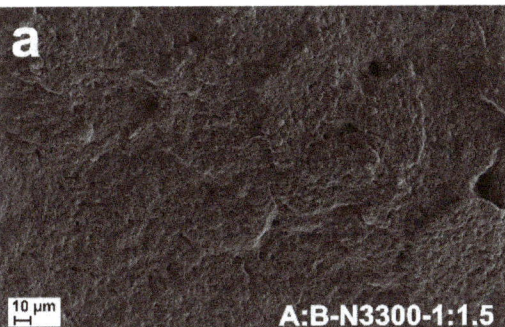 

Figure 12. SEM micrographs of two TISS paints with (**a**) binder completely covering the flakes (A:B-N3300-1:1.5) and (**b**) proper distribution of the components (A:B-N75-1:1.1).

4. Conclusions

In the present work, three different fluoropolymer binders L200, L9716 and L9721 were used for the preparation of black thickness-insensitive spectrally selective (TISS) paints which are, due to incorporation of metal flakes and pigment particles inside polymer binder, suitable for polymer solar absorbers. However, the adhesion of these paints on polymer substrates is usually inferior. To achieve good adhesion of the liquid paint and optimal spectrally selective properties of dry coating, which strongly depend on the distribution of the particles in the polymer binder, systematic research needs to be performed, and the rheological properties carefully determined. For this reason, our study focused on the rheological characterization of binders, pastes, and TISS liquid paints, together with the effect of two polyisocyanate hardeners. Moreover, the adhesion of dry coatings was evaluated with peel-off tests, while spectral selectivity was determined on the basis of solar absorptivity and thermal emissivity measurements.

The results showed that two binders exhibited Newtonian flow behavior, while the behavior of the binder L200 was slightly shear thinning. The addition of the same concentration of pigment particles to the binders changed the flow behavior of the binders towards significant shear thinning. The particle–particle interactions of pigment particles were the highest for the L9716 binder, since the shear thinning effect was the most significant for the paste with this binder. Strong agglomeration of the particles at rest was observed with high zero-shear viscosity, while de-agglomeration occurred when the shear increased. However, when the applied shear decreased, the re-agglomeration was slower, as the viscosity did not recover to its initial state.

For TISS paints, the results showed that the liquid paints exhibited shear thinning, time-dependent flow behavior with noticeable yield stress. Due to the higher viscosity of the binder L200, the TISS paint prepared with this binder exhibited the highest viscosity in the range of high shear stress. Three-step tests of the prepared TISS paints showed that for the paints TISS-L9716-A+B and TISS-L9721-A+B the recovery of the structure to its initial values was almost immediate; therefore, during the formation of the dry coating on the substrate, some inhomogeneities could be expected using these paints. On the other hand, the results of the recovery process—leveling of the paint with L200 binder (TISS-L200-A+B)—a homogeneous final dry coating on the substrate could be expected.

The comparison of the two catalysts used for the preparation of final liquid two-component TISS paints showed that the coatings with N75 catalyst enable the formation of a better and more homogeneous dry coating compared to the catalyst N3300. Moreover, the results showed that the rheological properties strongly depended on the type of catalyst, while the effect of the catalyst's concentration was almost negligible. A small increase in viscosity was observed when higher concentrations of catalyst were added; however, the shape of the viscosity curves remained similar.

The results of spectral selectivity showed that, compared to Desmodur N3300, the catalyst Desmodur N75 (A:B-N75-1:1.1) enabled slightly lower e_T values (e_T = 0.353) and better adhesion to the HDPE polymer surface. It was also observed that the spectral selectivity ($a_S - e_T$) with increasing concentration of catalyst decreased from 0.550 to 0.469 for N3300, while the spectral selectivity for the coatings with N75 catalyst was higher than 0.508 for all the paints prepared, regardless of the concentration of the hardener.

The obtained results confirmed that the rheological properties of liquid TISS paints are extremely important for achieving homogeneous dry solar coatings with good adhesion on polymer substrates. It was shown that only proper rheological properties of liquid paint enable the correct distribution of the metal flakes and black pigment in the binder during the application, which is crucial for the best functionality of dry coating, i.e., high solar absorptivity and low thermal emissivity.

Author Contributions: Conceptualization, L.S.P. and M.M.; methodology, M.M., M.K.G. and L.S.P.; formal analysis, M.M. and M.K.G.; investigation, M.M., M.K.G. and L.S.P.; writing—original draft preparation, L.S.P.; writing—review and editing, M.M., M.K.G. and L.S.P.; visualization, M.M., M.K.G. and L.S.P.; supervision, L.S.P.; project administration, L.S.P. All authors have read and agreed to the published version of the manuscript.

Funding: The authors acknowledge the financial support from the Slovenian Research Agency (research core funding No. P2-0264 and P2-0393).

Institutional Review Board Statement: Not applicable.

Informed Consent Statement: Not applicable.

Data Availability Statement: Not applicable.

Conflicts of Interest: The authors declare no conflict of interest. The funders had no role in the design of the study; in the collection, analyses, or interpretation of data; in the writing of the manuscript, or in the decision to publish the results.

References

1. Carlsson, B.; Persson, H.; Meir, M.; Rekstad, J. A total cost perspective on use of polymeric materials in solar collectors—Importance of environmental performance on suitability. *Appl. Energy* **2014**, *125*, 10–20. [CrossRef]
2. Meir, M.; Murtnes, E.; Dursun, A.; Rekstad, J. Polymeric Solar Collectors or Heat Pump?—Lessons Learned from Passive Houses in Oslo. *Energy Procedia* **2014**, *48*, 914–923. [CrossRef]
3. Kahlen, S.; Wallner, G. Aging Behavior of Polymeric Materials for Solar Thermal Absorber Applications. In *ISES Solar World Congress 2007, ISES 2007*; Springer: Berlin/Heidelberg, Germany, 2009; Volume 1, pp. 519–523. ISBN 978-3-540-75996-6.
4. Schulz, J.; Kent, D.; Crimi, T.; Glockling, J.L.D.; Hull, T.R. A Critical Appraisal of the UK's Regulatory Regime for Combustible Façades. *Fire Technol.* **2021**, *57*, 261–290. [CrossRef]

5. Štirn, Ž.; Čolović, M.; Vasiljević, J.; Šobak, M.; Žitko, G.; Čelan Korošin, N.; Simončič, B.; Jerman, I. Effect of bridged DOPO/polyurethane nanocomposites on solar absorber coatings with reduced flammability. *Sol. Energy* **2022**, *231*, 104–114. [CrossRef]
6. Orel, B.; Spreizer, H.; Vuk, A.; Fir, M.; Merlini, D.; Vodlan, M.; Köhl, M. Selective paint coatings for coloured solar absorbers: Polyurethane thickness insensitive spectrally selective (TISS) paints (Part II). *Sol. Energy Mater. Sol. Cells* **2007**, *91*, 108–119. [CrossRef]
7. Sung, L.-P.; Nadal, M.E.; McKnight, M.E.; Marx, E.; Laurenti, B. Optical reflectance of metallic coatings: Effect of aluminum flake orientation. *J. Coat. Technol.* **2002**, *74*, 55–63. [CrossRef]
8. Gunde, M.K.; Orel, Z.C.; Hutchins, M.G. The influence of paint dispersion parameters on the spectral selectivity of black-pigmented coatings. *Sol. Energy Mater. Sol. Cells* **2003**, *80*, 239–245. [CrossRef]
9. Kirchner, E.; Houweling, J. Measuring flake orientation for metallic coatings. *Prog. Org. Coat.* **2009**, *64*, 287–293. [CrossRef]
10. Slemenik Perše, L.; Mihelčič, M.; Orel, B. Rheological and optical properties of solar absorbing paints with POSS-treated pigments. *Mater. Chem. Phys.* **2015**, *149*, 368–377. [CrossRef]
11. Perše, L.S.; Bizjak, A.; Orel, B. The role of rheological properties and spraying parameters on the spectral selectivity of Thickness Insensitive Spectrally Selective (TISS) paint coatings. *Sol. Energy Mater. Sol. Cells* **2013**, *110*, 115–125. [CrossRef]
12. Shen, Z.; Rajabi-Abhari, A.; Oh, K.; Lee, S.; Chen, J.; He, M.; Lee, H.L. The Effect of a Polymer-Stabilized Latex Cobinder on the Optical and Strength Properties of Pigment Coating Layers. *Polymers* **2021**, *13*, 568. [CrossRef] [PubMed]
13. Oh, K.; Lee, J.-H.; Im, W.; Rajabi Abhari, A.; Lee, H.L. Role of Cellulose Nanofibrils in Structure Formation of Pigment Coating Layers. *Ind. Eng. Chem. Res.* **2017**, *56*, 9569–9577. [CrossRef]
14. Cohu, O.; Magnin, A. Rheology and flow of paints in roll coating processes. *Surf. Coat. Int.* **1997**, *80*, 102–107. [CrossRef]
15. Mihelčič, M.; Gaberšček, M.; Salzano de Luna, M.; Lavorgna, M.; Giuliani, C.; Di Carlo, G.; Surca, A.K. Effect of silsesquioxane addition on the protective performance of fluoropolymer coatings for bronze surfaces. *Mater. Des.* **2019**, *178*, 107860. [CrossRef]
16. Hutchins, M.G. Spectrally selective materials for efficient visible, solar and thermal radiation control. In *Solar Thermal Technologies for Buildings*; Santamouris, M., Ed.; James & James: London, UK, 2003; ISBN 9781315074467.
17. Chen, A.T.; Wojcik, R.T. Polyurethane coatings for metal and plastic substrates. *Met. Finish.* **2010**, *108*, 108–121. [CrossRef]

Article

The Effect of PVP on Thermal, Mechanical, and Dielectric Properties in PVDF-HFP/PVP Thin Film

Urška Gradišar Centa [1,*], Mohor Mihelčič [1], Vid Bobnar [2], Maja Remškar [2] and Lidija Slemenik Perše [1]

[1] Faculty of Mechanical Engineering, University of Ljubljana, Aškerčeva 6, 1000 Ljubljana, Slovenia
[2] Institut Jozef Stefan, Jamova 39, 1000 Ljubljana, Slovenia
* Correspondence: urska.gradisarcenta@fs.uni-lj.si

Abstract: In this research, the influences of the addition of PVP to PVDF-HFP polymers and the preparation of thin films using a solvent casting method were studied. The PVDF-HFP and polymer blend PVDF-HFP/PVP thin films with a nanostructured surface were investigated using scanning electron microscopy, differential scanning calorimetry, nanoindentation, and dielectric spectroscopy. The results showed that the PVP formed a dispersed phase (the poorer conductive islands) in the PVDF-HFP polymer matrix, which reduced its mechanical properties. The crystallinity of PVDF-HFP polymer decreased with the addition of PVP by 7.4%, but the PVP induced the formation of the polar β-phase of PVDF-HFP. Therefore, an improved dielectric response is expected, but it was not significantly improved even though the polar β-phase was detected. The contrasting effect was attributed to less conductive PVP islands on the surface of the PVDF-HFP/PVP polymer blend, which decreased its conductivity.

Keywords: polymer blend; PVDF-HFP; PVP; mechanical properties; crystallization

1. Introduction

Polymer composites, based on poly(vinylidene fluoride-co-hexafluoropropylene) (PVDF-HFP) polymer, can be employed in many electronics, i.e., sensor systems [1,2], actuators [3], nanogenerators [4], microelectronic components, and medical applications [5], due to their cost-effective, mechanically flexible, and high-temperature stability. The chemically inert polymer PVDF-HFP has been used in numerous scientific fields, e.g., as membranes [6], polymer electrolyte for batteries [7,8], electrospinning produced fibres [9,10], as supercapacitors [11], and so on. The wide range of usage of PVDF-HFP polymers is a consequence of its ferroelectric, piezoelectric, and pyroelectric properties [12,13], good mechanical strength [14], thermal stability of up to 143 °C [6,15,16], and low degree of crystallinity (about 50%) [17]. For a thin film of PVDF-HFP/PVP, it has already been proven that it can show antimicrobial activities against *Gram-positive bacteria L. monocytogenes* after 3 and 6 h of exposure, but a reduction in colony forming units (CFU) was observed for *P. anomala* and *A. flavus* after 6 h [18].

The semi-crystalline properties of PVDF-HFP polymer are strongly dependent on its preparation conditions, solvents, and additives, which affects the degree of crystallinity and the presence of individual phases (α, β, γ, δ, and ε). The polar γ-phase occurrence increases with crystallization temperatures and time [19]. Still, the non-polar α-phase is the most common and is usually present after melt crystallization at temperatures below 160 °C. The most polar β-phase of PVDF-HFP is formed by mechanical deformation of α-phase PVDF-HFP film or by doping the polymer with fillers, most commonly with organic nanoparticles [20] or clay minerals [21].

As we already showed in our previous research [18], the presence of polar β-phase was induced by the addition of amorphous water-soluble polyvinylpyrrolidone (PVP) polymers. PVP has a weakly basic nature; therefore, it is capable of forming complexes not only with

anionic reagents but also with ionic polymers [22]. For polymer blend PVDF/PMMA, it has been shown that 25% of β-phase was present after in situ blend crystallizations from ethanol with the addition of 1 wt.% of PMMA. In this study, it has been indicated that a good mixing ability of polymers results from the presence of hydrogen bonding, which increases the movement resistance of PVDF chain segments to the crystal front [23]. Therefore, the PVDF chain is forced to form its extended all-trans conformation that leads to crystallization in the β-phase. In another study, the polyaniline nanorods were dispersed in PVP and then incorporated into PVDF and, therefore, enabled the formation of dPANI@PVP/PVDF nanocomposites with a relatively low dielectric loss while maintaining a sufficiently high dielectric constant [24]. Moreover, the percolative nanocomposite Ag@PVP/PVDF, which is based on PVDF with the addition of silver core–shell nanoparticles, coated with the 5–10 nm layer of PVP for better dispersion into the matrix, and possessing low dielectric loss and high permittivity, has been prepared [25]. In the literature, some studies also described that the surface modification of ferroelectric particles (e.g., $BaTiO_3$) with PVP polymers, exhibiting fine core–shell structure, which was homogeneously dispersed in the PVDF matrix; in this way, new materials for energy storage have been prepared. With 40 vol% of $BaTiO_3$ loading, the prepared materials exhibited the largest dielectric constant of 65 at 25 °C and 1 kHz [26].

In recent years, many studies have been performed on polymers and their composites using the nanoindentation technique [27]. Nanoindentation measurements proved that the incorporation of various organic, inorganic, or organic-inorganic fillers help improve or modulate the mechanical properties of PVDF [28–30].

However, only a few studies were performed on PVDF copolymer (PVDF-HFP) films when determining their mechanical properties. The research of Yuennan and Muensit [31] showed that the addition of magnesium chloride hexahydrate ($MgCl_2 \cdot 6H_2O$) fillers to PVDF-HFP acted as a plasticizer and improved the flexibility of the film, which is advantageous in piezoelectric applications.

In this article, we study the effect of the addition of PVP polymers with polar pyrrolidinone substituents on the degree of interactions with the polar polymer matrix PVDF-HFP. Furthermore, we also investigated how the formation of polar β-phase affects mechanical and dielectric properties.

2. Materials and Methods

The PVDF-HFP, obtained from Sigma Aldrich (St. Louis, MO, USA), was dissolved in dimethylformamide (DMF), Carlo Erba Reagents (Cornaredo, Italy), by mixing for 4 h on a magnetic stirrer with 400 rpm at 80 °C. After achieving a homogenous solution, the thin film of PVDF-HFP was formed on a Teflon plate by solution casting and drying for 2 h at 80 °C.

Polymer solution PVDF-HFP/PVP was obtained by dissolving PVDF-HFP and PVP Sigma Aldrich (St. Louis, MO, USA) in DMF separately for 4 h on a magnetic stirrer with 400 rpm at 80 °C. The final PVDF-HFP/PVP solution was prepared by adding the PVP solution into the PVDF-HFP one in mass ratio 25:75 by mixing both components on a magnetic stirrer for additional 2 h. The thin film of PVDF-HFP/PVP was prepared in the same way and at the same conditions as explained above for the PVDF-HFP film.

The morphology of formed thin films was investigated with a scanning electron microscope using a field-emission gun Supra 36 VP, Carl Zeiss (Braunschweig, Germany). The accelerated voltage was 1.5 kV, and the working distance 4.3 mm and the figures were formed from the signal of secondary electrons. The thin films were placed on an adhesive carbon tape and sputtered with 10 nm of carbon to ensure better sample conductivity. Thermal properties of polymers and polymer blend were determined by using differential scanning calorimetry—DSC (Q2500, TA Instruments, New Castle, DE, USA). Heat–cool–reheat tests were performed according to standard ISO 11,357 in the temperature range from −80 °C to 250 °C for PVP, from −80 °C to 300 °C for PVDF-HFP, and from 40 °C to 300 °C for polymer blend PVDF-HFP/PVP with heating and cooling rates of 10 °C/min under an inert (nitrogen) atmosphere. To determine the values of the glass transition, melting and crystallization temperatures, and enthalpies, the TRIOS software was used. The degree of crystallinity of PVDF-HFP polymer was calculated using the following equation:

$$\chi_c = \frac{\Delta H_f}{\Delta H_f^0} \cdot 100\% \qquad (1)$$

where X_c is the degree of crystallinity, ΔH_f is the fusion enthalpy (calculated from DSC curve), and ΔH_f^0 is the fusion enthalpy for 100% crystallinity (for PVDF-HFP polymer, the value of 104.7 J/g was used [6]). The degree of crystallinity for polymer blend PVDF-HFP/PVP was calculated with the same equation, where the values of crystallinity were multiplied with weight fraction coefficient $1/\varphi$, where φ represents the weight fraction of the PVDF-HFP as the crystallized component (in our case 0.75).

The XRD diffractogram of the PVDF-HFP/PVP polymer blend was performed on a D4 Endeavor diffractometer (Bruker Corporation, Billerica, MA, USA) at room temperature using a quartz monochromator Cu Kα radiation source (λ = 0.1541 nm) with a Sol-X dispersive detector in the range of 2θ from 5° to 65° (step size 0.04°). The XRD diffractogram has been smoothed in the computer program Origin. Optical images of spherulites were taken with a polarized light optical Microscope Axioskop 2, Carl Zeiss (Oberkochen, Germany) at the magnification 200×.

Mechanical properties were characterized with nanoindentation techniques on the surface of the tested thin film samples. Nanoindentation is a useful non-destructive technique for determining mechanical properties such as elastic moduli and hardness at the nano-scale of many materials in bulk or as a coating deposited on the surface [32,33]. All tests were performed on a Nanoindenter G200 XP instrument manufactured by Agilent Technologies, Inc (Santa Clara, CA, USA). Continuous Stiffness Measurement (CSM) was performed using a standard three-sided pyramidal Berkovich probe with the tip oscillation frequency of 45 Hz and 2 nm harmonic amplitude. For the characterization of the samples, several sheets were cut and placed into a holder (diameter 8 mm, height 1 mm), melted at 140 °C for 10 min, and cooled down in air. Thirty-six indents were performed onto two samples for each coating with a 200 μm distance between adjacent indentations to exclude interaction effects. The highest depth of the indents was 2500 nm; however, for the calculation of the elastic modulus and hardness, values at the depths of 1000 and 2000 nm were used. For analyzing the results, the Poisson ratio of 0.33 was used for PVDF-HFP [34]. All measurements were conducted at room temperature.

The contact modulus (CM) method was used to determine the dynamic properties (storage and loss modulus) of films using a 100 μm flat-punch tip. The viscoelastic properties were determined at various frequencies from 45 Hz to 1 Hz. A pre-compression of 5 μm was used to make sure that the tip is in contact with the surface. Fifteen tests were performed on each sample, and averaged values are provided as a result.

Dielectric properties of PVDF-HFP and PVDF-HFP/PVP thin films were measured with a Precision LCR Meter (HP 4284A) using the amplitude of the probing AC electric signal of 1 V. Two-layer electrodes were sputtered on the thin films: 10 nm of chromium for better adhesion and 100 nm of gold. The complex dielectric constant $\varepsilon^* = \varepsilon' - i\varepsilon''$ was measured during cooling in the temperature range between 400 K and 150 K at a cooling rate of 1 K/min. The temperature of the samples was controlled by a lock-in bridge technique with a platinum resistor Pt100 and was stabilized within ±0.01 K. The real part of the complex AC conductivity $\sigma^* = \sigma' + i\sigma''$ was calculated via equation $\sigma' = 2\pi\nu\varepsilon_0\varepsilon''$, where ε_0 is the permittivity of free space.

3. Results

3.1. Morphology

The surface topography of the transparent thin films of PVDF-HFP and polymer blend PVDF-HFP/PVP is shown in Figure 1. On the surface of the PVDF-HFP thin film (Figure 1A), some random spherical structures are visible. The addition of the PVP polymer into the PVDF-HFP (PVDF-HFP/PVP polymer blend, Figure 1B) led to the formation of a nanostructured surface with spherical structures of an average diameter of 200 to 500 nm.

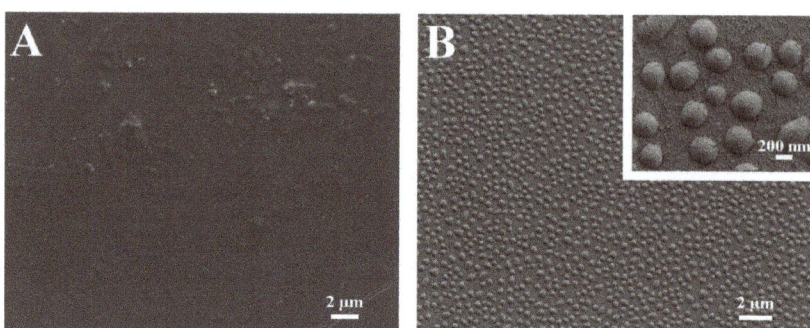

Figure 1. SEM images of the surface: (**A**) PVDF-HFP thin film and (**B**) PVDF-HFP/PVP thin film.

3.2. Thermal Properties

The DSC curves that indicate the thermal properties of PVP and both thin films are presented in the Figure 2. The values of the glass transition temperature (T_g), melting temperature (T_m), and melting enthalpy (ΔH) were determined from the re-heating curve (Figure 2A), while crystallization temperatures and enthalpies were determined from the cooling curve (Figure 2B). The obtained values of temperatures of phase transitions and fusion enthalpy are summarized in Table 1. The PVP is an amorphous polymer with an average molecular weight of 40 kDa and exhibits only the glass transition temperature at 164.8 °C. For PVDF-HFP thin films, the crystallinity was 28.9%, while the addition of 25% of the PVP polymer (polymer blend PVDF-HFP/PVP) decreased the crystallinity to 21.6%.

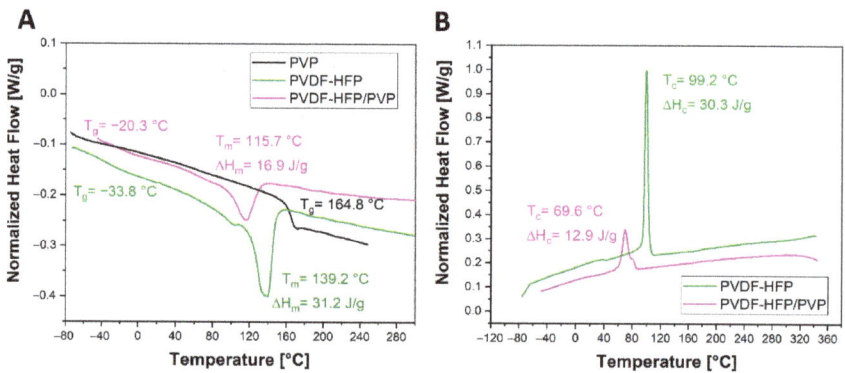

Figure 2. (**A**) DSC curves for PVP, PVDF-HFP polymer, and PVDF-HFP/PVP polymer blend at second heating; (**B**) DSC crystallization curves for PVDF-HFP polymer and PVDF-HFP/PVP polymer blend.

Table 1. Parameters of the DSC curves.

Thin Films	T_g (°C)	T_m (°C)	ΔH_m (J/g)	T_c (°C)	ΔH_c (J/g)
PVP	164.8	-	-	-	-
PVDF-HFP	−33.8	139.2	31.2	99.2	30.3
PVDF-HFP/PVP	−20.3	115.7	16.9	69.6	12.9

PVP is a hygroscopic amorphous polymer and, therefore, a broad endothermic peak ranging from 80 to 120 °C was observed in the DSC curve (the first heating cycle is not shown). This peak was attributed to the presence of a residual moisture [35]. In polymers, water acts as a plasticizer [36], which can be observed in the DSC thermogram as a shift in phase transition temperatures to lower temperatures. Moreover, the literature reports that the presence of water in PVP polymer affects the increasing mobility of the polymer blend [37], which could also increase the dielectric response of such thin films.

3.3. Crystal Structures

The XRD diffractogram of the PVDF-HFP/PVP polymer blend is presented in Figure 3A. It can be clearly seen that the multi-phase crystal structure in polymer blend was observed. The XRD diffractogram of PVDF-HFP/PVP polymer blends, presented according to our previous publication, show a broad peak centered at 2θ = 20°, which was attributed to the presence of the β-phase, as a result of the sum of the diffraction in (110) and (200) planes [37]. The next, less intensive peak, centered at 26.6°, was attributed to the diffraction of the crystal plane (021) and corresponds to the presence of mixed phases: α and γ-phase [37]. The predominant crystalline phase of PVDF-HFP in the polymer blend was attributed to the γ phase, because the most intensive peak was observed at about 2θ = 40°. The spherulites, formatted during the crystallization, were observed by a polarizable light optical microscope, and they are shown in Figure 3B. A greater share of the small and medium crystal grains was attributed to the presence of α and β phase of PVDF, whereas larger crystal grains are present in the γ-phase [37].

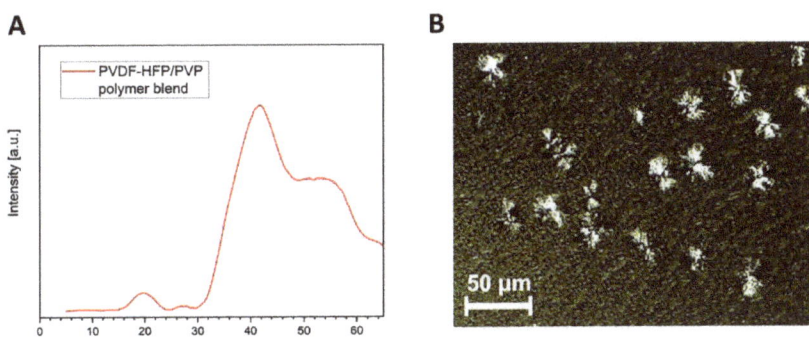

Figure 3. (**A**) XRD diffractogram of PVDF-HFP/PVP polymer blend thin film; (**B**) optical figure of spherulites on PVDF-HFP/PVP thin film.

3.4. Mechanical Properties

The mechanical properties of PVDF-HFP and PVDF-HFP/PVP were determined using a nanoindentation technique, and the results are presented in the Figure 4. The results show that the addition of PVP to PVDF-HFP decreased the elastic modulus and hardness of the thin film. The elastic modulus of the PVDF-HFP/PVP film was 1.61 ± 0.01 GPa and 1.99 ± 0.02 GPa of the film without PVP. A larger difference between both samples was observed by the determination of surface hardness, which was for PVDF-HFP/PVP almost twice lower compared to PVDF-HFP.

Figure 5A shows a variation in the storage modulus (E′), loss modulus (E″), and loss factor (tanδ) with the modulation of frequency at a depth of 5 μm for PVDF-HFP and PVDF-HFP/PVP, respectively. For both thin films, the values of storage modulus were found to be two orders of magnitude higher compared to the loss modulus. However, the addition of PVP to PVDF-HFP (PVDF-HFP/PVP thin film) decreased both the E′ and E″ modulus. The E′ for both thin films decreased for about 0.2 GPa with the decreasing frequency, while the E″ was almost independent upon frequency changes. The loss factor, which represents the ratio of E′/E″, exhibited the same trend for both thin films, with lower values for PVDF-HFP/PVP (Figure 5B).

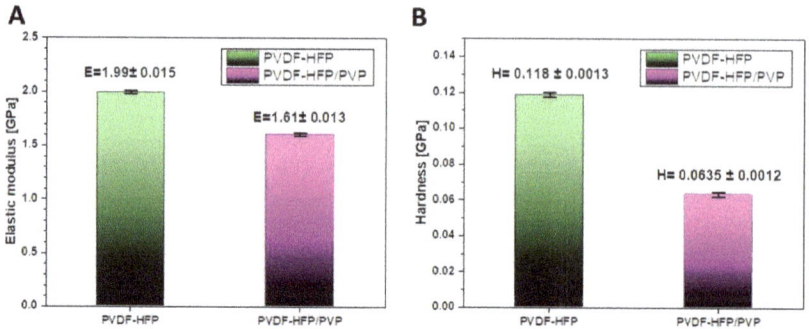

Figure 4. Results of nanoindentation: (**A**) elastic modulus of PVDF-HFP and PVDF-HFP/PVP; (**B**) hardness of PVDF-HFP and PVDF-HFP/PVP.

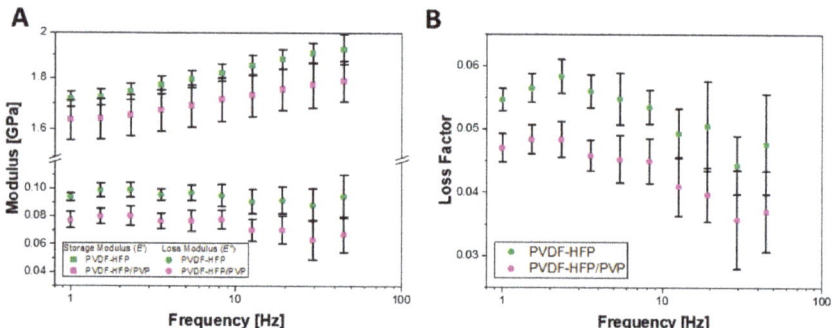

Figure 5. Results of nanoindentation—frequency dependent: (**A**) storage and loss modulus of PVDF-HFP and PVDF-HFP/PVP; (**B**) loss factor of PVDF-HFP and PVDF-HFP/PVP.

3.5. Dielectric Properties

The results of dielectric measurements are presented in Figure 6. The real and imaginary parts of the complex dielectric constant ε* are shown along with the real part of the complex AC conductivity at four frequencies (1, 10, 100 kHz, and 1 MHz) in the temperature range between −123 °C and 127 °C (150 and 400 K). Imaginary part ε'' represents dielectric losses or, in another words, the electrical conductivity of the system. The dielectric response of the PVDF-HFP thin film shows dielectric relaxation in the temperature range between −73 °C and 27 °C (200 and 300 K) due to the dynamic transition from the solid glass state to the soft rubber phase occurring in the amorphous part of the PVDF-HFP polymer [38]. The results show that, at room temperature, the dielectric constant of PVDF-HFP was about 10 and 9 for the polymer blend of PVDF-HFP/PVP, respectively. The characteristic dynamic peaks in the graph of dielectric losses (ε'') on the temperature scale occured at the same temperature (24 °C or 297 K) in both thin films. The addition of PVP polymer did not cause any significant changes in the values of ε', ε'', and σ' or in the structure of PVDF-HFP polymer. We observed that the electrical conductivity slightly decreased after the addition of the amorphous PVP polymer. The addition of PVP polymer simultaneously reduced the proportion of crystalline phases in the sample (which consequently reduces the relaxation intensity [38], i.e., the relaxation peaks in ε'' and the relaxation strength in ε'), but on the other hand, it triggered the crystallization of PVDF-HFP polymers into the ferroelectric polar phase and stabilized it; no significant change in the dielectric response was observed.

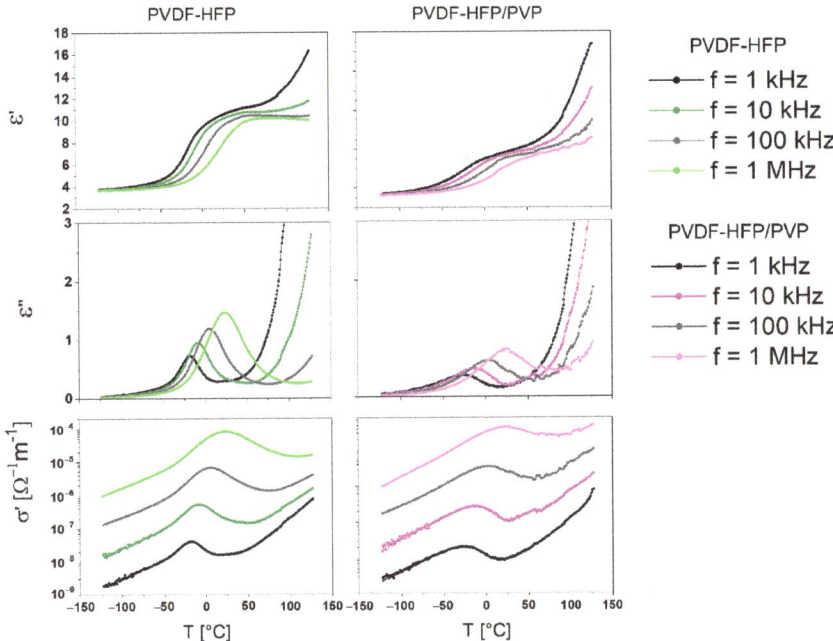

Figure 6. The real (ε') and imaginary (ε'') parts of the complex dielectric constant and the real part (σ') of the complex AC conductivity vs. temperature at different frequencies for thin films of PVDF-HFP and PVDF-HFP/PVP.

4. Discussion

Some studies in the literature have already reported that the PVP polymer enhances thermal stability, crosslinking, and mechanical strength if added to the composite because pyrrolidone groups can enable a homogeneous incorporation of organic and inorganic salts [39]. However, according to our knowledge, no research studies on thin films involving the PVDF-HFP polymer have been conducted. The goal of our research was, therefore, to study such films and characterize their physical properties, i.e., optical, thermal, mechanical, and dielectric properties. Physical properties are influenced also by the crystalline phase in the polymorphous matrix polymer PVDF-HFP. Technologically, the most interesting phase is the polar β-phase, which exhibits pyro, piezoelectric, and dielectric properties. It has already been shown that the predominance of the β-phase could be determined from the temperature of the solvent evaporation; the majority of β-phase was formed at temperatures below 70 °C. Between 70 and 110 °C, a mixture of α- and β-phase was formed, and above 110 °C, the γ-phase became dominant [19]. It has already been proven in our previous research that although the temperature of the crystallization process of PVDF-HFP thin films was 80 °C, Raman spectroscopy confirmed the presence of the β-phase in the PVDF-HFP/PVP polymer blend. On the other hand, the XRD diffractogram revealed that the γ-phase was dominant as the most intensive peak was observed at about 40°. In the XRD diffractogram (Figure 3A), the peaks of β-phase were also present, but the intensity was lower compared to the γ-phase. However, the β-phase was not detected in pure PVDF-HFP thin films prepared by crystallization at the same conditions [18]. It has been reported that the interphases between the crystalline and amorphous phases maintain the conformational characteristics of the crystalline regions [19]; therefore, we expect the improvement of the mechanical and dielectric response of the PVDF-HFP/PVP polymer blend. In particular, this expectation is because Li et al. expected an interaction between the polar pyrrolidinone substituents of PVP and the polar polymer matrix (PVDF-HFP) and, thus, an improvement in dielectric properties [25].

As already mentioned, the crystallization of PVDF-HFP thin films was performed at 80 °C. The results, already published in [21], showed that after crystallization, the film exhibited a rich multi-phase structure, including the presence of the polar β-phase. As already mentioned, the crystallization of PVDF-HFP thin film was performed at 80 °C. The results, already published in [21], showed that after crystallization, the film exhibited a rich multiphase structure, including the presence of the polar β-phase. However, the β-phase in the composite, prepared by "drawing at low temperatures", did not exhibited piezoelectric properties due to the random orientation of dipoles [40]. It turned out that in decreased electric fields, the α-phase was the most polarizable, slightly less β-phase, while due to the lowest remanent polarization, the γ-phase was the least polarizable 37. The loss of polarisation is attributed to the greater mobility of polymer molecular chains [41]. However, at the γ-phase of the PVDF, the crystal grains grow side by side, and a little block space between the adjacent crystal grains can be observed [37]. In our case, where the multiphase structure was observed, the crystal grains had different sizes. The larger light spherulites represented the γ-phase (Figure 3B).

After the addition of PVP, the degree of crystallinity in the formed PVDF-HFP/PVP film decreased, but we detected that the addition of a PVP polymer affected the PVDF-HFP crystal structure, while it crystallized in polar β- and γ-phases [18]. For the presence of the β-phase, of the formation of hydrogen bonds between hydroxyl groups (–OH) and carbonfluorine (–CF) is desirable, but the critical role for the polar molecular conformation of PVDF-HFP is to possess -OH groups [42]. In the following, the more intensive peak at 839 cm^{-1} of the PVDF-HFP/PVP thin film, detected with Raman spectroscopy, confirmed that the addition of PVP resulted in the stabilization of the polar β-phase of the PVDF-HFP polymer [43].

However, the addition of PVP polymer deteriorates the mechanical properties of the PVDF-HFP/PVP thin film compared to the PVDF-HFP thin film. The elastic modulus of the PVDF-HFP/PVP polymer blend decreased by around 20%. At higher contents of PVP polymer, stronger linkages may occur between the components of the polymer blends and the polymeric chains, which can reduce the motion of molecules [44]. A higher amount of PVP accumulated in the dispersed phase (PVP islands on the surface, Figure 1B) in the PVDF-HFP/PVP thin film, which decreases the mechanical properties of films [44]. The hardness of the polymer blend reduced for around 50%.

Due to the confirmed presence of the polar β-phase [18], we expected stronger dielectric responses from the films, prepared by the addition of PVP. However, the micrographs showed that nanostructured surface of the PVDF-HFP/PVP thin film contained PVP island structures (diameter between 200 and 500 nm, height 150 nm) with poorer conductivity. It turned out that these islands overshadowed the increase in dielectric responses and decreased the conductivity of thin PVDF-HFP/PVP films.

5. Conclusions

A polymer blend based on inert PVDF-HFP with the addition of PVP was successfully prepared and characterized by its thermal, mechanical, and dielectric properties. The addition of 25% of PVP polymer slightly decreased the elastic modulus of the polymer blend and reduced the thin film's hardness by half. Despite the detection of the ferroelectric β-phase of PVDF-HFP polymer, the polymer blend PVDF-HFP/PVP did not exhibit an improved dielectric response, because the PVP was present in the dispersed phase, which led to the formation of less conductive islands. However, the dielectric properties of thin films improved in a moisture environment, since their conductivity increased. With the presented results, we can conclude that the PVDF-HFP/PVP polymer blend can be further used in conductive biocompatible components in the field of biomedicine, sensors, and smart scaffolds. For future studies, crystallization parameters will be optimized in a manner in which large amounts of β phase will be detected. After that, the polarity of the thin film will be checked and the nanogenerators will be possibly formed.

Author Contributions: Conceptualization, U.G.C.; methodology, M.M. and U.G.C.; formal analysis, U.G.C., M.M. and V.B.; investigation, U.G.C., M.M. and L.S.P.; writing—original draft preparation, U.G.C.; writing—review and editing, U.G.C., M.M., L.S.P. and M.R.; visualization, U.G.C. and M.M.; supervision, M.R. and L.S.P.; project administration, U.G.C. All authors have read and agreed to the published version of the manuscript.

Funding: The authors acknowledge the financial support from the Slovenian Research Agency (research core funding Nos. P2-0264, P1-0125 and P1-0099) and the APC was funded by the Slovenian Research Agency (research core funding No. P2-0264).

Institutional Review Board Statement: Not applicable.

Informed Consent Statement: Not applicable.

Data Availability Statement: Not applicable.

Conflicts of Interest: The authors declare no conflict of interest. The funders had no role in the design of the study; in the collection, analyses, or interpretation of data; in the writing of the manuscript; or in the decision to publish the results.

References

1. Daneshkhah, A.; Shrestha, S.; Agarwal, M.; Varahramyan, K. Poly(Vinylidene Fluoride-Hexafluoropropylene) Composite Sensors for Volatile Organic Compounds Detection in Breath. *Sens. Actuators B Chem.* **2015**, *221*, 635–643. [CrossRef]
2. Karthieka, R.R.; Prakash, T. Dose-Dependent X-Ray Sensing Behaviour of Cs3Bi2I9: PVDF-HFP Nanocomposites. *Phys. E Low-Dimens. Syst. Nanostructures* **2021**, *133*, 114823. [CrossRef]
3. Terasawa, N.; Ono, N.; Hayakawa, Y.; Mukai, K.; Koga, T.; Higashi, N.; Asaka, K. Effect of Hexafluoropropylene on the Performance of Poly(Vinylidene Fluoride) Polymer Actuators Based on Single-Walled Carbon Nanotube–Ionic Liquid Gel. *Sens. Actuators B Chem.* **2011**, *160*, 161–167. [CrossRef]
4. Fu, H.; Jin, Y.; Ou, H.; Huang, P.; Liu, C.; Luo, Y.; Xiao, Z. High-Performance Ag Nanowires/PEDOT:PSS Composite Electrodes for PVDF-HFP Piezoelectric Nanogenerators. *J. Mater. Sci. Mater. Electron.* **2021**, *32*, 21178–21187. [CrossRef]
5. Azimi, S.; Golabchi, A.; Nekookar, A.; Rabbani, S.; Amiri, M.H.; Asadi, K.; Abolhasani, M.M. Self-Powered Cardiac Pacemaker by Piezoelectric Polymer Nanogenerator Implant. *Nano Energy* **2021**, *83*, 105781. [CrossRef]
6. Cao, J.H.; Zhu, B.K.; Xu, Y.Y. Structure and Ionic Conductivity of Porous Polymer Electrolytes Based on PVDF-HFP Copolymer Membranes. *J. Memb. Sci.* **2006**, *281*, 446–453. [CrossRef]
7. Miao, R.; Liu, B.; Zhu, Z.; Liu, Y.; Li, J.; Wang, X.; Li, Q. PVDF-HFP-Based Porous Polymer Electrolyte Membranes for Lithium-Ion Batteries. *J. Power Sources* **2008**, *184*, 420–426. [CrossRef]
8. Manuel Stephan, A.; Saito, Y. Ionic Conductivity and Diffusion Coefficient Studies of PVdF–HFP Polymer Electrolytes Prepared Using Phase Inversion Technique. *Solid State Ionics* **2002**, *148*, 475–481. [CrossRef]
9. Filip, P.; Zelenkova, J.; Peer, P. Electrospinning of a Copolymer PVDF-Co-HFP Solved in DMF/Acetone: Explicit Relations among Viscosity, Polymer Concentration, DMF/Acetone Ratio and Mean Nanofiber Diameter. *Polymers* **2021**, *13*, 3418. [CrossRef]
10. Parangusan, H.; Ponnamma, D.; Al-Maadeed, M.A.A. Stretchable Electrospun PVDF-HFP/Co-ZnO Nanofibers as Piezoelectric Nanogenerators. *Sci. Rep.* **2018**, *8*, 754. [CrossRef]
11. Tripathi, S.K.; Kumar, A.; Hashmi, S.A. Electrochemical Redox Supercapacitors Using PVdF-HFP Based Gel Electrolytes and Polypyrrole as Conducting Polymer Electrode. *Solid State Ionics* **2006**, *177*, 2979–2985. [CrossRef]
12. Lovinger, A.J. Annealing of Poly (Vinylidene Fluoride) and Formation of a Fifth Phase. *Macromolecules* **1982**, *15*, 40–44. [CrossRef]
13. Martins, P.; Lopes, A.C.; Lanceros-Mendez, S. Electroactive Phases of Poly (Vinylidene Fluoride): Determination, Processing and Applications. *Prog. Polym. Sci.* **2014**, *39*, 683–706. [CrossRef]
14. Wang, X.; Xiao, C.; Liu, H.; Huang, Q.; Hao, J.; Fu, H. Poly(Vinylidene Fluoride-Hexafluoropropylene) Porous Membrane with Controllable Structure and Applications in Efficient Oil/Water Separation. *Materials* **2018**, *11*, 443. [CrossRef] [PubMed]
15. Sundaram, N.T.K.; Subramania, A. Microstructure of PVdF-Co-HFP Based Electrolyte Prepared by Preferential Polymer Dissolution Process. *J. Memb. Sci.* **2007**, *289*, 1–6. [CrossRef]
16. Malmonge, L.F.; Malmonge, J.A.; Sakamoto, W.K. Study of Pyroelectric Activity of PZT/PVDF-HFP Composite. *Mater. Res.* **2003**, *6*, 469–473. [CrossRef]
17. Laxmayyaguddi, Y.; Mydur, N.; Shankar Pawar, A.; Hebri, V.; Vandana, M.; Sanjeev, G.; Hundekal, D. Modified Thermal, Dielectric, and Electrical Conductivity of PVDF-HFP/LiClO4 Polymer Electrolyte Films by 8 MeV Electron Beam Irradiation. *ACS Omega* **2018**, *3*, 14188–14200. [CrossRef]
18. Centa, U.G.; Sterniša, M.; Višić, B.; Federl, Ž.; Možina, S.S.; Remškar, M. Novel Nanostructured And Antimicrobial PVDF-HFP/PVP/MoO 3 Composite. *Surf. Innov.* **2020**, *9*, 256–266. [CrossRef]

19. Gregorio, R.; Ueno, E.M. Effect of Crystalline Phase, Orientation and Temperature on the Dielectric Properties of Poly (Vinylidene Fluoride) (PVDF). *J. Mater. Sci.* **1999**, *34*, 4489–4500. [CrossRef]
20. Chakhchaoui, N.; Farhan, R.; Eddiai, A.; Meddad, M.; Cherkaoui, O.; Mazroui, M.; Boughaleb, Y.; Van Langenhove, L. Improvement of the Electroactive β-Phase Nucleation and Piezoelectric Properties of PVDF-HFP Thin Films Influenced by TiO_2 Nanoparticles. *Mater. Today Proc.* **2021**, *39*, 1148–1152. [CrossRef]
21. Zhang, S.; Tong, W.; Wang, J.; Wang, W.; Wang, Z.; Zhang, Y. Modified Sepiolite/PVDF-HFP Composite Film with Enhanced Piezoelectric and Dielectric Properties. *J. Appl. Polym. Sci.* **2020**, *137*, 48412. [CrossRef]
22. Shawky, A.I.; Megat Mohd Noor, M.J.; Nasef, M.M.; Khayet, M.; Nallappan, M.; Ujang, Z. Enhancing Antimicrobial Properties of Poly(Vinylidene Fluoride)/Hexafluoropropylene Copolymer Membrane by Electron Beam Induced Grafting of: N -Vinyl-2-Pyrrolidone and Iodine Immobilization. *RSC Adv.* **2016**, *6*, 42461–42473. [CrossRef]
23. Zhao, X.; Cheng, J.; Zhang, J.; Chen, S.; Wang, X. Crystallization Behavior of PVDF/PMMA Blends Prepared by in Situ Polymerization from DMF and Ethanol. *J. Mater. Sci.* **2012**, *47*, 3720–3728. [CrossRef]
24. Yu, S.; Qin, F.; Wang, G. Improving the Dielectric Properties of Poly(Vinylidene Fluoride) Composites by Using Poly(Vinyl Pyrrolidone)-Encapsulated Polyaniline Nanorods. *J. Mater. Chem. C* **2016**, *4*, 1504–1510. [CrossRef]
25. Li, H.; Yang, J.; Fang, D.; Liu, L.; Chen, J.; Jiang, M.; Xiong, C. Preparation and Dielectric Properties of Silver@poly(Vinyl Pyrrolidone)/Poly (Vinylidene Fluoride) Composites. *Micro Nano Lett.* **2016**, *11*, 746–748. [CrossRef]
26. Fu, J.; Hou, Y.; Zheng, M.; Wei, Q.; Zhu, M.; Yan, H. Improving Dielectric Properties of PVDF Composites by Employing Surface Modified Strong Polarized $BaTiO_3$ Particles Derived by Molten Salt Method. *ACS Appl. Mater. Interfaces* **2015**, *7*, 24480–24491. [CrossRef]
27. Díez-Pascual, A.M.; Gómez-Fatou, M.A.; Ania, F.; Flores, A. Nanoindentation in Polymer Nanocomposites. *Prog. Mater. Sci.* **2015**, *67*, 1–94. [CrossRef]
28. Zeng, F.; Liu, Y.; Sun, Y.; Hu, E.; Zhou, Y. Nanoindentation, Nanoscratch, and Nanotensile Testing of Poly (Vinylidene Fluoride)-polyhedral Oligomeric Silsesquioxane Nanocomposites. *J. Polym. Sci. Part B Polym. Phys.* **2012**, *50*, 1597–1611. [CrossRef]
29. Flyagina, I.S.; Mahdi, E.M.; Titov, K.; Tan, J.-C. Thermo-Mechanical Properties of Mixed-Matrix Membranes Encompassing Zeolitic Imidazolate Framework-90 and Polyvinylidine Difluoride: ZIF-90/PVDF Nanocomposites. *APL Mater.* **2017**, *5*, 86104. [CrossRef]
30. Liu, D.; Wang, J.; Peng, W.; Wang, H.; Ren, H. An Organic Dielectric Filler Polyethylene Glycol-Polyaniline Block Copolymer with Low-Density Used in PVDF-Based Composites. *Compos. Sci. Technol.* **2022**, *221*, 109300. [CrossRef]
31. Yuennan, J.; Muensit, N. Preparation and Characterization of Flexible PVDF-HFP Film for Piezoelectric Applications. In Proceedings of the IOP Conference Series: Materials Science and Engineering, Ulaanbaatar, Mongolia, 10–13 September 2020; IOP Publishing: Bristol, UK, 2020; Volume 715, p. 12107.
32. Oliver, W.C.; Pharr, G.M. An Improved Technique for Determining Hardness and Elastic Modulus Using Load and Displacement Sensing Indentation Experiments. *J. Mater. Res.* **1992**, *7*, 1564–1583. [CrossRef]
33. Oliver, W.C.; Pharr, G.M. Measurement of Hardness and Elastic Modulus by Instrumented Indentation: Advances in Understanding and Refinements to Methodology. *J. Mater. Res.* **2004**, *19*, 3. [CrossRef]
34. Shir Mohammadi, M.; Hammerquist, C.; Simonsen, J.; Nairn, J.A. The Fracture Toughness of Polymer Cellulose Nanocomposites Using the Essential Work of Fracture Method. *J. Mater. Sci.* **2016**, *51*, 8916–8927. [CrossRef]
35. Yadav, P.S.; Kumar, V.; Singh, U.P.; Bhat, H.R.; Mazumder, B. Physicochemical Characterization and in Vitro Dissolution Studies of Solid Dispersions of Ketoprofen with PVP K30 and D-Mannitol. *Saudi Pharm. J.* **2013**, *21*, 77–84. [CrossRef] [PubMed]
36. Lehmkemper, K.; Kyeremateng, S.O.; Heinzerling, O.; Degenhardt, M.; Sadowski, G. Impact of Polymer Type and Relative Humidity on the Long-Term Physical Stability of Amorphous Solid Dispersions. *Mol. Pharm.* **2017**, *14*, 4374–4386. [CrossRef] [PubMed]
37. Zhao, Y.; Yang, W.; Zhou, Y.; Chen, Y.; Cao, X.; Yang, Y.; Xu, J.; Jiang, Y. Effect of Crystalline Phase on the Dielectric and Energy Storage Properties of Poly (Vinylidene Fluoride). *J. Mater. Sci. Mater. Electron.* **2016**, *27*, 7280–7286. [CrossRef]
38. Taylor, P.; Furukawa, T. Phase Transitions: A Multinational Ferroelectric Properties of Vinylidene Fluoride Copolymers. *Phase Transit.* **1989**, *18*, 143–211.
39. Basha, S.K.S.; Sundari, G.S.; Kumar, K.V.; Reddy, K.V.B.; Rao, M.C. Electrical Conduction Behaviour of Pvp Based Composite Polymer Electrolytes. *Rasayan J. Chem.* **2017**, *10*, 279–285.
40. Ruan, L.; Yao, X.; Chang, Y.; Zhou, L.; Qin, G.; Zhang, X. Properties and Applications of the β Phase Poly (Vinylidene Fluoride). *Polymers* **2018**, *10*, 228. [CrossRef]
41. Vasic, N.; Steinmetz, J.; Görke, M.; Sinapius, M.; Hühne, C.; Garnweitner, G. Phase Transitions of Polarised PVDF Films in a Standard Curing Process for Composites. *Polymers* **2021**, *13*, 3900. [CrossRef]
42. Yuan, D.; Li, Z.; Thitsartarn, W.; Fan, X.; Sun, J.; Li, H.; He, C. β Phase PVDF-Hfp Induced by Mesoporous SiO_2 Nanorods: Synthesis and Formation Mechanism. *J. Mater. Chem. C* **2015**, *3*, 3708–3713. [CrossRef]
43. Singh, P.; Borkar, H.; Singh, B.P.; Singh, V.N.; Kumar, A. Ferroelectric Polymer-Ceramic Composite Thick Films for Energy Storage Applications. *AIP Adv.* **2014**, *4*, 087117. [CrossRef]
44. Salih, S.I.; Jabur, A.R.; Mohammed, T.A. The Effect of PVP Addition on the Mechanical Properties of Ternary Polymer Blends. In Proceedings of the IOP Conference Series: Materials Science and Engineering, Kerbala, Iraq, 26–27 March 2018; IOP Publishing: Bristol, UK, 2018; Volume 433, p. 12071.

Article

In-Depth Rheological Characterization of Tungsten Sol-Gel Inks for Inkjet Printing

Urša Opara Krašovec [1], Tjaša Vidmar [1], Marta Klanjšek Gunde [2], Romana Cerc Korošec [3] and Lidija Slemenik Perše [4,*]

[1] Faculty of Electrical Engineering, University of Ljubljana, Tržaška 25, 1000 Ljubljana, Slovenia; ursa.opara@fe.uni-lj.si (U.O.K.); tjasavidmar2@gmail.com (T.V.)
[2] National Institute of Chemistry, Hajdrihova 19, 1000 Ljubljana, Slovenia; marta.k.gunde@ki.si
[3] Faculty of Chemistry and Chemical Engineering, University of Ljubljana, Večna pot 113, 1000 Ljubljana, Slovenia; romana.cerc-korosec@fkkt.uni-lj.si
[4] Faculty of Mechanical Engineering, University of Ljubljana, Aškerčeva cesta 6, 1000 Ljubljana, Slovenia
* Correspondence: lidija.slemenik.perse@fs.uni-lj.si

Abstract: The inkjet printing of the functional materials prepared by the sol-gel route is gaining the attention for the production of the variety of the applications not limited to the printed boards, displays, smart labels, smart packaging, sensors and solar cells. However, due to the gelation process associated with the changes from Newtonian to non-Newtonian fluid the inkjet printing of the sol-gel inks is extremely complex. In this study we reveal in-depth rheological characterization of the WO_3 sols in which we simulate the conditions of the inkjet printing process at different temperature of the cartridge (20–60 °C) by analyzing the structural and rheological changes taking place during the gelation of the tungsten oxide (WO_3) ink. The results provide the information on the stability of the sol and a better insight on the effects of the temperature on the gelation time. Moreover, the information on the temperature and the time window at which the inkjet printing of the sol-gel inks could be performed without clogging were obtained. The WO_3 ink was stable in a beaker and exhibited Newtonian flow behavior at room temperature over 3 weeks, while the gelation time decreased exponentially with increasing temperature down to 0.55 h at 60 °C.

Keywords: rheology; inkjet printing; tungsten oxide; sol-gel

Citation: Opara Krašovec, U.; Vidmar, T.; Klanjšek Gunde, M.; Cerc Korošec, R.; Slemenik Perše, L. In-Depth Rheological Characterization of Tungsten Sol-Gel Inks for Inkjet Printing. *Coatings* **2022**, *12*, 112. https://doi.org/10.3390/coatings12020112

Academic Editor: Michelina Catauro

Received: 30 December 2021
Accepted: 17 January 2022
Published: 19 January 2022

Publisher's Note: MDPI stays neutral with regard to jurisdictional claims in published maps and institutional affiliations.

Copyright: © 2022 by the authors. Licensee MDPI, Basel, Switzerland. This article is an open access article distributed under the terms and conditions of the Creative Commons Attribution (CC BY) license (https:// creativecommons.org/licenses/by/ 4.0/).

1. Introduction

Functional materials are used in wide-ranging industrial fields including, but not limited to solar cells, energy storage devices, displays, smart windows, catalysts for chemical reactions and sensors [1]. Among them the tungsten oxide (WO_3) represents an inorganic transition metal oxide with chromogenic and semiconductor properties, which enable its applicability in numerous applications mentioned above [2–4]. The WO_3 layers can be produced with complex and expensive techniques (physical vapor deposition—PVD, chemical vapor deposition—CVD and electrodeposition) or by less expensive application from a solution phase (sol-gel, mixture of powder and different solvents) [5]. A variety of the sol-gel chromogenic devices enabling optical modulation of the interior light in the buildings [6,7], such as electrochromic [7,8], photoelectrochromic [9,10], photochromic [11–13] and gasochromic [14] have been prepared with the dip-coated WO_3 layers from the corresponding sols. This publication focuses on the in-depth study of the rheological characteristics of the sol-gel derived WO_3 inks suitable for inkjet printing [15].

The sol-gel process enables the fabrication of the variety of the functional materials [16]. The process involves the conversion of small molecules (monomers) into a colloidal solution (sol) that transforms to an integrated network (gel) in which the solvent is trapped. Further drying of the gel leads to the formation of a solvent free xerogel while annealing of the xerogel results in the powder. On the other hand, the printing is becoming more and more linked with the applications such as printed boards, displays, smart labels, smart packaging, and various printed electronic such as different sensors, solar cells, etc., which all require the availability of the functional materials in ink or paste form required for the planned printing technique [17–20]. Therefore, nowadays, functional sol-gel materials are entering the printing production, but the area is still new and underexplored. Every ink or paste should exhibit proper rheological properties, which are essential for every specific deposition process. The inkjet printing requires low viscosity of the inks (1–30 mPa.s) and surface tension around 30 mN/m, the ink typically contains a complex mixture of many solvents with high and low boiling points [18,20]. This allows proper drop generation, avoids nozzle clogging and provides optimal printing results. However, due to irreversible behavior of sol-gel material, inkjet printing of functional sol-gel materials is even more complex and requires a careful gelation control of inks, with necessary fine tuning of their rheological properties [15].

In 1997, Atkinson and coworkers reported one of the first examples of inkjet printouts using sol-gel inks [21]. Since then, the research interest in using sol-gel materials for inkjet printing production of metal oxides has increased, which is demonstrated by numerous scientific publications over the past decade [19,22–25]. Furthermore, the printability of the tungsten sol using inkjet printing has been successfully demonstrated [15]. The tungsten oxide—WO_3 printouts with the thickness of around 300 nm were realized of very good optical quality and enabled the realization of the electrochromic devices [15]. The tungsten sol used as functional ink has been modified by using 2-propoxy propanol to match the inkjet printing requirements for proper drop generation (jetting characteristics), smooth ink transfer through printer and uniformity of deposited films [15].

Moreover, inkjet printing has complex drying behavior, therefore an appropriative solvent system, temperature modulation of printer vacuum plates (up to 60 °C) and also cartridge temperature (up to 70 °C) should be carefully chosen for each individual system and substrate in order to form uniform deposited films [18,20]. From this perspective, it is of paramount importance to perform a rheological study of inks by simulating the conditions in printing process. An in-depth study of the sol rheology could enable better insight of the sol-gel inks limitations as well as the control of the sol-gel material stability (sol-gel transition) which is required for continuous inkjet printing without clogging. The inks should have Newtonian behavior, which means that viscosity of fluid or ink is constant with applied shear rates. In the case of the sol-gel inks the transition of the sol to gel occurs therefore the ink changes from Newtonian to non-Newtonian fluid. In this regard printing of functional sol-gel materials is even more complex and requires a careful gelation control of the ink, with necessary understanding of its rheological properties.

To our knowledge, we are the first to reveal in-depth rheological characterization of the WO_3 sols for inkjet printing in which we simulate the conditions of the printing process. Publication by Karimi-Nazarabad et al. [26] describes the rheological properties of nanofluids of tungsten oxide nanoparticles in ethylene glycol and glycerol. The authors don't report on practical usage of studied nanofluids, neither on the rheological characteristics of the gelation process of the samples. Moreover, D. Tripkovic et al. [27] demonstrated tailoring of $BaTiO_4$ sol-gel inks for inkjet printing. The most relevant publication [28], describes the study of rheological properties of TiO_2 sol for direct write assembly in planar and 3D configuration.

The aim of this study is to characterize the sol to gel transition of the WO_3 inks at various temperatures that the ink could be exposed to during the inkjet printing. The results were obtained by coupling two measurement techniques, IR spectroscopy and rheological characterization. Rheological study enabled the insight into the gelation process, while the IR spectroscopy shed the light on the changes of the chemical structure of the WO_3 sols taking place during the transformation of the sol to the gel.

2. Materials and Methods

2.1. Preparation of the WO₃ Sol

Step one involved the synthesis of a WO$_3$ sol. First, peroxo-tungstic acid (PTA) was synthesized by reacting 5 g of tungsten monocrystalline powder (mean particle size < 1 micron, purity 99.95%, Sigma-Aldrich GmbH, Schneldorf, Germany) with 20 mL of hydrogen peroxide (30%, Belinka). This reaction is strongly exothermic. Sols where then prepared by adding solvent to the PTA solution at 120 °C. The addition of alcohol resulted in the formation of the W-ether that polymerizes to peroxopolytungstic acid (P-PTA) [29]. Solvent for inkjet printing ink should have boiling point higher than 100 °C, therefore WO$_3$ sol used in this study as ink was prepared with a mixture of two solvents with different boiling points, i.e., 2-propanol (ACS reagent, Sigma-Aldrich GmbH, Schneldorf, Germany) and 2-propoxy ethanol (Puriss, Sigma-Aldrich, Schneldorf, Germany). The ink appeared slightly orange and contained 27.7 mmol of tungsten per 30 mL of sol. Figure 1 presents the WO$_3$ sol and gel state—Figure 1a, the WO$_3$ xerogel dried at RT—Figure 1b and the WO$_3$ powder annealed for 1 h at 450 °C—Figure 1c. More details on the preparation of the sols could, also the ones based solely on ethanol and 2-propanol could be found in Ref [15].

 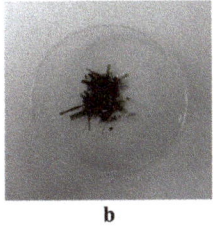 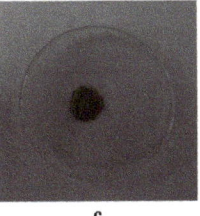

Figure 1. WO$_3$ sol and gel: (**a**), xerogel (**b**) and powder (**c**).

2.2. Thermogravimmetric Analysis of the WO₃ Xerogles

Thermogravimetric (TG) measurements of WO$_3$ sols, prepared with different solvents, were performed using Mettler Toledo TGA/DSC1 instrument (Mettler Toledo, Schwerzenbach, Switzerland). Samples were heated from room temperature to 500 °C with the hating rate of 5 K min^{-1} under the dynamic air flow (100 mL min^{-1}). 75 µL of the prepared sol was pipetted into 150 µL platinum crucible and dried under air for 24 h. After drying, masses of the samples used for analyses were around 15 mg. The blank curve was subtracted.

Dynamic TG curves of all three samples are shown in Figure 2. The course of thermal decomposition is similar for xerogels B and C, where ethanol and 2-propanol were used as a solvent while the mass-loss curve has a different course in the case of xerogel A, where the solvent was a mixture of 2-propoxy ethanol and 2-propanol. We ascribe the first step, which takes place from room temperature to approximately 200 °C, to solvent evaporation. In this step mass-loss rate is higher in the case of xerogel B and C. Up to 185 °C, xerogel B losses 6.0% of the initial mass, while xerogel C 8.10% and xerogel A 4.5%, respectively. The second and the third step, where condensation reactions continue, are partially overlapped. For xerogel B mass loss is additional 3.0% in a temperature range from 270 °C to 350 °C and from 350 °C to 430 °C another 0.5%. Similar behavior is observed for xerogel C except that the second step begins at higher temperature, i.e., 325 °C. Successive reactions are much more overlapped in the case of xerogel A. Second step occurs in a temperature range from 185 °C to approximately 270 °C with a mass loss of 3.8% and turns then to the third step with a slower mass loss. Mass for the sample A is not constant even at 500 °C.

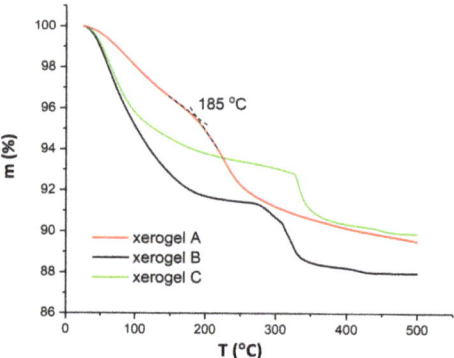

Figure 2. TG measurement of the WO$_3$ xerogels prepared with different solvents: xerogel A with 2-propoxy ethanol and 2-propanol; xerogel B with ethanol and xerogel C with 2-propanol.

2.3. XRD Analysis of the WO$_3$ Xerogles and Powders

The XRD measurements of the xerogels and powders were performed using a PW 1710 Philips X-ray diffractometer (Philips, Almelo, Netherlands). The XRD spectra of the WO$_3$ xerogels prepared by either a mixture of 2-propoxy ethanol and 2-propanol or solely with 2-propanol are presented in Figure 3a, while the corresponding powders obtained after annealing of the xerogel at 450 °C for 1 h are shown in Figure 3b. The results confirmed that both WO$_3$ xerogels, regardless on the solvent used for the sol preparation are amorphous while annealing of the xerogels leads to the crystallization. The analysis of the XRD spectra of the WO$_3$ powders reveals the presence of the monoclinic phase, which is well in agreement with our previous results for the sample C [12].

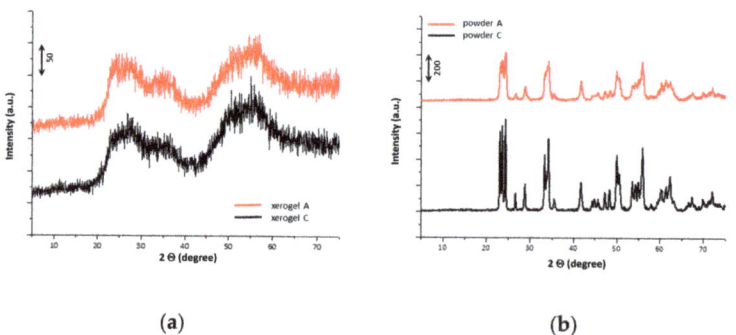

(a)　　　　　　　　　　　　(b)

Figure 3. XRD patterns of the WO$_3$ xerogel (a) and corresponding powders obtained after annealing of the xerogels for 1 h at 450 °C (b). Xerogel A and powder A denotes WO$_3$ prepared with 2-propoxy ethanol and 2-propanol. Xerogel C and powder C denotes WO$_3$ prepared with 2-propanol.

2.4. Rheological Characterization

Rheological measurements were performed with a rotational controlled rate rheometer (Physica MCR302, Anton Paar, Graz, Austria), equipped with a cone and plate sensor system (CP 50/2°). Temperature of the measurements was controlled with a Peltier HOOD (Anton Paar, Graz, Austria).

All samples were tested under rotational and oscillatory shear conditions. Rotational flow tests were performed with a triangular method by changing the shear rate from 0–1000–0 s^{-1}. Oscillatory stress sweep tests at constant frequency of oscillation (1 Hz) were used to determine the linear viscoelastic range (LVR). Frequency tests were performed at constant small deformation in LVR by decreasing the frequency from 20–0.01 Hz.

2.5. IR Spectroscopic Measurements

The IR spectra of the sols, xerogels and gels have been taken by FT-IR Perkin Elmer System 2000 spectrometer (PerkinElmer, Waltham, MA, USA). The samples have been deposited as thin layers on the double side polished Si-resin.

3. Results and Discussion

3.1. Structural Analysis of the Sols and Gels

A characteristic of a WO_3 sol prepared by the peroxo sol-gel route is the presence of the peroxopolytungstic acid (P-PTA) clusters. The structure of the P-PTA was assessed in 1991 by Nanba et al. [29]. It is complex and consists of two edge-sharing 3-membered (W_3O_{13}) rings, located above and below the corner-shared 6-membered (WO_7 pentagonal bipyramids) ring. The contact between these species is established via the H-bonds of water placed between them. The P-PTA structures transform during the sol preparation, drying and gelation process to the network of the tungsten polyhedra (WO_6) connected via corners and edges. There are numerous factors influencing this transformation, among them are also the temperature and the alcohol used for the sol preparation.

For the IR spectrum of the WO_3 sols a variety of the W-O bond oscillations are characteristic and the bands could be assigned to: the terminal bond, i.e., double bond between tungsten and oxygen, $\nu(W=O)$ at 980 cm^{-1}, single W-O bond of tungsten polyhedra connected via corners, $\nu(W-O-W)$ at 630–650 cm^{-1} or via the edges, $\nu(W-O-W)$ at 700–720 cm^{-1}. In addition, the peroxo bonds are characterized by the absorption peaks of the W-O-O-W and W-O-O oscillations at 800–830 cm^{-1} and 560 cm^{-1}, respectively. The presence of water in the structure is evident from the broad band in the range 3000–3500 cm^{-1} (ν(O-H)) and a band at 1630 cm^{-1} ($\delta(H_2O)$) [30–33].

Figure 4 shows the IR spectra of the freshly deposited WO_3 sols based on different alcohols, the xerogel of the sol that was dried in a recording chamber of the FTIR spectrophotometer and the WO_3 gel. For the comparison of the influence of different solvents on the structure of the sols and gel formation we have analyzed beside the printable WO_3 ink prepared by 2-propoxy ethanol (Figure 4a) also the IR spectra of the WO_3 sols based on ethanol (Figure 4b) or 2-propanol (Figure 4c). It should be mentioned, that the WO_3—ethanol and WO_3—2-propanol sols have been found inappropriate for the inkjet printing [15] but are very suitable for dip-coating deposition of the active transparent WO_3 layers used in chromogenic devices [7,9–12,14].

To compare how the temperature of the WO_3 ink influences the cross-linking of the tungsten polyhedra and gel formation we have taken the IR spectra after the gelation of the WO_3 sol in a rheometer at the temperatures (20 °C, 30 °C, 40 °C, 50 °C and 60 °C) at which the rheological properties of the sols were examined (Figure 5).

In the IR spectra of the wet sols prepared by different solvents (Figure 4) the bands characteristic for the –CH groups (2840–3000 cm^{-1}) of the alcohol are present. The comparison of the IR spectra of the wet and dried sols-xerogels shows a noticeable difference in the intensity of the –CH bands. The results confirm that the alcohol entirely evaporates during drying of the sols at room temperature when the sols are prepared by ethanol or 2-propanol (Figure 4b,c), while it remains present when the 2-propoxy ethanol (Figure 4a) is used.

The IR spectra of the sols (Figure 4) show the presence of all the peaks characteristic for the P-PTA structure. However, the intensity of the peaks, characteristic for the crosslinking of the tungsten polyhedral, differs among the studied sols. The highest intensity of the band typical for peroxo groups (810 cm^{-1} and 560 cm^{-1}) is characteristic for the fresh sol and the sol dried at RT (xerogel) prepared by 2-propoxy-ethanol (Figure 4a) which still contains some alcohol. While the intensity of the peroxo groups in the xerogel is much smaller in the case of ethanol and 2-propanol based WO_3 xerogel that are solvent (alcohol) free. From these we conclude that the decomposition of the peroxo groups present in the P-PTA structure is associated with drying and evaporation of the solvent from the sol during the xerogel formation. The evaporation process is fast and complete for the sols prepared by ethanol and 2-propanol (Figure 4b,c). A slower decomposition of the peroxo groups has been found in the case of the sol containing 2-propoxy-ethanol (Figure 4a). The slower evaporation could have resulted also in slower gelation process, but the rheological studies showed the opposite. The rheological studies showed that the gelation is even faster in case of the WO$_3$ sol prepared by 2-propoxy-ethanol. The reason for the faster gelation can be attributed to a different way of cross-linking of the WO$_3$ polyhedra when different alcohols are used for the sol preparation which has been confirmed by the IR spectra analysis.

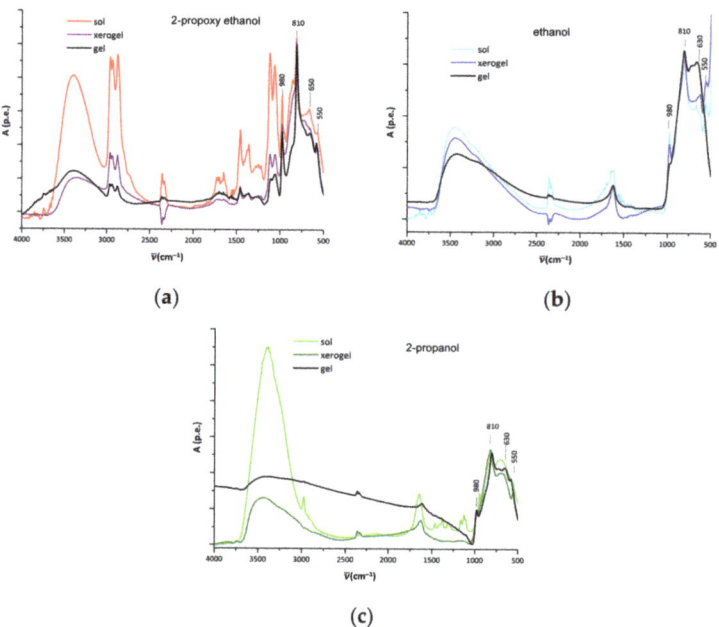

Figure 4. The IR spectra of the WO$_3$ sol, xerogel and gel prepared by different alcohols, (**a**) 2-propoxy ethanol, (**b**) ethanol and (**c**) 2-propanol.

In addition to the IR spectra of dried sols, xerogel the spectra of the gels (Figure 4) confirmed that only in the case of the sol containing the 2-propoxy ethanol the alcohol remains trapped in the gel structure (Figure 4a). The results demonstrate a strong influence of the solvent on the cross-linking of the tungsten polyhedra. The analysis of the IR spectra of the gels prepared by ethanol and by 2-propanol reveals that the intensity of the band attributed to the terminal W = O bond (980 cm^{-1}) strongly decreases during gelation, while the skeletal W-O bonds typical for the connection of the WO$_6$ polyhedra at 630–650 cm^{-1} and 700–720 cm^{-1} increase in the intensity (Figure 4b,c). The results confirm also the lowest intensity of the terminal W = O bonds for the gels prepared with ethanol (Figure 4b) which implies that the strongest cross-linking of the tungsten polyhedra took place in the gel formed from the ethanol-based sol.

On the other hand, in the IR spectra of the gel formed from the sol based on 2-propoxy ethanol a peak characteristic for the terminal double W = O bond remains intensive as well as the bands characteristic for the peroxo groups (Figure 4a). This leads to the conclusion that in the WO$_3$ gel based on 2-propoxy ethanol the P-PTA structure remains present to some degree, while the cross-linking of the tungsten polyhedra is hindered. In addition, the results confirmed the presence of the –CH bonds (peak at 2840–3000 cm^{-1}) characteristic for the 2-propoxy ethanol meaning that the solvent remained trapped in the cross-linked WO$_3$ gel structure (Figure 4a).

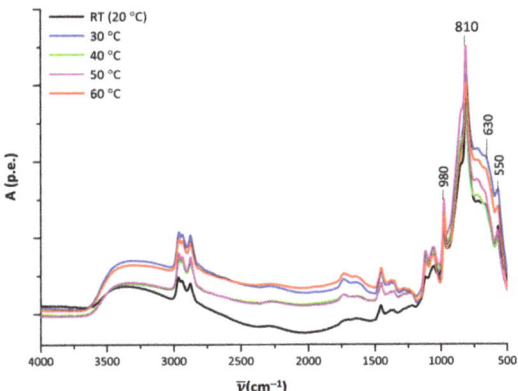

Figure 5. The IR spectra taken after the gelation of the WO$_3$ sol (2-propoxy ethanol) in the rheometer at the temperatures (20 °C, 30 °C, 40 °C, 50 °C and 60 °C) at which the rheological properties of the sols were examined.

To summarize, the IR spectra analysis of the gels showed that the alcohol used for the WO$_3$ sol preparation strongly influences the cross-linking process taking place in the gel formation. The WO$_3$ gel based on 2-propoxy ethanol has the solvent kept in the gel structure, the structure of the P-PTA is to some extent preserved and the bonding of the tungsten polyhedra is not complete which results in a weaker and softer WO$_3$ gels compared to the gels formed from the ethanol and 2-propanol tungsten sols (Figure 4).

In a further IR analysis, we followed the gelation of the WO$_3$ sol based on 2-propoxy ethanol at different temperatures that the WO$_3$ ink could be exposed during the inkjet processing while adjusting the printing parameters. The IR spectra of the WO$_3$ gels were obtained after the gelation of the WO$_3$ sol in a rheometer at 20, 30, 40, 50 and 60 °C (Figure 5). In the IR spectra of the gel a high intensity peaks characteristic for the peroxo bonds (peaks at 810 and 550 cm^{-1}) were observed regardless of the temperature at which gelation took place. In addition, no significant difference of the skeletal W-O modes typical for corner or edge shared tungsten polyhedra has been noticed that might suggest different cross-linking mechanisms taking place at different temperature during the gel formation. Moreover, regardless of the temperature at which the gel has been formed the solvent, 2-propoxy ethanol remains trapped in the WO$_3$ gel structure (Figure 5). The IR analysis shows no significant influence of the temperature, in the range between 20 and 60 °C, on the chemical structure of the WO$_3$ gel.

3.2. Viscosity of WO$_3$ Sols

At all temperatures examined (20, 30, 40, 50 and 60 °C), the prepared WO$_3$ sol exhibited Newtonian viscosity, which exponentially decreased with increasing temperature (Figure 6). However, after gelation process was finished, the viscosity of all gels, formed at different temperatures, was in the same range, regardless of the temperature at which the gelation took place. The formed gels exhibited similar complex viscosity with much higher values (η *~400 Pa.s) compared to the initial sols (η *~0.001–0.01 Pa.s) (Figure 6). This is well in agreement with the results of the IR analysis of the gels taken after the rheology study at different temperature (Figure 5). The dependencies of the viscosities on the temperature were for the initial sols as well as for the formed gels expressed by Arrhenius model:

$$\eta = \eta_0 e^{\left(\frac{E}{RT}\right)} \quad (1)$$

where T is temperature, η_0 is a material constant, E is the activation energy and R is the universal gas constant [34].

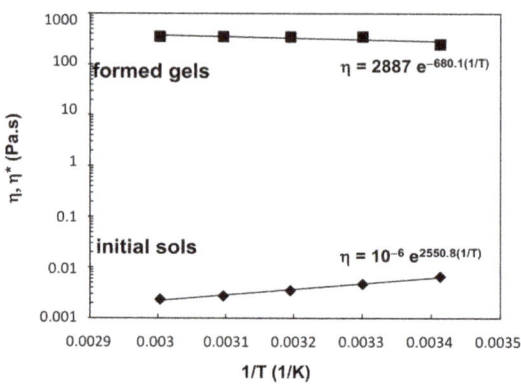

Figure 6. Temperature dependence of the dynamic viscosity of freshly prepared WO_3 sol and complex viscosity of gels, formed at different temperatures.

3.3. In Situ Rheological Characterization of WO_3 Gelation Process

A detailed insight into the progress of the rheological properties of WO_3 sols during gelation process was obtained with in situ oscillatory test in the linear viscoelastic range at deformation so small that no destruction of the structure could occur. For inkjet printing we used material deposit system—Dimatix Materials Printer DMP2800 [20]. It enables temperature modulation of printer vacuum plates (up to 60 °C) and cartridge temperature (up to 70 °C), which should be tailored for each material or ink. For inkjet printing of the sol-gel WO_3 ink the most optimal temperature of printer plates was 35 °C, which warmed ink in cartridge up to 26 °C. Moreover, to predict stability of inkjet process we studied temperature dependence of the WO_3 sol-gel characteristics. In situ oscillatory test in the linear viscoelastic range was used for 5 different temperatures: 20 °C, 30 °C, 40 °C, 50 °C and 60 °C, respectively. The progress of viscoelastic properties, i.e., elastic G' and viscous modulus G'', during gelation process of WO_3 sol was similar, especially for the temperatures from 30 °C to 60 °C (Figure 7a). Initial sols exhibited "liquid-like" behavior with low consistency and only viscous contribution G'' detected. As the gelation process started the loss modulus G'' commenced to increase continuously, while the storage modulus G' suddenly appeared after a certain time and rose sharply until it intersected and exceeded the loss modulus G''. The time at which the both moduli reached the same value indicated the sol to gel transition point, which is often referred to as a "gel-point" [34]. From this point onward, the elastic behavior G' dominated and the behavior of the sample became "solid-like". Both moduli leveled off as the reaction came to completion; moreover, at the end of gelation the viscous modulus was too low to be detected. As a result, the formed gel became brittle without any viscous effects. The gels, formed at 30 °C to 60 °C exhibited similar values of G' and G'' at the end of sol-gel process, while the gel, formed at 20 °C exhibited softer gel structure with lower values of G' and G''; moreover, the values of G'' were not negligible. We can conclude that the gelation time depended strongly on the temperature, to which the sol was exposed. Higher temperature leaded to faster gelation process, i.e., for the sol, which was exposed to 60 °C (Figure 7a), the gel formed after ~0.55 h, while the transition from sol to gel at 20 °C was observed not earlier as after

~10 h (Figure 7b). The final values of G' were for the gels, formed at temperatures from 30 °to 60 °C in the range of 3000 Pa, while G'' was negligible. On the other hand, the gel, formed at 20 °C exhibited the G' below 1000 Pa, while the G'' was in the range of 100 Pa.

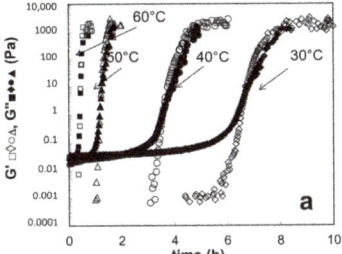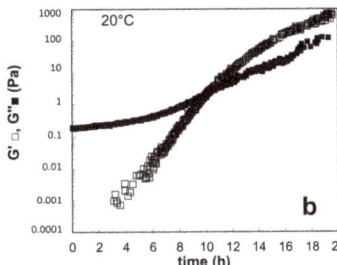

Figure 7. Dependence of dynamic moduli G' and G'' on time of the WO$_3$ gelation process at different temperatures: (**a**) 30 °C, 40 °C, 50 °C and (**b**) 20 °C. The tests were performed in linear viscoelastic range. Solid symbols represent viscous modulus G'', hollow symbols represent elastic modulus G'.

To evaluate different parameters, important for the sol-gel process, the experimental asymmetric dependences of phase shift angle δ on the time of gelation could be the best fitted with five-parameter logistic function [35]:

$$\delta = \delta_1 - \frac{(\delta_1 - \delta_2)}{\left(1 + \left(\frac{t}{t_g}\right)^{-W}\right)^s} \tag{2}$$

where δ_1 and δ_2 are the highest (initial) and the lowest (the end) values of phase shift angle, respectively, t_g is the time of sol-gel transition, while s jointly with W controls the rate of approach to the δ_2 asymptote. The values of the five parameters, obtained by fitting the experimental data are presented in Table 1, while the experimental data of phase shift angle δ and calculated values obtained with the above equation are presented in Figure 8. The results show excellent agreement of the experimental data with the predicted values for all five temperatures, used in the presented study. The five-parameter logistic model enabled the exact determination of the time at which sol to gel transition occurs (t_g) together with precise viscoelastic properties of formed gels (δ_{min}). Moreover, the model can be used for accurate prediction of total rate and time of sol-gel process (W, s).

Table 1. The parameters of the five-parameter logistic model (Equation (1)) for the gelation process at different temperatures.

	20 °C	30 °C	40 °C	50 °C	60 °C
δ_{max}	90.0	90.0	90.0	90.0	90.0
δ_{min1}	7.4	0.0028	0.0028	0.0023	0.0027
t_g (h)	10.06	6.2	3.1	1.5	0.55
W	7.8	26.4	30.18	24.73	18.56
s	0.92	1.43	1.32	1.09	1.06

Gelation times (t_g), determined from the Equation (1) are organized as a dependence of the temperature (Table 1), to which the initial sol was exposed. The dependence (Figure 9) is very good fitted with the exponential function, which shows that the time of gelation exponentially decreased with increasing temperature of the gelation process.

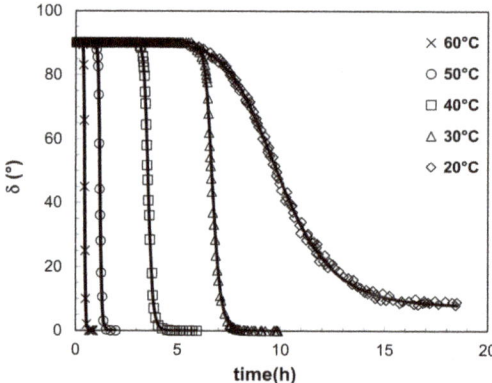

Figure 8. Changing of phase angle d with time of gelation process at different temperatures. Dependence of dynamic moduli G′ and G″ on time of the WO$_3$ gelation at different temperatures.

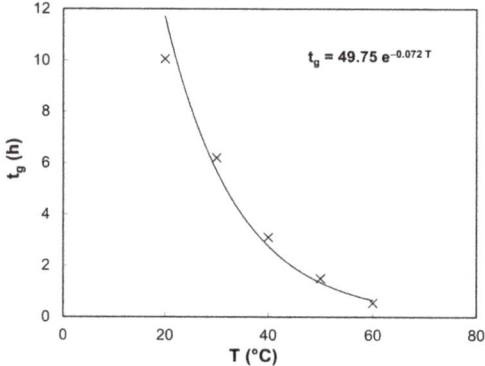

Figure 9. Dependence of gelation time on temperature of the gelation process of the WO$_3$ ink.

3.4. Ex-Situ Rheological Characterization of WO$_3$ Gelation Process

In addition to the in-situ rheological characterization the ex-situ rheological tests were performed for the sol, exposed to the room temperature for several days. Moreover, we followed the sol stability, prepared with different solvents (ethanol, isopropanol and a mixture of isopropanol and 2-propoxy ethanol) stored in a 100 mL glass bottle in the refrigerator. Our observation was that sol-gel transition occurs in 22 days, when using the mixture of isopropanol and 2-propoxy ethanol, 6 months, when using isopropanol and 10 months when using ethanol. It should be noted here that the transition from sol to gel in the bulk of the beaker (sample volume ~100 mL) occurred much later compared to the transition, which occurred in the sensor system of the rheometer and inkjet cartridge content, where the whole volume of the sample was considerably lower (cca 1.6 mL). The stability of the sols could be significantly prolonged by keeping the inks at lower temperature, for example when kept in freezer (below −15 °C) the sols remain stable over 1 year.

In the ex-situ characterization various rheological tests were performed under destructive and non-destructive shear conditions. First, the flow behavior was followed with flow tests under destructive shear conditions (Figure 10). One day after the sol preparation, the sol (Figure 10, initial) exhibited Newtonian flow behavior with constant viscosity of 0.0054 Pa.s (at T = 23 °C). 14 days after the preparation the sol maintained the Newtonian character, while the value of the viscosity slightly increased to 0.0065 Pa.s (at T = 23 °C).

Higher increase of the viscosity and the first deviation from Newtonian flow behavior was observed on the 21st day after the preparation. The next day (22nd day) the viscosity again slightly increased with similar flow behavior, while 24th day after the preparation the flow behavior of the solution significantly changed to pronounced shear thinning flow behavior, where the viscosity decreased for almost three decades as the shear rate increased from 0.1 to 1000 s^{-1}. A constant decreasing of the viscosity curve from the low to the high shear rates indicates that these solutions showed some structure stability at rest [34], while hysteresis loop during the decreasing of the shear rate towards initial value (0.1 s^{-1}) indicates time-dependent flow properties during structure recovery.

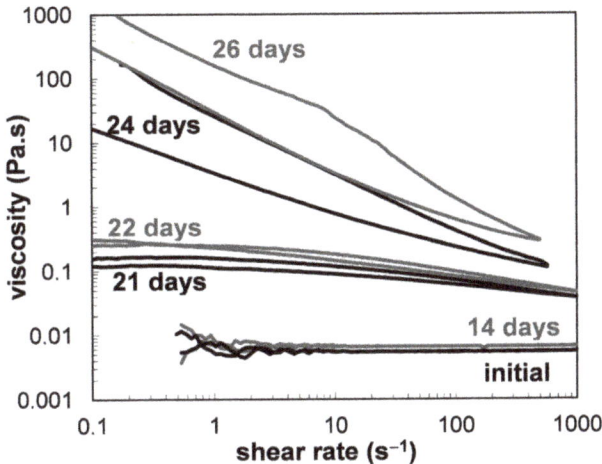

Figure 10. Flow curves (the dependence of viscosity on shear rate) for WO_3 sols at T = 23 °C.

The rheological properties of WO_3 sols were, at different times during the gelation process, followed also with non-destructive oscillation tests in the range of linear viscoelastic response (Figure 11). At the beginning, immediately after the preparation, the sol exhibited Newtonian flow behavior as only viscous contribution G'' was present, which linearly increased with the frequency of oscillation. As it was observed during flow tests, at 21st day some changes were observed also with frequency tests. Newtonian character of the sol changes to viscoelastic liquid as the dynamic storage modulus G' occurred with lower values compared to the loss modulus G''; moreover, the moduli exhibited similar, linear dependences with frequency of oscillation. The next day (22nd day) the values of the moduli increased and their dependences on the frequency changed. The G' equaled G'', and both moduli depended on the frequency of oscillation by the order of 0.45. Such behavior is characteristic for weak gels, which resemble the strong gels in their mechanical behavior, particularly at low frequencies, but as the deformation increases, their networks undergo a progressive breakdown into smaller clusters. As a consequence, the system can flow with flow properties typical of a disperse system [36]. Gel was formed after 26 days, when the sample exhibited "solid-like" behavior with much higher values of storage modulus G' compared to the loss modulus G''; moreover, the moduli were frequency independent, i.e., $G' \sim G'' \sim \omega_0$. Such behavior is characteristic for strong gels [37], which are usually, due to lack of viscous contribution, also very hard and brittle. Under the conditions of small deformation, strong gels manifest the typical behavior of viscoelastic solids and, above a critical deformation value, they rupture rather than flow [37]. At other temperatures examined, the sol to gel transition was similar to the one, explained in Figure 11; except that the time of gelation process was shorter.

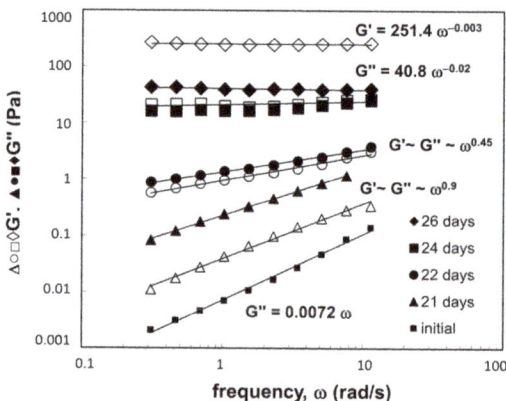

Figure 11. Dependence of dynamic moduli G′ and G″ on the frequency of oscillation in the range of linear viscoelastic response of the WO_3 ink. Solid symbols represent viscous modulus G″, hollow symbols represent elastic modulus G′.

For chemical, i.e., covalently cross-linked gels formed in our study, linear viscoelastic behavior has been extensively investigated in the vicinity of the sol-gel transition point. Winter and Chambon [38,39] showed that at critical gel point dynamic moduli follow a simple power law: $G'(\omega) \approx G''(\omega) \sim \omega_n$, where n depends on the particular gelation mechanism. In our work, the transition from sol to gel was observed for the sol at 23 °C (Figure 11), where after 22 days the G′ and G″ depended on the frequency by the order of 0.5 ($G' \sim G'' \sim \omega^{0.5}$).

At the end of gelation process the initial Newtonian structure of the sol changed to solid-like gel at all temperatures examined. The formed gels were examined with oscillatory tests in linear viscoelastic range to evaluate the influence of the temperature on the structure of formed gels. For the sake of clarity only gels, formed at temperatures 60 °C, 40 °C and 20 °C, respectively, are presented in Figure 12. The results indicate that all formed gels exhibited similar dependences of G′ and G″ on the frequency of oscillation. The differences in the structure can be observed regarding consistency of the gels. Higher temperature of the gelation process led to faster gelation process and higher consistency of formed gel. Thus, the gel, formed at the highest temperature exhibited the highest consistency with the most "solid-like" structure, while the gel, formed at 20 °C, exhibited softer gel structure with lower consistency.

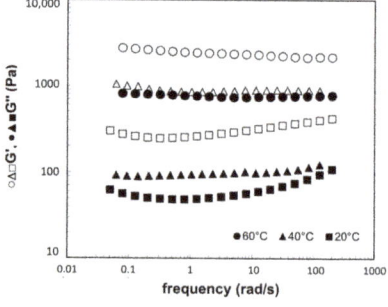

Figure 12. Dependence of dynamic moduli G′ and G″ on the frequency of oscillation in the range of linear viscoelastic response for gels, formed at 60 °C, 40 °C and 20 °C, respectively. Solid symbols represent viscous modulus G″, hollow symbols represent elastic modulus G′.

4. Conclusions

The IR analysis of the WO$_3$ sols showed the difference of the cross-linking of the peroxopolytungstic acid (P-PTA) clusters for the sols based on different alcohols. The most pronounced cross-linking of the tungsten polyhedral is found for the ethanol, followed by the 2-propanol based sol. In the WO$_3$ sol based on 2-propoxy ethanol the cross-linking of the P-PTA is hindered which is demonstrated by the strong presence of terminal W = O bonds and the peroxo groups in the IR spectra of the corresponding sol. The drying of the 2-propoxy ethanol based WO$_3$ sol at room temperature was not complete as is the case for ethanol and 2-propanol. The slower evaporation could have slowed down the gelation process, but the rheological studies show that the gelation was even faster in case of the WO$_3$ sol prepared by 2-propoxy-ethanol. The reason for the faster gelation can be attributed to a different way of cross-linking of the tungsten polyhedra through corners and edges when different alcohols are used. The IR spectra analysis of the gels shows that the WO$_3$ gel based on 2-propoxy ethanol has the solvent kept in the gel structure, the structure of the P-PTA clusters is to some extent preserved and the cross-linking of the tungsten polyhedra is not complete which results in a weaker and softer WO$_3$ gel compared to the gels formed from the ethanol and 2-propanol tungsten sols.

The IR analysis of the 2-propoxy ethanol WO$_3$ gels performed after the gelation of the WO$_3$ sol in a rheometer at 20 °C, 30 °C, 40 °C, 50 °C and 60 °C showed no significant difference on the chemical structure of the WO$_3$ gel which is well in accordance with the rheological studies confirming similar progress of viscoelastic properties, i.e., elastic G' and viscous modulus G'', during gelation process of WO$_3$ sol. Initial sols exhibit "liquid-like" behavior with low consistency and only viscous contribution G'' while during gelation process the elastic modulus G' suddenly appeared after a certain time and rise sharply until it intersected and exceeded the loss modulus G''. The gelation time decreases exponentially with the temperature, the gelation of the sol exposed to 60 °C was completed in 0.55 h, while the transition from sol to gel at 20 °C was observed not earlier as after ~10 h. To evaluate the parameters, important for the sol-gel process, the experimental asymmetric dependences of phase shift angle δ on the time of gelation were fitted with five-parameter logistic function [35]. The results show excellent agreement of the experimental data with the predicted values for all five temperatures which enabled the exact determination of the time at which sol to gel transition occurred (t_g), the precise viscoelastic properties of formed gels (δ_{min}) and the accurate prediction of total rate and time of sol-gel process (W, s).

The ex-situ rheological characterization of WO$_3$ sols prepared with different alcohols was performed to study the stability of the sol during ageing at RT. The results show that the fastest gelation for the 2-propoxy ethanol-based sol (22 days), while the 2-propanol and ethanol remain stable up to 6 and 10 months, respectively. The fresh sol (1 day after preparation) exhibited Newtonian flow behavior with constant viscosity of 0.0054 Pa.s, after 14 days the sol maintained the Newtonian character, while the viscosity slightly increased to 0.0065 Pa.s. The first deviation from Newtonian flow behavior was observed on the 21st day after the preparation. On the 22nd day the G' equaled G'', and both moduli depend on the frequency of oscillation by the order of 0.45 which is characteristic for weak gels. Gel was formed after 26 days, when the sample exhibited "solid-like" behavior with much higher values of storage modulus G' compared to the loss modulus G''; moreover, the moduli were frequency independent, i.e., G'~G''~ω_0. Such behavior is characteristic for strong gels [37], which are usually, due to lack of viscous contribution, also very hard and brittle.

In overall the results of this study confirmed that in-depth rheological characterization linked with the IR spectroscopy of the sol-gel inks could provide the information on the stability of the sol and a better insight on how the temperature influences the gelation time. It provides the information on the temperature and the time window at which the continuous inkjet printing of the sol-gel inks could be performed without clogging. The WO$_3$ ink was stable in a beaker and have Newtonian flow behavior at room temperature over 3 weeks, while the gelation time decreases exponentially with the temperature down to 0.55 h at 60 °C.

Author Contributions: Conceptualization, U.O.K. and L.S.P.; methodology, T.V. and M.K.G.; formal analysis, U.O.K., L.S.P., T.V., R.C.K. and M.K.G.; investigation, U.O.K., L.S.P., T.V., R.C.K. and M.K.G.; writing—original draft preparation, U.O.K. and L.S.P.; writing—review and editing, U.O.K., L.S.P., T.V., R.C.K. and M.K.G.; supervision, L.S.P.; project administration, L.S.P. All authors have read and agreed to the published version of the manuscript.

Funding: The authors acknowledge the financial support from the Slovenian Research Agency (research core funding No. P2-0264, P2-0244 and P2-0393).

Institutional Review Board Statement: Not applicable.

Informed Consent Statement: Not applicable.

Data Availability Statement: The data presented in this study are available on request from the corresponding author.

Conflicts of Interest: The authors declare no conflict of interest. The funders had no role in the design of the study; in the collection, analyses, or interpretation of data; in the writing of the manuscript, or in the decision to publish the results.

References

1. Naito, M.; Yokoyama, T.; Hosokawa, K.; Nogi, K. *Nanoparticle Technology Handbook*, 3rd ed.; Elsevier: Amsterdam, The Netherlands, 2018.
2. Deb, S.K. Opportunities and challenges in science and technology of WO_3 for electrochromic and related applications. *Sol. Energ. Mat. Sol. Cells* **2008**, *92*, 245–258. [CrossRef]
3. Monk, P.; Mortimer, R.; Rosseinsky, D. *Electrochromism and Electrochromic Devices*; Cambridge: New York, NY, USA, 2007; pp. 139–151.
4. Zhang, M.; Yang, C.; Zhang, Z.; Tian, W.; Hui, B.; Zhang, J.; Zhang, K. Tungsten oxide polymorphs and their multifunctional applications. *Adv. Colloid Interface Sci.* **2022**, *300*, 102596. [CrossRef] [PubMed]
5. Cezarina, C.M.; Hassel, A.W. Review on the versatility of tungsten oxide coatings. *Phys. Status Solidi A* **2019**, *216*, 1–16.
6. Klanjšek Gunde, M.; Opara Krašovec, U.; Platzer, W. Color rendering properties of interior lighting influenced by a switchable window. *J. Opt. Soc. Am. A* **2005**, *22*, 416–423. [CrossRef] [PubMed]
7. Opara Krašovec, U.; Šurca Vuk, A.; Orel, B. Comparative studies of all sol–gel electrochromic windows employing various counter-electrodes. *Sol. Energy Mater. Sol. Cells* **2002**, *73*, 21–37. [CrossRef]
8. Cremonesi, A.; Bersani, D.; Lottici, P.P.; Djaoued, Y.; Ashrit, P.V. WO_3 thin films by sol–gel for electrochromic applications. *J. Non-Cryst. Solids* **2004**, *45&346*, 500–504. [CrossRef]
9. Hauch, A.G.; Opara Krašovec, U.; Orel, B. Comparison of photoelectrochromic devices with different layer configurations. *J. Electrochem. Soc.* **2002**, *149*, H159–H163. [CrossRef]
10. Opara Krašovec, U.; Georg, A.; Georg, A.; Wittwer, V.; Luther, J.; Topič, M. Performance of a solid-state photoelectrochromic device. *Sol. Energy Mater. Sol. Cells* **2004**, *84*, 369–380. [CrossRef]
11. Hočevar, M.; Opara Krašovec, U. A photochromic single glass pane. *Sol. Energy Mater. Sol. Cells* **2018**, *186*, 111–114. [CrossRef]
12. Hočevar, M.; Opara Krašovec, U. Cubic WO_3 stabilized by inclusion of Ti: Applicable in photochromic glazing. *Sol. Energy Mater. Sol. Cells* **2016**, *154*, 57–641. [CrossRef]
13. Dao, T.T.; Park, S.; Hong, S.; Sarwar, S.; Van Tran, H.; Lee, S.I.; Park, H.S.; Song, S.H.; Nguyen, H.D.; Lee, K.-K.; et al. Novel flexible photochromic device with unprecedented fast-bleaching kinetic via platinum decoration on WO_3 layer. *Sol. Energy Mater. Sol. Cells* **2021**, *231*, 111316. [CrossRef]
14. Opara Krašovec, U.; Orel, B.; Georg, A.; Wittwer, V. The gasochromic properties of sol-gel WO_3 films with sputtered Pt catalyst. *Sol. Energy* **2000**, *68*, 541–551. [CrossRef]
15. Vidmar, T.; Topič, M.; Dzik, P.; Opara Krašovec, U. Inkjet printing of sol-gel derived tungsten oxide inks. *Sol. Energ. Mater. Sol. Cells* **2014**, *125*, 87–95. [CrossRef]
16. Brinker, J.C.; Scherer, G.W. *Sol-Gel Science: The Physics and Chemistry of Sol–Gel Processing*; Academic Press: Cambridge, MA, USA, 1999. [CrossRef]
17. Cui, Z. *Printed Eletronics: Materials, Technologies and Applications*; John Wiley and Sons: Hoboken, NJ, USA, 2016.
18. Zapka, W. *Handook of Industrial Inkjet Printing: A Full System Approach*; John Willey and Sons: Hoboken, NJ, USA, 2017.
19. Yang, P.; Jin Fan, H. Inkjet and extrusion printing for electrochemical energy storage: A mini review. *Adv. Mater. Technol.* **2020**, *5*, 2000217. [CrossRef]
20. Dimatix. *Materials Printer DMP-2800 Series: User Manual*; Doc. # PM000040 Rev. 05; FUJIFILM Dimatix, Inc.: Tokyo, Japan, 2010.

21. Atkinson, A.; Doorbar, J.; Hudd, A.; Segal, D.L.; White, P.J. Continuous ink-jet printing using sol-gel "Ceramic" inks. *JSST 1* **1997**, *8*, 1093–1097. [CrossRef]
22. Liu, X.; Tarn, T.J.; Huang, F.; Fan, J. Recent advances in inkjet printing synthesis of functional metal oxides. *Particuology* **2015**, *19*, 1–13. [CrossRef]
23. Žitňan, M.; Müller, L.; Zub, K.; Schubert, U.S.; Galusek, D.; Wondraczek, L. Low-cost inkjet printing of thin-film mullite structures. *Int. J. Appl. Glass Sci.* **2022**, *13*, 135–142. [CrossRef]
24. Homola, T.; Ďurašová, Z.; Shekargoftar, M.; Souček, P.; Dzik, P. Optimization of TiO_2 mesoporous photoanodes prepared by inkjet printing and low-temperature plasma processing. *Plasma Chem. Plasma Process.* **2020**, *40*, 1311–1330. [CrossRef]
25. Kassem, O.; Saadaoui, M.; Rieu, M.; Sao-Joao, S.; Viricelle, J.-P. Synthesis and inkjet printing of sol–gel derived tin oxide ink for flexible gas sensing application. *J. Mater. Sci.* **2018**, *53*, 12750–12761. [CrossRef]
26. Karimi-Nazarabad, M.; Goharshadi, E.K.; Entezari, M.H. Rheological properties of the nanofluids of tungsten oxide nanoparticles in ethylene glycol and glycerol. *Microfluid. Nanofluid.* **2015**, *19*, 1191–1202. [CrossRef]
27. Tripkovic, D.; Vukmirovic, J.; Bajac, B.; Samardzic, N.; Djurdjic, E.; Stojanovic, G.; Srdica, V.V. Inkjet patterning of in situ sol–gel derived barium titanate thin films. *Ceram. Int.* **2016**, *42*, 1840–1846. [CrossRef]
28. Duoss, E.B.; Twardowski, M.; Lewis, J.A. Sol-gel inks for direct-write assembly of functional oxides. *Adv. Mater.* **2007**, *19*, 3485–3489. [CrossRef]
29. Nanba, T.; Takano, S.; Yasui, I.; Kudo, T. Structural study of peroxopolytungstic acid prepared from metallic tungsten and hydrogen peroxide. *J. Solid State Chem.* **1991**, *90*, 47–53. [CrossRef]
30. Daniel, M.F.; Desbat, B.; Lassegues, J.C.; Gerand, B.; Figlartz, M. Infrared and Raman study of WO_3 tungsten trioxides and WO_3 x H_2O tungsten trioxide hydrates. *J. Solid State Chem.* **1987**, *67*, 235–247. [CrossRef]
31. Grošelj, N.; Gaberšček, M.; Opara Krašovec, U.; Orel, B.; Dražič, G.; Judeinstein, P. Electrical and IR spectroscopic studies of peroxopolytungstic acid/organic–inorganic hybrid gels. *Solid State Ion.* **1999**, *125*, 125–133. [CrossRef]
32. Opara Krašovec, U.; Ješe, R.; Orel, B.; Grdadolnik, J.; Dražič, G. Structural, vibrational and gasochromic properties of porous WO_3 films templated with a sol-gel organic-inorganic hybrid. *Monatsefte Chem.* **2002**, *133*, 1115–1133. [CrossRef]
33. Opara Krašovec, U.; Šurca Vuk, A.; Orel, B. IR Spectroscopic studies of charged–discharged crystalline WO_3 films. *Electrochim. Acta* **2001**, *46*, 1921–1929. [CrossRef]
34. Metzger, T.G. *The Rheology Handbook*, 4th ed.; Vincentz Network: Hannover, Germany, 2014.
35. Gottschalk, P.G.; Dunn, J.R. The five-parameter logistic: A characterization and comparison with the four-parameter logistic. *Anal. Biochem.* **2005**, *343*, 54–65. [CrossRef]
36. Lapasin, R.; Pricl, S. *Rheology of Industrial Polysaccharides: Theory and Applications*; Blackie Academic & Professional: London, UK, 1995.
37. Larson, R.G. *The Structure and Rheology of Complex Fluids*; Oxford University Press: New York, NY, USA, 1999.
38. Winter, H.H.; Chambon, F. Analysis of linear viscoelasticity of a crosslinking polymer at the gel point. *J. Rheol.* **1986**, *30*, 367. [CrossRef]
39. Chambon, F.; Petrovic, Z.S.; MacKnight, W.J.; Winter, H.H. Rheology of model polyurethanes at the gel point. *Macromolecules* **1986**, *19*, 2146–2149. [CrossRef]

Review

Integration of Antifouling and Anti-Cavitation Coatings on Propellers: A Review

Jingying Zhang [1,2,†], Weihua Qin [1,2,†], Wenrui Chen [1,2], Zenghui Feng [1,2], Dongheng Wu [1,2], Lanxuan Liu [1,2] and Yang Wang [1,2,*]

1. Wuhan Research Institute of Materials Protection, Wuhan 430030, China; zhangjy288@163.com (J.Z.); qinweihua0@163.com (W.Q.); andrewcwr@whut.edu.cn (W.C.); 13419694843@163.com (Z.F.); 18804050206@163.com (D.W.); liulanxuan@rimp.com.cn (L.L.)
2. State Key Laboratory of Special Surface Protection Materials and Application Technology, Wuhan 430030, China
* Correspondence: wangyang@rimp.com.cn
† These authors contributed equally to this work.

Abstract: The performance of an entire ship is increasingly impacted by propellers, which are the essential components of a ship's propulsion system that have growing significance in a variety of aspects. Consequently, it has been a hot research topic and a challenge to develop high-performance antifouling and anti-cavitation coatings due to the issue of marine biofouling and cavitation faced by propellers in high-intensity service. While there is an overwhelming number of publications on antifouling and anti-cavitation coatings, a limited number of papers focus on integrated protective coatings on propellers. In this paper, we evaluated the development of antifouling and anti-cavitation coatings for ship propellers in the marine environment as well as their current status of research. These coatings include self-polishing antifouling coatings, fouling-releasing antifouling coatings, and biomimetic antifouling coatings for static seawater anti-biofouling, as well as anti-cavitation organic coatings and anti-cavitation inorganic coatings for dynamic seawater anti-cavitation. This review also focuses both on the domestic and international research progress status of integrated antifouling and anti-cavitation coatings for propellers. It also provides research directions for the future development of integrated antifouling and anti-cavitation coatings on propellers.

Keywords: propeller; antifouling coatings; anti-cavitation coatings; integration technology

1. Introduction

Propellers have been around for more than two hundred years, i.e., since the 19th century. Propellers have consistently been the preferred option for ship propulsion due to their notable efficiency and commendable hydrodynamic characteristics. Consequently, the overall performance and efficiency of an entire vessel are inherently influenced by the state of the propellers [1]. The presence of microorganisms and proteins in stagnant seawater poses a significant challenge in the maritime industry, as it leads to the attachment of marine organisms onto propellers and other materials, resulting in biofouling issues [2]. Simultaneously, the prolonged and rapid rotation of the propeller induces the development of cavitation on the material surface, leading to fatigue-related degradation of the propeller blades. Both of these factors have emerged as the primary determinants influencing the ship's performance, as depicted in Figure 1. The ship's propulsion system is expected to experience significant degradation due to the presence of biofouling and cavitation. These factors will lead to changes in the propeller's surface morphology, disrupt its dynamic balance, and result in reduced efficiency and range, accompanied by increased energy consumption. At present, the effectiveness of protective coatings, both within national borders and beyond international contexts, is below the desired standard. The significant challenges of seawater scouring and cavitation present considerable problems in the pursuit of long-lasting protection, rendering it susceptible to succumbing to these deleterious effects. Therefore, an urgent need for the creation of a comprehensive protective coating for ship propellers that demonstrates resistance to both dynamic cavitation and static fouling has arisen.

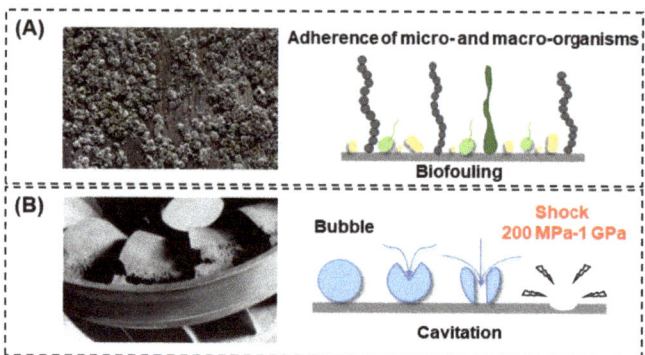

Figure 1. Ship propeller biofouling and cavitation erosion phenomenon and its mechanism diagram. (**A**) The phenomenon of marine biofouling and its mechanism; (**B**) The phenomenon of cavitation erosion and its mechanisms.

Marine biofouling refers to the phenomenon wherein submerged objects, such as marine propellers, pumps, turbines, and various components of the hull located beneath the waterline, become encrusted with biological organisms [3]. The attachment, propagation, and accumulation of marine organisms on the surface of these objects result in surface corrosion and material degradation [4]. The organisms responsible for this occurrence are referred to as marine fouling organisms, which can result in economic losses for human activities and livelihoods [5]. Additionally, this can result in unavoidable economic disadvantages in terms of human productivity and livelihood.

Based on statistical data, the global count of marine fouling organisms is estimated to range between 4000 and 5000 species [6]. Among these organisms, barnacles, mussels, oysters, and other similar species have been identified as the most detrimental along the coastal regions of China. The presence of marine fouling organisms has been found to have a direct impact on ship navigation by increasing the roughness of ship hull surfaces [7]. The presence of these substances has the potential to harm the protective coating designed to prevent corrosion on the hull, expedite the corrosion process on the underlying metal surface, escalate maintenance expenses, and result in significant financial setbacks. According to reports, the aggregate economic burden resulting from marine fouling organisms continues to surpass a staggering sum of 150 billion dollars annually. The proliferation and attachment of fouling organisms can result in an increase in weight and a decrease in the speed of a ship's hull [8]. Moreover, the presence of microorganisms adhered to ships can potentially exert an influence on the surrounding ecosystem [9].

The intricate process of marine biofouling has been found to be a dynamic and continuous phenomenon that can be divided into four distinct phases, as illustrated in Figure 2. A wide variety of organic components, including protein molecules, polysaccharides, and esters, immediately cling to the surface of the material in the early phase after burying it in saltwater. This phenomenon is attributed to several physical mechanisms, such as Brownian motion, van der Waals forces, electrostatic forces, and hydrogen bonding. Consequently, a provisional film is formed [10,11]. The subsequent phase involves the prompt adherence of microorganisms, specifically bacteria, facilitated by the adsorption of the conditioned film onto the metal substrate's surface. This process leads to the development of a biofilm, characterized by the production of metabolic secretions by the microorganisms and the entrapment of the polymer material itself [12,13]. In the third phase, the organisms responsible for pollution release various substances such as proteins, polysaccharides, nucleic acids, and other compounds. These materials serve the dual purpose of capturing additional polluting species and providing them with essential nutrients to facilitate their growth. The fourth stage is characterized by the attachment and accumulation of minute organisms, leading to the settlement and growth of larvae from larger marine invertebrates and other organisms on the surface, ultimately resulting in substantial biofouling. The fouling process described above is generally applicable to the majority of marine organisms responsible for fouling in contemporary times, although it is not universally consistent. For instance, certain types of algae spores and barnacle larvae have the ability to directly adhere to material surfaces without the need for biofilms [14]. Furthermore, the marine environment exhibits a diverse array of marine fouling organisms, alongside its inherent complexity and variability. Several factors within the marine environment can affect the adhesion of marine fouling organisms to propeller surfaces. These factors include temperature, current velocity, shear stress, pH of seawater, and salinity. Each of these factors exerts a distinct influence on the fouling phenomenon that manifests on the surface of propellers.

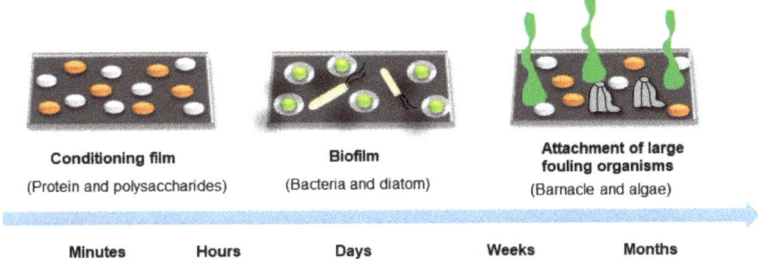

Figure 2. Diagram illustrating the marine biofouling process [15].

Cavitation is a unique form of erosion–corrosion that almost every propeller suffers from. The phenomenon of cavitation was initially observed and postulated in 1887 by the esteemed British scientist S.W. Barnaby during an investigation into the propeller efficiency of maritime vessels [16]. Currently, it is widely accepted that there are two factors contributing to the material degradation resulting from cavitation, as indicated in [17]. The initial cause can be attributed to the mechanical force that is generated upon the bursting of the bubble. The impact pressure wave generated by the collapse of the bubble at its center induces surface alterations in the nearby materials, leading to consequential material degradation [18]. On the other hand, the bubble undergoes collapse when subjected to deformation, and this deformation intensifies as the pressure increases, thereby facilitating the formation of a high-velocity microjet. These repeated impacts can lead to local plastic deformation of the materials and mass removal [19]. Furthermore, as a consequence of the influence of thermodynamics, when the bubble undergoes collapse within a region of elevated pressure, the vapor contained within the bubble, characterized by low pressure, swiftly condenses. This condensation process results in the liberation of a substantial quantity of heat, thereby inducing thermal damage of high temperature to the material [20]. However, propeller cavitation in a real marine environment involves the cooperative effects of mechanical-related corrosion and electrochemical corrosion, which act synergistically to accelerate the corrosion processes and therefore can be called erosion–corrosion. Cavitation is a prevalent and significant concern in hydraulic systems, encompassing various components such as turbine blades, valves, propellers, pipelines, and others. This phenomenon poses substantial financial losses and safety risks within industries such as ship water conservancy [21]. Based on the findings of the British Ship Research Association (BSRA), it has been observed that despite the relatively small surface area and volume of the propeller, the combined effects of biofouling and cavitation contribute to nearly one-third of the total loss [22]. Consequently, it is crucial to the utilization of integrated technology for the purpose of propeller surface antifouling and anti-cavitation.

2. Antifouling Coatings

The phenomenon of hull biofouling is characterized by its intricate and multifaceted nature. Various aspects such as different sections of the hull, depth of immersion, speed of the current, and additional factors play a significant role in determining the attachment behavior of marine fouling organisms. Organisms exhibiting robust adhesion tend to cause fouling on propellers as a consequence of prolonged exposure to seawater and the rapid rotational motion they experience. In a broad sense, fouling is primarily affected by the surface properties of the material, including the substrate's surface energy, wettability, and microstructure, among other factors. One of the most prevalent and efficacious methods for accomplishing this objective involves the utilization of antifouling coatings. Consequently, the deliberate alteration of the surface structure represents a more straightforward approach to managing fouling. Based on the antifouling properties exhibited by coatings, it is possible to categorize common antifouling coatings into three main types: self-polishing antifouling coatings, fouling release antifouling coatings, and bionic antifouling coatings. The application of antifouling coatings on propellers has the potential to mitigate energy dissipation, minimize frictional resistance between the propeller surface and saltwater, and restrict the attachment of marine fouling organisms [23]. As a result, the propellers of the ship will exhibit enhanced operational efficiency and durability, thereby more effectively fulfilling diverse requirements and minimizing financial losses.

2.1. Self-Polishing Antifouling Coatings

Self-polishing antifouling coatings function through the gradually hydrolyzing polymers, resulting in the formation of a peel layer on the external surface of the coating. The presence of this layer serves as a deterrent for marine fouling organisms, impeding their ability to adhere to surfaces such as ship propellers [24,25]. During the early 1970s, the development of organotin self-polishing antifouling coatings took place, utilizing analogs of butyltin (TBT) [26]. The primary function of this coating principle is to apply a coating onto a resin material based on acrylate, facilitating the hydrolysis of ester bonds and subsequently releasing the antifouling effect of TBT. Figure 3A provides an illustration of this technique. Simultaneously, the incorporation of an antifouling agent, such as cuprous oxide, within the paint film enables the release of copper ions. The combined effect of these ions enhances the coating's ability to resist fouling from a wide range of organisms. The process commonly known as "self-polishing" involves the application of a coating onto a surface base material, which is subjected to a continuous flow of seawater. This constant exposure to seawater results in the ionization of dissolved salts and a continuous alteration of the surface, leading to its self-polishing effect.

Upon its initial implementation, the coating quickly gained dominance in the antifouling coating industry [27]. Nevertheless, the persistent utilization of organotin antifouling coatings has revealed that tributyltin (TBT), despite its remarkable efficacy in preventing fouling, exhibits significant toxicity, particularly towards fish and shellfish. This toxicity poses a substantial threat to marine organisms, thereby endangering the marine ecosystem and potentially leading to species extinction [26]. In 2001, the International Maritime Organization (IMO) issued a declaration stating that the utilization of TBT antifouling coatings on commercial vessels worldwide would be banned, effective from 2008. Consequently, there has been a shift in research attention towards the advancement of environmentally friendly antifouling coatings. Subsequently, scholars employed organic copper, organic zinc, organic silicon, and various other elements [28–30] as alternatives to organotin in the formulation of antifouling coatings that are both tin-free and possess reduced toxicity. The grafts mentioned above exhibit a lower ionic antifouling capacity compared to organotin. Therefore, the inclusion of a copper antifouling agent is typically necessary during the preparation process to ensure a satisfactory antifouling effect [31].

The researchers have integrated acrylic acid and polyurethane copolymers that are capable of undergoing hydrolysis in seawater into the coatings, with the aim of developing environmentally sustainable self-polishing coatings that exhibit strong antifouling characteristics [25]. Xuezhi Jiang and colleagues [32,33] from Wuhan University of Technology employed acrylic acid chloride as a chemical modifier for glyphosate, subsequently undergoing a sequence of polymerization reactions to produce polyacrylic acid resins featuring glyphosate side-linked branches. These resins were then subjected to fouling bio-inhibition experiments. The findings indicated that the polyacrylic acid resin, which featured glyphosate side-chained branches, exhibited an attachment inhibition rate of 41% for barnacle Venus larvae. Moreover, the highest observed attachment inhibition rate of 46.9% was achieved for Rhodophyta crescentica. The team led by Professor Xia Li at Ocean University of China [34] conducted a study where they introduced indole derivatives (NPI) as side chains into acrylic resins through the process of free radical polymerization. The researchers then evaluated the resulting resins for their self-polishing and antifouling properties. The findings of the study indicate that the copolymer exhibited significant antifouling characteristics due to the synergistic effect of the acrylate's self-polishing capability and the antifouling property of the indole derivative, as depicted in Figure 3B. Boron acrylate polymers were synthesized by Professor Rongrong Chen's team at the Harbin Institute of Technology [35]. Pyridine-diphenyl-borane with hydrolysis function was employed as the side chain in the synthesis process. The experimental findings indicated that these polymers exhibited enhanced antifouling properties against diatoms. Furthermore, the polymers demonstrated improved antifouling effects in suspension experiments conducted in the Yellow Sea of China. The research team led by Professor Chunfeng Ma from the South China University of Technology [36] synthesized a polyurethane with main chain degradability using N-2,4,6 trichlorophenylmaleimide (TCPM) through the integration of the mercapto-alkene reaction and condensation reaction. The subject of investigation involved conducting hydrolysis experiments to examine the release of TCPM in relation to the degradation of the polyurethane main chain. The experimental results demonstrated that TCPM was indeed released during this procedure. Simultaneously, the degradation rate of the material would exhibit a decrease as the quantity of TCPM content increases, thereby promoting the enhancement of antifouling material durability. The antifouling capability of a polyurethane material has been demonstrated through experiments conducted on marine pegboards. The observed phenomenon can be attributed to the regulated rate at which the material undergoes surface self-renewal, as well as the release of antifouling chemicals that occurs as a consequence of its degradation.

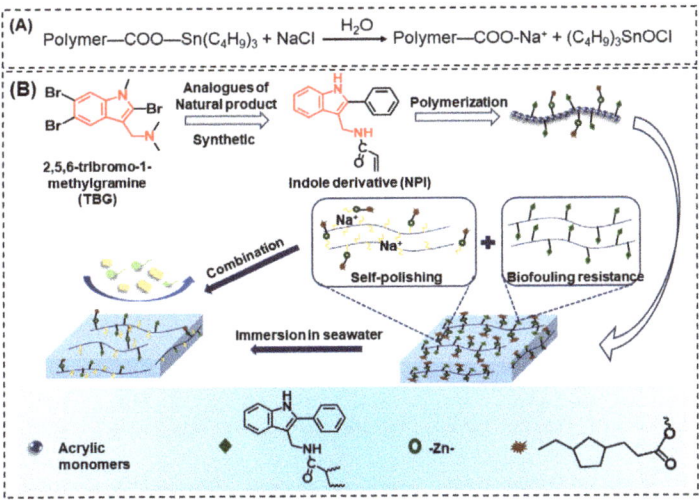

Figure 3. (A) Hydrolysis mechanism of organotin self-polishing resins; (B) Schematic diagram of the antifouling mechanism of NPI and acrylate resin copolymer [34] with permission (Copyright © 2023, Elsevier).

2.2. Fouling Release Antifouling Coatings

The efficacy of fouling-release antifouling coatings is primarily attributed to the inherent ability of marine fouling organisms to adhere to the surface of the coating and subsequently accumulate fouling materials with ease. Conventional fouling-release antifouling coatings possess low surface energy, thereby facilitating the removal of fouling through mechanical cleaning methods or minimal application of shear force. The aforementioned coatings exhibit consistent fouling-release properties and demonstrate enduring antifouling effects, as depicted in Figure 4A [37,38]. The efficacy of fouling-release antifouling coatings is typically influenced by various factors, including the thickness of the coating, its modulus of elasticity, surface roughness, and other parameters, as well as the material's hydrophobicity and low surface energy [39]. The antifouling performance of hydrophobic samples can be attributed to the inherent material properties and surface roughness, which can effectively trap air bubbles in the interstices. This, in turn, reduces the contact between protein organisms and the sample surface [40]. Fouling-release antifouling coatings have garnered significant attention in recent years due to their exceptional environmental compatibility [41,42]. In marine contexts, the implementation of this coating has the potential to mitigate the navigational resistance of ships by virtue of its smooth surface, thereby resulting in reduced economic consumption, such as fuel usage, and the enhanced longevity of ships, particularly with regard to propellers.

Currently, the two most commonly employed substances for fouling-release coatings are organ-silicone polymers and organ-fluorine polymers. This is primarily attributed to the limited strength of the interactions between chain segments containing silicone or fluorine and protein biomolecules, which renders them inadequate for the formation of biofouling [43,44]. In a study conducted by Yarbrough et al. [45] at the University of North Carolina at Chapel Hill, a set of perfluoro polymer brushes was fabricated on a glass substrate. The presence of functional groups, specifically perfluorooxymethylene, on the surface of the glass was found to significantly enhance its low-surface-energy properties. Consequently, the material exhibited improved resistance to protein adsorption and possessed fouling-release properties. Experimental investigations on the antifouling capabilities of the coatings demonstrated a remarkable 90% fouling release efficiency against microorganisms. The researchers led by Professor Chunfeng Ma from the South China University of Technology [46] developed a metal–ligand crosslinked organosilicon coating. The experimental results suggest that this coating exhibits enhanced self-healing and antifouling characteristics.

The aforementioned materials, which possess either hydrophilic or hydrophobic properties and have a single graft on their surface, still fail to achieve the intended antifouling effect. The development of a dirt-releasing coating has been achieved by researchers, which demonstrates the ability to prevent the attachment of fouling organisms and decrease the adhesion force between fouling and the surface of the material. This has been accomplished through the continuous optimization of the physical and chemical parameters of the coatings, as well as the rational design of the molecular structure. Furthermore, the coating exhibits effective protein adsorption prevention properties and possesses amphiphilic characteristics due to the presence of both hydrophilic and hydrophobic components. Additionally, it demonstrates certain capabilities for releasing dirt. The research team led by Professor Lingmin Yi at Zhejiang University of Technology [47] developed an amphiphilic polymer. This polymer is composed of a flexible fluorosilicon macromonomer, which serves as the hydrophobic component with a low surface energy, and an amphiphilic monomer, which acts as the hydrophilic component. This molecular structure is depicted in Figure 4B. The experimental results indicate that the coated surface effectively retained a substantial quantity of low-surface-energy fluoro silicone particles, even when immersed in water. A copolymer coating was synthesized by carefully adjusting the proportions of hydrophilic and hydrophobic surface segments, resulting in a coating with remarkable protein resistance. The researchers led by Junpeng Zhao from the South China University of Technology [48] synthesized amphiphilic polyurethane coatings with heterostructures by incorporating poly(ethylene oxide) and birch alcohol. The experimental findings demonstrated that these coatings exhibited favorable characteristics in terms of protein adsorption resistance and fouling release properties. In a study conducted by Wooley et al. (University of Washington, USA), a polymer brush was synthesized using hydrophilic PEG chains and crosslinked hyperbranched fluorinated units. The researchers then performed simulated experiments using bovine serum protein as the fouling material [49]. The findings of the study indicated that the amphiphilic polymer brush exhibited enhanced efficacy in preventing protein adhesion, superior performance in resisting fouling, and improved ability to release fouling. In a work by Martinelli et al. (University of Pisa, Italy), they conducted a study in which they synthesized polystyrene polymer brushes containing amphiphilic side chains (PEG-b-PTFE). These brushes exhibited distinct phase regions attributed to the hydrophilic and hydrophobic components present. Upon exposure to water, the conformation of the brushes underwent changes, resulting in an increase in surface roughness and subsequent enhancement of inhomogeneity [50]. The results of the investigation into the degree of adhesion that various microorganisms and proteins showed to the surface suggested that the coated surface had a more significant antifouling effect.

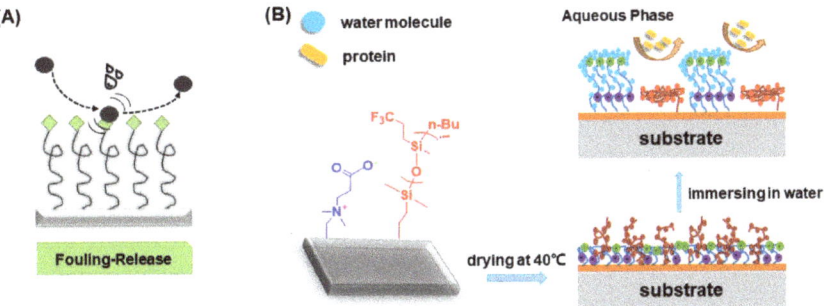

Figure 4. (**A**) Fouling release coating mechanism [51]; (**B**) Antifouling mechanism of amphiphilic copolymer [47] with permission (Copyright © 2023, American Chemical Society).

2.3. Bionic Antifouling Coatings

Numerous organisms in the natural environment possess remarkable antifouling characteristics owing to their distinct surface morphology and properties, rendering them highly resistant to the attachment of other organisms. As a result, when developing antifouling coatings, researchers frequently take inspiration from biological mechanisms. Presently, the prevailing body of research pertaining to biomimetic antifouling coatings can be categorized into two distinct groups. The initial approach involves the utilization of naturally derived antifouling agents, including terpene sugars emitted by certain marine organisms [52] and halofuranones [53], as well as extracts from terrestrial plants such as capsaicin [54] and carvacrol [55], all of which possess specific antifouling characteristics. One approach involves employing micro- and nano-construction techniques to replicate the biological surface properties of lotus leaves and sharks, as depicted in Figure 5A.

In the field of marine antifouling, there is an expectation that environmentally friendly antifouling agents will utilize natural compounds with comparable efficacy to effectively combat fouling. Nevertheless, there exist certain limitations in the present scenario, including intricate extraction processes and inadequate stability [55]. Consequently, it becomes imperative to establish a harmonious combination of natural antifouling agents and resin carriers to guarantee their stability and regulated release. The research team led by Jinggang Gai from Sichuan University [56] employed guanidine groups as antifouling agents in the fabrication of filter membranes. The results indicated that the membranes exhibited exceptional antifouling properties along with a notable selectivity in permeability. The research team led by Professor Peiyuan Qian from The Hong Kong University of Science and Technology [57,58] developed a butenolide-based antifouling agent. This agent was prepared by utilizing polycaprolactone-based (PCL) polyurethane as a carrier, enabling a gradual and regulated release of the antifouling agent due to the biodegradable properties of the resin. The experimental results, depicted in Figure 5B, demonstrated that the developed agent exhibited a certain level of antifouling efficacy against larvae of marine fouling organisms. A research team led by Professor Xiaoli Zhan from Zhejiang University [59] employed a polycondensation reaction to graft modified coumarin and eugenol onto polyurethane side chains. This process was utilized to develop a polydimethylsiloxane (PDMS)-based marine antifouling polyurethane coating with smart properties. Based on the results, the layer exhibited enhanced mechanical properties, antibacterial properties, and resistance against algae growth. Additionally, it demonstrated a notable marine antifouling performance lasting for a duration of 9 months. These findings suggest promising prospects for the application of this layer in the field of marine antifouling.

The surfaces of shark skin and lotus leaves exhibit robust biofouling inhibition and self-cleaning properties, effectively preventing the attachment of fouling organisms such as barnacles and algae to the underlying substrate material. A collaborative effort between researchers from the University of Florida, USA, and the University of Birmingham, UK, resulted in the development of a microstructured antifouling coating inspired by shark skin [60]. This coating demonstrates a significant reduction of 85% in the attachment rate of macroalgae spores while also exhibiting a favorable effect on drag reduction. In their study, Zheng et al. conducted research at the Wuhan University of Technology to fabricate a polyurethane surface with lotus leaf characteristics [61]. This was achieved by utilizing the natural lotus leaf as a template through the replica molding method. The resulting surface exhibited a notable decrease in protein adsorption compared to its pre-construction state, resembling the lotus leaf's distinctive properties. In the realm of antifouling surface modification strategies, chemical surface modification is employed alongside physical alterations in the surface microstructure. The prevailing technique employed in current research is polymer brushing, characterized by a substantial polymer density, strong adherence to the water layer, and the ability to readily incorporate functional groups with anti-adhesion properties. The presence of the polymer creates a physical barrier that effectively maintains a predetermined distance between dirt particles and the surface of the material, thereby resulting in the attainment of antifouling properties. Qian Ye et al. [62] conducted a study at Northwestern Polytechnical University where they utilized 3D printing technology to create a surface resembling shark skin, which was then modified with a poly ionic liquid brush. The researchers used bovine serum protein as a model for fouling in their experiments, as depicted in Figure 5C. The findings indicated that the surface exhibited a superior anti-protein adhesion efficacy. The initial discovery of the remarkable anti-protein adhesion properties of polyethylene glycol (PEG) derivatives was made by Prime et al. [63] at Harvard University. The anti-protein adhesion properties of novel copolymers of poly (ethylene glycol) methacrylate (AEM-PEG) on glass substrates were discovered by Lonov et al. [64] from the Max Planck Institute in Germany. This finding is illustrated in Figure 5D. Additionally, the researchers highlighted the copolymers' advantageous features, including their low cost, ease of use, and compatibility with aqueous solutions. In a study conducted by Philip et al. [65] at the University of Texas, it was observed that the presence of bovine serum protein as a simulated fouling agent on the substrate surface was reduced to a minimum when the polyethylene glycol (PEG) content was 45%. This finding can be attributed to the decrease in the size of the split-phase region, which occurs as the PEG content decreases. The dimensions of an object have an impact on the ability of proteins and marine organisms, including microorganisms, to adhere to the material's surface.

Micro- and nano-structured antifouling materials predominantly leverage their microstructure to hinder the attachment of fouling organisms without causing any detrimental effects on the marine ecosystem. This contrasts with traditional antifouling techniques, which entail discharging antifouling substances. The majority of micro- and nanostructures are predominantly composed of soft materials, including silicon wafers, PDMS, and similar substances. Currently, there is a scarcity of research pertaining to microstructured antifouling techniques specifically tailored for the hard metals employed in ship propellers. Moreover, the high method employed for such purposes is accompanied by certain limitations, including a complex procedural approach and substantial financial expenses. These factors collectively impede the widespread application of micro-nanostructures within the domain of ship propeller antifouling.

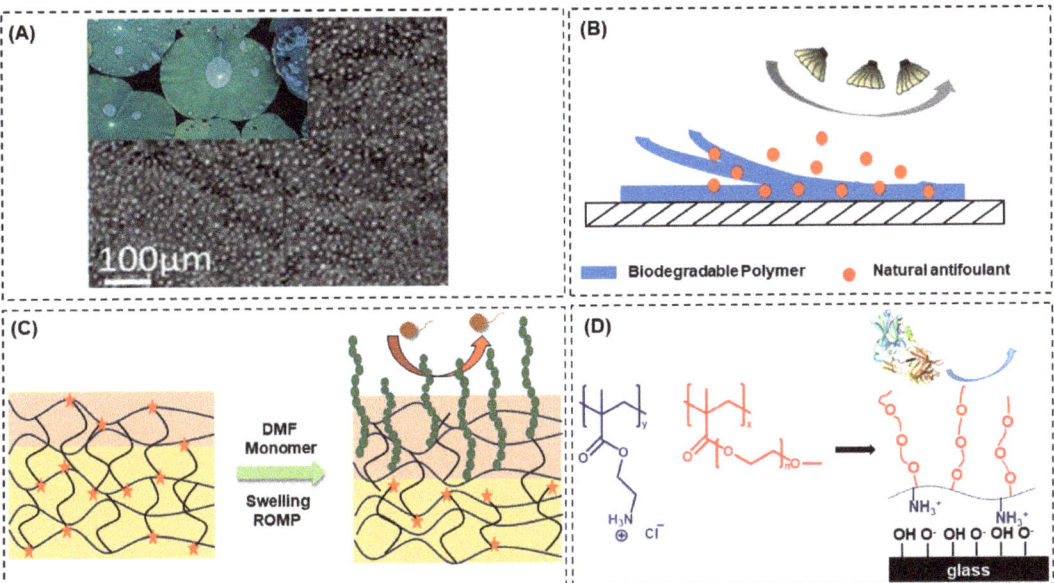

Figure 5. (A) Surface of lotus leaf and its SEM image [66] with permission (Copyright © 2023, American Chemical Society); (B) Mechanism diagram of natural antifouling agent to achieve antifouling performance using resin as a matrix [58] with permission (Copyright © 2023, American Chemical Society); (C) Mechanism of preparation of poly ionic liquid brush grafted imitation shark skin surface [62] with permission (Copyright © 2023,Elsevier); (D) Mechanism of anti-protein adhesion of linear PEG polymer brushes on glass substrates [64] with permission (Copyright © 2023, American Chemical Society).

3. Anti-Cavitation Coatings

The propeller, as the primary power system of a ship, exerts a substantial influence on the ship's operational lifespan owing to its prolonged and high-velocity operation, which gives rise to cavitation erosion. Currently, the mitigation of cavitation erosion in ship propellers is commonly addressed through two methods. Initially, the material undergoes surface modification treatment to enhance its anti-cavitation properties. Alternatively, research and development efforts are focused on exploring novel materials with superior characteristics in this regard. Nevertheless, the implementation of these modifications is limited due to the time-consuming and arduous nature of new material research and development. The second approach involves the application of material surface modification treatments, such as surface nitriding technology and surface coating technology [67]. These treatments have the potential to significantly mitigate cavitation damage in the overcurrent region by implementing anti-cavitation coatings. Surface coating technology has garnered significant attention due to its ability to provide long-term protection against cavitation-induced damage to materials [68]. Surface coating technology plays a significant role in mitigating cavitation effects. Based on the divergent characteristics of their constituent materials, the prevailing anti-cavitation coating technologies currently employed can be classified into two primary categories: organic coatings and inorganic coatings, predominantly comprising metal coatings. The enhancement of the material's resistance to cavitation in organic coatings is primarily achieved through the utilization of the material's elastic properties to effectively absorb the energy generated during the cavitation process. Firstly, it is important to note that the application of metal coatings serves to enhance the fatigue resistance of the material surface, thereby improving its ability to withstand cavitation-induced damage. This is achieved through the utilization of the material's inherent hardness, relatively elevated mechanical properties, and resistance to high temperatures. The approach has been extensively developed.

3.1. Cavitation-Resistant Organic Coatings

In recent years, there has been a growing interest among domestic and international researchers in organic coatings for ship propellers and other overcurrent components that are required to function in seawater for prolonged durations. These coatings possess unique properties that effectively minimize the risk of corrosion in seawater. Moreover, they offer additional advantages such as affordability, ease of operation, and the ability to easily regulate their molecular structure. In recent years, there has been a significant increase in scholarly attention, both domestically and internationally, towards this subject matter. Currently, the most extensively studied polymer organic coatings for anti-cavitation purposes encompass polyurethane elastomers and polyurea elastomers. And considering the need for environmental protection, they use waterborne polymers more often. However, the organic coatings have poor abrasion resistance and low binding force with the surface of materials. So it is very vulnerable to falling off under the frequent impact of a periodic alternating cavitation force, which has become a major obstacle to application.

Polyurethane is a polymer material that consists of a diisocyanate-coupled polyether soft segment and a macromolecule diisocyanate hard segment. The soft segment imparts elasticity and flexibility to the system, while the hard segment allows for reversible crosslinking. Polyurethane exhibits favorable mechanical properties, such as resistance to abrasion and corrosion, along with the advantages of high strength and robust toughness [69]. Simultaneously, it is worth noting that polyurethane elastomer exhibits a notable loss factor, thereby facilitating the dissipation of a substantial amount of energy upon encountering external vibrations or impacts. This property effectively mitigates external harm and renders it a superior material for anti-cavitation purposes [70,71]. During the cavitation process, the molecular chain segments of polyurethane undergo softening due to the repetitive impacts from the airflow and microjet. Simultaneously, motion deformation takes place to absorb the energy generated by these impacts, as depicted in Figure 6A.

The initial application of polyurethane as a coating to prevent cavitation occurred in the year 1996. In a study conducted by Liangmin Yu et al. [72], a polyurethane coating was applied to propellers and other components to assess its effectiveness in mitigating cavitation. The findings indicated that the coating exhibited superior anti-cavitation properties and adhesion. However, it was observed that the coating had a shorter service life. Hydrophobic polydimethylsiloxane-based polyurethanes (Si-PUx) were synthesized by Professor Rongrong Chen's research team at Harbin Engineering University [73]. The synthesis involved a polycondensation reaction utilizing hydroxypropyl polydimethylsiloxane (H-PDMS) and polytetramethylene glycol (PTMG) as the soft segments and 2,4-toluene diisocyanate, 1,4-butanediol, and triethanolamine as the hard segments. Based on the results obtained, it was observed that the cavitation resistance of Si-PUx coatings exhibited a consistent increase with the progressive augmentation of H-PDMS content. However, a decrease in the adhesion of Si-PUs coatings was observed as the level of H-PDMS increased. The figure depicted in Figure 6B illustrates the surface morphology of Si-PUx and epoxy coatings subjected to varying cavitation durations. It is noteworthy that the Si-PUx coating, which incorporated 12.5 wt% of H-PDMS, exhibited a remarkably low accumulated mass loss of merely 2.96 mg. Following a duration of 80 h of cavitation testing, the surface demonstrated exceptional resistance to cavitation as it remained free from any observable fractures or voids throughout the entire testing period. In their study,

Qinghua Dai et al. [74] employed a three-layer composite coating approach to develop a cavitation-resistant polyurethane coating. The top layer consisted of polyurethane, while the intermediate layer was composed of a combination of epoxy and polyurethane interpenetrating resin. The primer layer, on the other hand, was made of epoxy resin. This composite coating exhibited enhanced mechanical properties and demonstrated superior adhesion to multiple substrates. The coating exhibited minimal surface degradation following 200 h of cavitation experiments, indicating superior resistance to cavitation. The cavitation resistance of polyurethane coatings reinforced with carbon nanofibers (CNFs) in seawater was examined by Lee et al. [75] through the application of the ultrasonic vibration method. The findings of the study demonstrated that the polyurethane coating lacking cellulose nanofibers (CNFs) exhibited the least resistance to cavitation. This suggests that the incorporation of CNFs has the potential to enhance the coating's ability to withstand the impact pressure resulting from the rupture of cavitation bubbles. Meanwhile, the polyurethane coatings that were supplemented with fluorine exhibited exceptional resistance to cavitation. The observed phenomenon of enhanced resistance to cavitation in polyurethane may be attributed to the synergistic effect between CNF and fluorine, which influences the material's structural properties. In a study conducted by Ning Qiu et al. [67] at Zhejiang University, various coatings including epoxy, ceramic, and polyurethane were applied onto rigid alloy surfaces in order to evaluate their resistance against cavitation. The primary factors influencing the cavitation durability of the coatings were found to be the adhesion and thickness of the coatings. Furthermore, upon analyzing the degradation mechanism of the coatings, it was determined that polyurethane coatings exhibited a prolonged latent period, thereby enhancing the materials' resistance to cavitation.

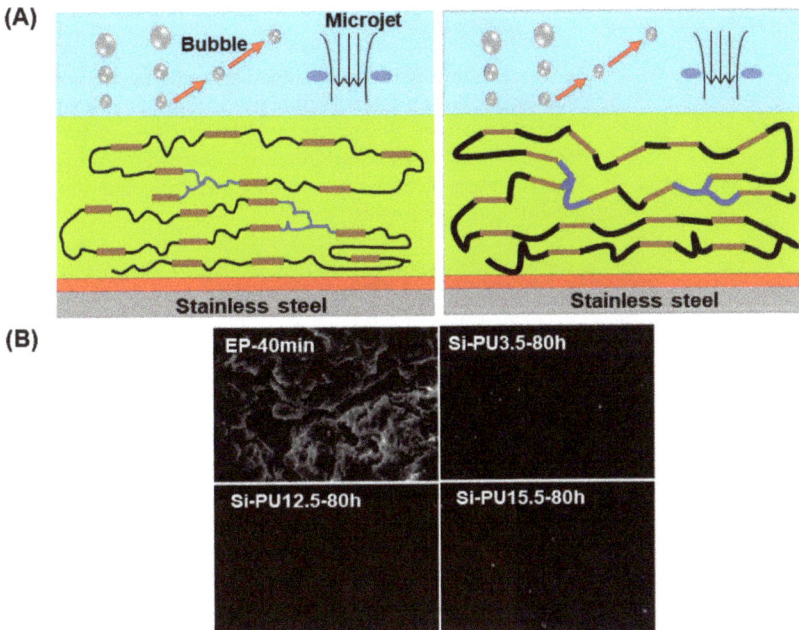

Figure 6. (**A**) Cavitation resistance mechanism of polyurethane elastomer coating [73]; (**B**) SEM image of Si-PUx and epoxy coatings at different cavitation times [73] with permission (Copyright © 2023, Wiley).

The emergence of polyurea as a corrosion-reducing coating with impact-resistant properties was initially observed in the 1980s [76]. Polyurea is a type of elastomeric substance that is produced through the chemical reaction between isocyanate groups and amino compounds. This material is characterized by its solvent-free nature, making it environmentally friendly. Additionally, it exhibits remarkable resistance to moisture and humidity, along with exceptional mechanical properties, including good corrosion and weather resistance. Furthermore, this material finds widespread application in various domains such as corrosion resistance, abrasion resistance, and waterproofing [76,77]. In their study, Shi Feng et al. [78] synthesized a range of polyurea anti-cavitation coatings by varying the concentration of MOCA, HDI trimer prepolymer, and polyaspartic ester. MOCA was employed as the chain extender in this two-component system. The findings indicate that the polyurea coatings produced exhibit favorable flexibility and notably superior resistance to cavitation compared to metal coatings. However, it is worth noting that these polyurea coatings have a relatively restricted lifespan. The researchers Marlin et al. [79] discovered that the failure of the coating was a result of the combined impact of locally transmitted impulse load and material heating. This conclusion was drawn based on their examination of the cavitation surfaces of the polyurea coatings. Furthermore, the resistance of the coating to cavitation is influenced by various factors, including the polyurea composition, the intensity of the cavitation field, the substrate material, the coating thickness, and several other contributing elements. Corrosion and temperature are also affected by these factors.

In comparison to polyurethanes, polymers featuring urea bonding exhibit enhanced responsiveness, thereby expediting the spraying procedure for polyurea coatings and simplifying their application on large-scale equipment such as ships. In the actual cavitation process, the use of polymer coatings may have some small polymer particles shedding into the ocean, but waterborne polymers do not have a significant impact on the marine environment and organisms, and most of them are environmentally friendly.

3.2. Cavitation-Resistant Inorganic Coatings

Inorganic coatings designed to mitigate cavitation on ship propellers encompass various options, such as copper alloy coatings, stainless steel coatings, nickel plating, and ceramic coatings. The predominant techniques employed in the modification of propeller surface metal anti-cavitation coatings consist of thermal spraying technology, surface laser modification technology, surface nitriding technology, and surface plasma modification technology, among others.

The phrase "thermal spraying technology" encompasses the procedure of elevating the temperature of the material intended for spraying until it reaches a molten or partially molten state. Subsequently, this material is expelled at a predetermined velocity to form a protective layer on a previously prepared substrate. The process exhibits traits such as operational flexibility, a high rate of material deposition, and the ability to control the thickness of the coating. Currently, the thermal spraying materials that are predominantly utilized encompass cobalt-based alloys, nickel-based alloys, and other similar materials [80]. In their study, Gu et al. [81] employed Cu95 as the coating material and employed thermal spraying technology to conduct propeller corrosion experiments. The findings of the study indicate that the coatings mentioned above exhibit exceptional resistance to cavitation and corrosion in seawater. Consequently, these coatings can be effectively employed as protective surface coatings for propellers made of manganese-copper alloy. Amarendra and colleagues [82] fabricated a thermal spray coating consisting of 70% nickel and 30% chromium through the utilization of the high-velocity oxygen fuel (HVOF) process. The researchers subsequently conducted an examination and comparison of the cavitation resistance between the thermal spray coating and the uncoated martensitic stainless steel (ss410). Furthermore, the researchers conducted experiments to assess the hardness, bending, and peeling characteristics of the specimens. The results indicated that, when subjected to identical conditions, the coated specimens exhibited significantly higher resistance to cavitation compared to the uncoated specimens. Moreover, the cavitation resistance of stainless steel 410 (ss410) can be significantly enhanced through the application of high-velocity oxygen fuel (HVOF) coating, which employs a combination of brittle and ductile erosion mechanisms. The aforementioned findings were revealed as a result of the analysis conducted on the specimens. In their study, Szala et al. [83] employed supersonic flame spraying as a technique to apply HVOF coatings of MCrAlY and NiCrMoFeCo onto a stainless steel substrate, specifically AISI310 (X15CrNi25-20). Subsequently, cavitation experiments were conducted using the vibration method. The results of the study indicate that the MCrAlY coating exhibited a comparatively reduced level of wear resistance in comparison to the NiCrMo coating. Meanwhile, the sliding wear resistance of the alloy demonstrates enhancement as the nickel content proportionately increases.

The technology of surface laser modification has its origins in the 1960s. It involves the utilization of high-energy density laser irradiation to treat the surface of metallic materials, resulting in the formation of an alloy layer with distinct properties. The introduction of alloy layers has the potential to alter the structural arrangement, hardness, density, and uniformity of the material, thereby enhancing its resistance to cavitation. Furthermore, the preservation of the material's bulkiness facilitates the development of the alloy system's microstructure and the formation of substable phases. The primary techniques encompassed within this method are laser surface alloying (LSA), laser cladding (LSC), laser melting condensation (LSM), and various other processes [84]. In their study, Lian Fen et al. [85] employed surface laser modification as a technique to create texturization on Ti6Al4V alloy. The grid-textured specimens exhibited superior strength and cavitation resistance compared to the straight-textured specimens, owing to their elevated surface hardness, broader hardness gradient, and even distribution of prominent high-hardness regions.

Laser surface alloying (LSA) is a technique that involves the use of laser light and solid-phase material to perform directional energy-beam-assisted surface alloying. The thermal phenomenon resulting from the interaction between the laser and solid-phase substances leads to the melting of the metal surface, including the alloying elements that have been introduced. Subsequently, rapid condensation occurs, resulting in the formation of a coating on the substrate's surface. The alloying elements can be incorporated using either the direct injection method or the pre-painted coating method [86]. In the study conducted by Yi [87], laser alloying technology was employed to fabricate high entropy alloyed coatings with diverse compositions on the surface of 304 stainless steel while ensuring controlled atmosphere conditions. Following a 5 h cavitation process, it was observed that the high entropy alloyed coating exhibited no material spalling on the surface of the sample material. Additionally, there was no apparent plastic deformation and a significant reduction in the cavitation phenomenon. These findings indicate that the coating demonstrates superior resistance to cavitation when compared to 304 stainless steel. Laser cladding (LSC) refers to the process of applying an alloy or composite material as a protective layer onto the surface of a substrate. The production of protective surface materials can greatly benefit from its extensive range of potential applications and value. To improve the mechanical properties and resistance to cavitation of 17-4PH stainless steel, Ding et al. [88] employed laser technology to implement two different surface treatments. One method entails the laser-induced solidification of tungsten-chromium-cobalt alloy powder, resulting in its hardening, whereas the other method involves the application of a laser coating onto the alloy surface. The findings indicate that the application of heat treatment and laser cladding techniques can potentially enhance the hardness, Young's modulus, resistance to plastic deformation, and resistance to cavitation of the alloy. Although the laser cladding treatment did not significantly enhance the hardness of the steel, it outperformed the laser heat treatment in terms of augmenting the steel's Young's modulus and resistance to cavitation. In a study conducted by Zhang Song et al. [89], a laser melting technique was employed to apply a NiCrBSi coating onto the surface of an aluminum alloy. The objective of this approach was to enhance the material's resistance against cavitation. The application of the laser melting solidification (LSM) technique results in an improved resistance to cavitation by promoting a more uniform distribution of applied force. Ren Yuhang [90] employed a combination of laser cladding technology and laser melting coagulation technology. Initially, a laser melting coagulation treatment was performed on the surface of stainless steel, followed by a subsequent treatment using laser melting cladding technology on NiTi alloy. This approach resulted in an enhancement of the material's resistance to cavitation.

The term "surface nitriding technology" encompasses various methods such as plasma nitriding and ion nitriding, which are employed to modify the surface properties of a material. These techniques aim to introduce a durable nitride coating that can effectively integrate with the substrate, thereby improving the material's resistance to cavitation. In a study conducted by Huang et al. [91] from Feng Chia University in Taiwan, it was observed that ion nitriding had a significant effect on enhancing the cavitation resistance of carbon steel. The cavitation resistance of 316L stainless steel in seawater was examined by Chong et al. [92] following ion nitriding at various temperatures. The plasma nitriding process was conducted with a N_2 to H_2 ratio of 1:4 at temperatures ranging from 400 to 500 °C for a duration of 10 h. An improvement in cavitation resistance was observed as the nitriding temperature was raised from 400 °C to 500 °C. When comparing the untreated samples to the treated ones, it was observed that the material exhibited a notable reduction in weight loss and damage rate within the temperature range of 400 to 500 °C. Additionally, the hardness, mechanical properties, and resistance to cavitation of the material exhibited a significant increase. In a study conducted by Szkodo et al. [93], stainless steel was subjected to nitriding treatment. The researchers observed that the surface layer of the nitrided stainless steel exhibited significantly enhanced cavitation resistance compared to the reference samples. In their study, Mitelea et al. [94] employed gas nitriding technology to fabricate coatings on the surface of aluminum alloys. This approach demonstrated a significant enhancement in the cavitation resistance of the alloys.

Currently, the predominant technique for surface plasma modification involves the introduction of metal ions, such as nitrogen (N) or boron (B), into the plasma to enhance surface hardness. Another commonly employed method is cathodic arc ion plating, which results in the deposition of a wear-resistant plating layer with high hardness. These modifications serve to enhance the material's resistance to cavitation. The cavitation resistance of cathodic arc ion plating is significantly influenced by the strength of the bond. Based on the findings of Krella et al. [95], it was observed that the TiN coating exhibits optimal bonding strength with the substrate at a temperature of 350 °C. Conversely, the CrN layer demonstrates superior bonding strength with the substrate at a temperature of 500 °C, resulting in reduced mass loss and improved resistance against cavitation.

In brief, the application of inorganic coatings has demonstrated the potential to mitigate material and economic detriments arising from cavitation, thereby enhancing anticavitation efficacy to a certain degree. However, in real-world scenarios encompassing corrosive settings like the ocean, as well as in critical components such as propellers necessitating prolonged operation at high velocities amidst overcurrent circumstances, metal coatings are susceptible to electrochemical corrosion, thereby diminishing their longevity to a certain extent. Simultaneously, the process requirements for these surface treatment technologies are comparatively stringent, thereby rendering their implementation more challenging. When comparing metal coatings to organic coatings, it is observed that organic coatings possess several advantages. These advantages include lower cost, increased operational flexibility, and the potential for certain organic coatings to exhibit elasticity. This elasticity enables the absorption of energy released during the cavitation process, thereby enhancing the material's resistance to cavitation. The issue of cavitation resistance has garnered significant interest in the realm of ship propeller protection in recent times. The utilization of organic coatings for this objective has witnessed a growing trend.

4. Integration of Antifouling and Anti-Cavitation Coatings

In the context of surfaces such as ship overflow parts and hydraulic turbine propeller blades, it is commonly observed that biofouling and cavitation corrosion occur concurrently, with an evident interplay between these phenomena. The presence of fouling organisms on a material leads to a decrease in its performance, as it promotes the formation of surface vacuoles. This, in turn, exacerbates the occurrence of cavitation-induced damage to the material. Additionally, the chloride ions in seawater can penetrate and destroy the surface of the materials and make them prone to localized corrosion such as pitting corrosion. The occurrence of cavitation erosion exacerbates the surface roughness of materials, such as propellers, thereby promoting the adhesion of marine fouling organisms. This creates a self-perpetuating cycle of fouling attachment [96].

In relation to the present market conditions, there is an expectation that the demand for antifouling coatings will exhibit a growth rate of 10% from 2021 to 2025, resulting in a market value of USD 1.96 billion. Notably, the Asia–Pacific region is projected to experience a substantial increase of 51%, equivalent to an 8.66% growth in 2021 alone. The demand for anti-cavitation coatings in the Asia–Pacific region accounts for 33% of the overall market demand. It is projected that the global demand for these coatings will experience a growth rate of 5% in the coming decade. By the year 2027, it is anticipated that the global market demand for anti-cavitation coatings will exceed USD 20 billion. Hence, the global imperative for the development of antifouling and anti-cavitation coatings is evident.

In 1987, Japanese scientists conducted an experiment involving the application of self-polishing antifouling coatings and silicone synthetic resins to propeller blades. Following 500 h of operation, the propeller blades exhibited minimal adherence of seafood and peeling of the paint film, thereby maintaining a notably clean condition [97]. The silicone coatings Intersleek® 700 and fluoropolymer coatings Intersleek® 900 were developed by the Netherlands International Paint Company (IP) as antifouling coatings with low surface energy properties. The application of a series of coatings was initially implemented on submarine propellers in 1995. Following a period of 12 months of utilization, it was observed that the coating remained in an exceptional condition. The Belzona2141 polyurethane resin coatings, manufactured by the British Belzona Company, are designed for application on various components such as propellers, turbines, valves, and other overflow equipment. The construction and operation of these devices are characterized by simplicity while also exhibiting a commendable degree of cavitation and corrosion resistance. The Metaline® series 700 two-component polyurethane coatings were manufactured by Germany's MetaLine Company. The findings indicated that the product exhibited improved resistance against fouling and cavitation simultaneously. Currently, there has been limited advancement in the incorporation of foreign antifouling and anti-cavitation coatings. This progress has primarily been observed in military applications, which do not encompass a significant portion of commercial and civil usage. Conversely, research on domestic antifouling and anti-cavitation integrated coatings is still in its early stages.

In the year 2011, scientists from China submitted a patent application with the objective of mitigating the corrosion of ship propellers and the accumulation of marine organisms. The initial step involved the removal of the oxide skin from the surface of copper alloy propellers to achieve a textured surface. Subsequently, a ceramic insulating coating composed of metal oxides was prepared using the thermal spraying technique. Following this, a metal antifouling coating was thermally sprayed onto the prepared ceramic coating, resulting in a composite coating that possesses both fouling and anti-cavitation characteristics. In their study, Weiwei Cong et al. [98] conducted research at the State Key Laboratory of Marine Coatings to develop a fouling release protective coating for propellers. The coating was prepared using an anticorrosive primer, elastic buffer paint, intermediate connecting paint, and antifouling topcoat. The researchers asserted that this coating exhibited both safety and environmental friendliness for construction personnel and the marine environment. Furthermore, they reported that the coating demonstrated exceptional efficacy in inhibiting the adhesion of marine fouling organisms and exhibited strong anti-cavitation performance. The assessment of the static antifouling efficacy was conducted at the Qingdao Zhongkang offshore test station, revealing that the surface condition remained favorable even after a duration of 36 months. In their study, Haocheng Yang et al. [99] conducted the synthesis of polyurethane (PU)/ZIF-8 (PHZ) composite coatings using a nanocomposite approach. The researchers utilized a multi-scale zeolite imidazolium skeleton material (ZIF-8) as a nano-filler, which is known for its environmentally friendly properties. The findings indicate that the ZIF-8 polyurethane coating exhibited superior antifouling and anti-cavitation properties at a particle size of 50 nm. The coverage of the selected fouling agent, small crescent-shaped rhododendron algae (Nitzschia clostridium), in the simulation experiment was found to be only 0.51%, as depicted in Figure 7A. The mass loss of the coating during the 30 h cavitation experiment amounted to a mere 9.9 milligrams. The incorporation of nanoparticles resulted in the improvement of the coating's hydrophobic properties, thermal stability, and mechanical characteristics. Nevertheless, it should be noted that ZIF-8 exhibits a diminished level of durability when exposed to saltwater, necessitating a gradual release of zinc ions over an extended duration in order to attain optimal antifouling efficacy. This, in turn, increases the probability of potential contamination in the secondary marine environment. Poly (dimethylsiloxane etherimide) (APT-PDMS) and poly (tetrahydrofuranediol) (PTMG) were utilized as raw materials in their study [100]. To introduce perfluorohexanediol (PFHT) into the polyurethane backbone, a two-step polymerization reaction was employed to synthesize a range of fluorinated isocyanate prepolymers (FIPs) with varying contents. This process resulted in the production of a fluoro silicone-containing polyurethane elastomer coating known as SFPU-x. The incorporation of fluorinated isocyanate prepolymers (FIPs) facilitates the formation of fluorinated polyurethane microdomains, thereby enabling the generation of microstructures with low surface energy. This is achieved through the enhancement of hard-segmented microzone structures within the coating. Based on the findings, it was observed that the SFPU-5 coating containing 5% FIP exhibited the most optimal level of internal microphase separation. The surface of this coating displayed a distinct microstructure, a significantly high water contact angle, and evident hydrophobic properties, as depicted in Figure 7B. These characteristics effectively hindered the adhesion of proteins and algae. The fouling simulation using bovine serum proteins resulted in a coverage of only 1.3% after a duration of 72 h. Additionally, the surface showed no discernible holes or cracks after 10 h of cavitation. The cumulative mass loss in deionized water and seawater was measured to be 2.7 mg and 2.9 mg, respectively, indicating a high level of cavitation resistance. Additionally, the investigation revealed that the microstructural characteristics of the coating surface have a significant impact on its ability to resist fouling and cavitation. The prevention of microorganism reproduction on the coating surface can be attributed to the presence of an inhomogeneous microstructure. This microstructure hinders the microorganisms' ability to reproduce through the "size-matching" effect and enrichment, thereby impeding their attachment and accumulation on the surface. The bubbles generated during the cavitation process undergo simultaneous collapse upon reaching the surface of the irregular microstructure. The region of the microstructure that is situated at its uppermost portion frequently experiences the lowest local pressure. Consequently, the area characterized by low external pressure is where the high internal pressure cavitation will ultimately collapse. The implosion force will be absorbed by the coating and transformed into both elastic and plastic deformation. This transformation leads to a reduction in the impact force, thereby preventing the occurrence of cracks and enhancing the quality of the loss, as depicted in Figure 7C.

For the sake of concision, it is clear that there is a sizable market need and substantial research relevance for the creation of an integrated coating that offers ship propellers both antifouling and anti-cavitation capabilities. However, the existing body of research on this topic is limited, necessitating a substantial number of studies to comprehensively understand and optimize the performance of propellers with the desired characteristics of "dynamic anti-cavitation and static fouling resistance". In light of the growing emphasis on environmental preservation, there has been a heightened interest in utilizing eco-friendly substances for various purposes. For instance, numerous antifouling agents found in nature, such as capsaicin, carvacrol, and halogenated furanone, have been identified as potential alternatives. Additionally, biomimetic materials, such as lotus leaves, sharkskin, and dolphins, have also garnered attention for their potential applications in this domain. They can be employed to fabricate high-quality integrated coatings that possess both antifouling and anti-cavitation properties. Furthermore, it is feasible to fabricate coatings that possess anti-cavitation properties and are functionalized with antifouling agents. This can be achieved through the process of gradient compounding, wherein antifouling functionalized nanoparticles are incorporated into the matrix. This approach can be explored from two research perspectives: modification of micro and nano functionalized fillers, as well as the design of the polymer chain segment structure.

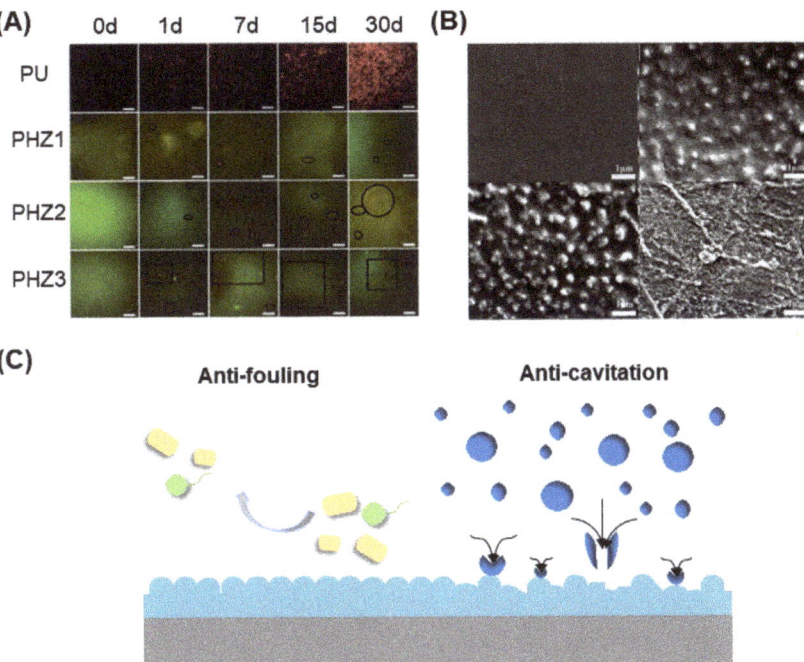

Figure 7. (A) Fluorescence microscope images of algae on PU, PHZ1, PHZ2, and PHZ3 coatings at different immersion times with a scale of 50 μm [99] with permission (Copyright © 2023, Elsevier); (B) SEM image of the microstructure of polyurethane-coated SFPU-x surface with a scale bar of 1 μm [100] with permission (Copyright © 2023, Elsevier); (C) Mechanism of antifouling and anti-cavitation of microphase separation structure.

5. Conclusion and Prospects

In brief, the global research community is actively engaged in investigating and addressing the challenges associated with ship propeller technology, with a particular focus on dynamic anti-cavitation and static fouling. The condition of the propeller has a direct influence on the ship's performance and lifespan, particularly as the maritime environment undergoes changes and the ship's service life extends. There is currently a disparity between domestic and international antifouling and anti-cavitation integrated technologies, hindering the full understanding and implementation of environmentally friendly coatings in this field. There exist numerous unresolved matters that necessitate the attention and ongoing investigation of scientific and technological professionals. These concerns primarily manifest in the subsequent domains.

Firstly, previous research has focused on addressing various issues related to the antifouling coating, such as enhancing its efficiency through functionalization, ensuring its environmental friendliness, and improving its strength and adhesion to counteract cavitation effects. Previous studies have proposed various solutions for addressing the issue at hand. These include the utilization of fouling release antifouling coatings and bionic antifouling coatings as alternatives to the environmentally harmful self-polishing antifouling coatings, with the aim of achieving enhanced antifouling efficacy. Simultaneously, the utilization of organic materials instead of inorganic ones can contribute to the mitigation of corrosion, particularly in seawater environments. This approach offers advantages over inorganic coatings, which are more susceptible to the flaws associated with electrochemical corrosion. This will potentially enhance the durability of the coating to a certain degree.

Furthermore, it is imperative for future researchers to prioritize the exploration of practical strategies aimed at comprehensively resolving the diverse challenges associated with integrated coatings for antifouling and anti-cavitation purposes. By employing rational molecular structure design and surface modification techniques, as well as incorporating functional fillers through compounding, it is possible to enhance the antifouling and anti-cavitation properties of ship propellers. Simultaneously, this approach can also strengthen the bond between the integrated coating and the underlying base material. This enables us to enhance and evaluate ship propeller protective coatings that possess the capability to deliver exceptional performance and extended durability in real-world marine conditions.

Author Contributions: Conceptualization, J.Z. and W.Q.; software, D.W. and Z.F.; investigation, W.C.; resources, L.L.; writing—original draft preparation, J.Z.; writing—review and editing, W.C. and J.Z.; visualization, J.Z.; supervision, Y.W.; project administration, W.Q. All authors have read and agreed to the published version of the manuscript.

Funding: This research received no external funding.

Institutional Review Board Statement: Not applicable.

Informed Consent Statement: Not applicable.

Data Availability Statement: Available in the paper.

Conflicts of Interest: The authors declare no conflict of interest.

References

1. Bertram, V. Chapter 2—Propellers. In *Practical Ship Hydrodynamics*, 2nd ed.; Butterworth-Heinemann: Oxford, UK, 2012; pp. 41–72.
2. Liang, L.; Pang, Y.; Tang, Y.; Zhang, H.; Liu, H.; Liu, Y. Combined wear of slurry erosion, cavitation erosion, and corrosion on the simulated ship surface. *Adv. Mech. Eng.* **2019**, *11*, 1687814019834450. [CrossRef]
3. Syamsundar, C.; Usha, S.R.I.P. Computational Analysis of Cavitation Structures on a Ship Propeller. In Proceedings of the International Conference on Emerging Trends in Engineering, Hyderabad, India, 22–23 March 2019.
4. Flemming, H.C.; Murthy, P.S.; Venkatesan, R.; Murthy, S.P. ; Cooksey. *Marine and Industrial Biofouling*; Springer: Duisburg, Germany, 2009; pp. 3–12.
5. DüRR, S.; Watson, D. Biofouling and Antifouling in Aquaculture. *Biofouling* **2010**, *12*, 267–287.
6. Anderson, C.D.; Hunter, J.E. International Conference on Ship and Shipping Research. In Proceedings of the NAV2000 Conference Proceedings, Venice, Italy, 19–22 September 2000.
7. Carchen, A.; Atlar, M.; Turkmen, S.; Pazouki, K.; Murphy, A.J. Ship performance monitoring dedicated to biofouling analysis: Development on a small size research catamaran. *Appl. Ocean Res.* **2019**, *89*, 224–236. [CrossRef]
8. Bixler, G.D.; Bhushan, B. Biofouling: Lessons from nature. *Phil. Trans. R. Soc. A* **2012**, *370*, 2381–2417. [CrossRef] [PubMed]
9. Pettengill, J.B.; Wendt, D.E.; Schug, M.D.; Hadfield, M.G. Biofouling likely serves as a major mode of dispersal for the polychaete tubeworm Hydroides elegans as inferred from microsatellite loci. *Biofouling* **2007**, *23*, 161–169. [CrossRef] [PubMed]
10. Yang, Y.; Wikieł, A.J.; Dall'agnol, L.T.; Eloy, P.; Genet, M.J.; Moura, J.J.G.; Sand, W.; Dupont-Gillain, C.C.; Rouxhet, P.G. Proteins dominate in the surface layers formed on materials exposed to extracellular polymeric substances from bacterial cultures. *Biofouling* **2016**, *32*, 95–108. [CrossRef] [PubMed]
11. Rosenhahn, A.; Schilp, S.; Kreuzer, H.J.; Grunze, M. The role of "inert" surface chemistry in marine biofouling prevention. *Phys. Chem. Chem. Phys.* **2010**, *12*, 4275–4286. [CrossRef] [PubMed]
12. Qian, P.Y.; Lau, S.C.K.; Dahms, H.U.; Dobretsov, S.; Harder, T. Marine Biofilms as Mediators of Colonization by Marine Macroorganisms: Implications for Antifouling and Aquaculture. *Mar. Biotechnol.* **2007**, *9*, 399–410. [CrossRef]
13. Hamilton, W.A. Microbially Influenced Corrosion as a Model System for the Study of Metal Microbe Interactions: A Unifying Electron Transfer Hypothesis. *Biofouling* **2003**, *19*, 65–76. [CrossRef]
14. Roberts, D.E.; Rittschof, D.; Holm, E.R.; Schmidt, A.R. Factors influencing initial larval settlement: Temporal, spatial and surface molecular components. *J. Exp. Mar. Biol. Ecol.* **1991**, *150*, 203–221. [CrossRef]
15. Liu, D.; Shu, H.; Zhou, J.; Bai, X.; Cao, P. Research Progress on New Environmentally Friendly Antifouling Coatings in Marine Settings: A Review. *Biomimetics* **2023**, *8*, 200. [CrossRef] [PubMed]

16. Barnaby, S.W. *Marine Propellers; Being a Course of Three Lectures Delivered at the Royal Naval College*; E. & F.N. Spon: London, UK, 1887.
17. Kazasidis, M.; Yin, S.; Cassidy, J.; Volkov-Husović, T.; Vlahović, M.; Martinović, S.; Kyriakopoulou, E.; Lupoi, R. Microstructure and cavitation erosion performance of nickel-Inconel 718 composite coatings produced with cold spray. *Surf. Coat. Technol.* **2020**, *382*, 125195. [CrossRef]
18. Brennen, C.E. *Fundamentals of Multiphase Flow*; Cambridge University Press: Cambridge, UK, 2005.
19. Moholkar, V.S.; Kumar, P.S.; Pandit, A.B. Hydrodynamic cavitation for sonochemical effects. *Ultrason. Sonochem.* **1999**, *6*, 53–65. [CrossRef] [PubMed]
20. Zhang, H.; Ren, X.; Luo, C.; Tong, Y.; Larson, E.A.; Lu, Z.; Gu, J. Study on transient characteristics and influencing of temperature on cavitation bubbles in various environments. *Optik* **2019**, *187*, 25–33. [CrossRef]
21. Wood, R.J.K. Erosion–corrosion interactions and their effect on marine and offshore materials. *Wear* **2006**, *261*, 1012–1023. [CrossRef]
22. Jing, Y.J.; Dachuan, H.E. Research Progress on Corrosion and Prevention Measure of Marine Propeller. *Corros. Sci. Prot. Technol.* **2019**, *31*, 443–448.
23. Li, K.; Zhuo, X.F.; Guan, F.; Qian, Z.; Zhang, M.; Duan, J.; Hou, B. Progress on Materials and Protection Technologies for Marine Propeller. *J. Chin. Soc. Corros. Prot.* **2017**, *37*, 495–503.
24. Kiil, S.; Weinell, C.E.; Pedersen, M.S.; Dam-Johansen, K. Mathematical Modelling of a Self-Polishing Antifouling Paint Exposed to Seawater: A Parameter Study. *Chem. Eng. Res. Des.* **2002**, *80*, 45–52. [CrossRef]
25. Kiil, S.; Weinell, C.E.; Pedersen, M.S.; Dam-Johansen, K. Analysis of self-polishing antifouling paints using rotary experiments and mathematical modeling. *Ind. Eng. Chem. Res.* **2001**, *40*, 3906–3920. [CrossRef]
26. De Mora, S.J.; Stewart, C.; Phillips, D. Sources and rate of degradation of tri(n-butyl)tin in marine sediments near Auckland, New Zealand. *Mar. Pollut. Bull.* **1995**, *30*, 50–57. [CrossRef]
27. Yebra, D.M.; Kiil, S.; Dam-johansen, K.; Weinell, C. Reaction rate estimation of controlled-release antifouling paint binders: Rosin-based systems. *Prog. Org. Coat.* **2005**, *53*, 256–275. [CrossRef]
28. Omae, I. General aspects of tin-free antifouling paints. *Chem. Rev.* **2003**, *103*, 3431–3448. [CrossRef] [PubMed]
29. Yonehara, Y.; Yamashita, H.; Kawamura, C.; Itoh, K. A new antifouling paint based on a zinc acrylate copolymer. *Prog. Org. Coat.* **2001**, *42*, 150–158. [CrossRef]
30. Omae, I. Organotin antifouling paints and their alternatives. *Appl. Organomet. Chem.* **2003**, *17*, 81–105. [CrossRef]
31. Ye, Z.; Chen, S.S.; Ma, C.F.; Wu, J.; Zhang, G. Development of Novel Environment-friendly Antifouling Materials. *Surf. Technol.* **2017**, *46*, 62–70.
32. Jiang, X.Z.; Huang, C.S.; Huang, Z.X.; Ren, R.; Wang, J. Study on Synthesis of Glyphosate—Acrylate Resinand Its Inhibition Toxicity on Microbes. *Paint Coat. Ind.* **2012**, *42*, 55–58.
33. Jiang, X.Z. The Synthesis of Glyphosate acrylic resin And Their Antifouling Properties. Master's Thesis, Wuhan University of Technology, Wuhan, China, 2012.
34. Feng, K.; Ni, C.; Yu, L.; Zhou, W.; Li, X. Synthesis and evaluation of acrylate resins suspending indole derivative structure in the side chain for marine antifouling. *Colloids Surf. B* **2019**, *184*, 110518. [CrossRef]
35. Li, Y.; Chen, R.; Feng, Y.; Liu, L.; Sun, X.; Tang, L.; Takahashi, K.; Wang, J. Antifouling behavior of self-renewal acrylate boron polymers with pyridine-diphenylborane side chains. *New J. Chem.* **2018**, *42*, 19908–19916. [CrossRef]
36. Ma, J.; Ma, C.; Yang, Y.; Xu, W.; Zhang, G. Biodegradable Polyurethane Carrying Antifoulants for Inhibition of Marine Biofouling. *Ind. Eng. Chem. Res.* **2014**, *53*, 12753–12759. [CrossRef]
37. Patterson, A.; Wenning, B.M.; Rizis, G.; Calabrese, D.R.; Finlay, J.; Franco, S.C.; Zuckermann, R.; Clare, A.; Kramer, E.; Ober, C.; et al. Role of backbone chemistry and monomer sequence in amphiphilic oligopeptide- and oligopeptoid-functionalized PDMS- and PEO-based block copolymers for marine antifouling and fouling release coatings. *Macromolecules* **2017**, *50*, 2656–2667. [CrossRef]
38. Zhao, Z.; Ni, H.; Han, Z.; Jiang, T.; Xu, Y.; Lu, X.; Ye, P. Effect of Surface Compositional Heterogeneities and Microphase Segregation of Fluorinated Amphiphilic Copolymers on Antifouling Performance. *ACS Appl. Mater. Interfaces* **2013**, *5*, 7808–7818. [CrossRef] [PubMed]
39. Brady, R. Fouling-release Coatings for Warships. *Def. Sci. J.* **2005**, *55*, 75–81. [CrossRef]
40. Truong, V.K.; Webb, H.K.; Fadeeva, E.; Chichkov, B.N.; Wu, A.H.F.; Lamb, R.; Wang, J.Y.; Crawford, R.J.; Ivanova, E.P. Air-directed attachment of coccoid bacteria to the surface of superhydrophobic lotus-like titanium. *Biofouling* **2012**, *28*, 539–550. [CrossRef] [PubMed]
41. Noguer, A.C.; Olsen, S.M.; Hvilsted, S.; Kill, S. Diffusion of surface-active amphiphiles in silicone-based fouling-release coatings. *Prog. Org. Coat.* **2017**, *106*, 77–86. [CrossRef]
42. Galhenage, T.P.; Hoffman, D.; Silbert, S.D.; Stafslien, S.J.; Daniels, J.; Miljkovic, T.; Finlay, J.A.; Franco, S.C.; Clare, A.S.; Brian T Nedved, B.T.; et al. Fouling-Release Performance of Silicone Oil-Modified Siloxane-Polyurethane Coatings. *ACS Appl. Mater. Interfaces* **2016**, *8*, 29025–29036. [CrossRef] [PubMed]
43. Wood, C.D.; Truby, K.; Stein, J.; Wiebe, D.; Holm, E.; Wendt, D.; Smith, C.; Kavanagh, C.; Montemarano, J.; Swain, G.; et al. Temporal and spatial variations in macrofouling of silicone fouling-release coatings. *Biofouling* **2000**, *16*, 311–322. [CrossRef]

44. Lejars, M.; Margaillan, A.; Bressy, C. Fouling release coatings: A nontoxic alternative to biocidal antifouling coatings. *Chem. Rev.* **2012**, *112*, 4347–4390. [CrossRef]
45. Yarbrough, J.C.; Rolland, J.P.; Desimone, J.; Callow, M.E.; Finlay, J.A.; Callow, J.A. Contact Angle Analysis, Surface Dynamics, and Biofouling Characteristics of Cross-Linkable, Random Perfluoropolyether-Based Graft Terpolymers. *Macromolecules* **2006**, *39*, 2521–2528. [CrossRef]
46. Hu, P.; Xie, Q.; Ma, C.; Zhang, G. Fouling resistant silicone coating with self-healing induced by metal coordination. *Chem. Eng. J.* **2021**, *406*, 126870. [CrossRef]
47. Wang, T.; Zhang, J.; Cai, Y.; Xu, L.; Yi, L. Protein-Resistant Amphiphilic Copolymers Containing Fluorosiloxane Side Chains with Controllable Length. *ACS Appl. Polym. Mater.* **2022**, *4*, 7903–7910. [CrossRef]
48. Chen, Y.; Song, Q.; Zhao, J.; Gong, X.; Schlaad, H.; Zhang, G. Betulin-Constituted Multiblock Amphiphiles for Broad-Spectrum Protein Resistance. *ACS Appl. Mater. Interfaces* **2018**, *10*, 6593–6600. [CrossRef]
49. Gudipati, C.S.; Finlay, J.A.; Callow, J.A.; Callow, M.E.; Wooley, K.L. The antifouling and fouling-release performance of hyperbranched fluoropolymer (HBFP)-poly(ethylene glycol) (PEG) composite coatings evaluated by adsorption of biomacromolecules and the green fouling alga Ulva. *Langmuir* **2005**, *21*, 3044–3053. [CrossRef] [PubMed]
50. Martinelli, E.; Agostini, S.; Galli, G.; Chiellini, E.; Glisenti, A.; Pettitt, M.A.; Callow, M.E.; Callow, J.A.; Graf, K.; Bartels, F.W. Nanostructured films of amphiphilic fluorinated block copolymers for fouling release application. *Langmuir* **2008**, *24*, 13138–13147. [CrossRef] [PubMed]
51. Maan, A.M.C.; Hofman, A.H.; Vos, W.M.; Kamperman, M. Recent Developments and Practical Feasibility of Polymer-Based Antifouling Coatings. *Adv. Funct. Mater.* **2020**, *30*, 2000936. [CrossRef]
52. Prieto, I.M.; Paola, A.; Pérez, M.; García, M.; Blustein, G.; Schejter, L.; Palermo, J.A. Antifouling Diterpenoids from the Sponge Dendrilla antarctica. *Chem. Biodivers.* **2022**, *19*, e2021006182022. [CrossRef] [PubMed]
53. De Nys, R.; Steinberg, P.D.; Willemsen, P.; Dworjanyn, S.A.; Gabelish, C.L.; King, R.J. Broad spectrum effects of secondary metabolites from the red alga delisea pulchra in antifouling assays. *Biofouling* **1995**, *8*, 259–271. [CrossRef]
54. Hao, X.; Chen, S.; Qin, D.; Zhang, M.; Li, W.; Fan, J.; Wang, C.; Dong, M.; Zhang, J.; Cheng, F.; et al. Antifouling and antibacterial behaviors of capsaicin-based pH responsive smart coatings in marine environments. *Mater. Sci. Eng. C* **2020**, *108*, 110361. [CrossRef] [PubMed]
55. Vitali, A.; Stringaro, A.; Colone, M.; Muntiu, A.; Angiolella, L. Antifungal Carvacrol Loaded Chitosan Nanoparticles. *Antibiotics* **2021**, *11*, 11. [CrossRef]
56. Zhang, H.L.; Gao, Y.B.; Gai, J.G. Guanidinium-functionalized nanofiltration membranes integrating anti-fouling and antimicrobial effects. *J. Mater. Chem. A* **2018**, *6*, 6442–6454. [CrossRef]
57. Qian, P.Y.; Wong, Y.H.; Zhang, Y. Changes in the proteome and phosphoproteome expression in the bryozoan Bugula neritina larvae in response to the antifouling agent butenolide. *Proteomics* **2010**, *10*, 3435–3446. [CrossRef]
58. Ma, C.; Zhang, W.; Zhang, G.; Qian, P. Environmentally Friendly Antifouling Coatings Based on Biodegradable Polymer and Natural Antifoulant. *ACS Sustain. Chem. Eng.* **2017**, *5*, 6304–6309. [CrossRef]
59. Tong, Z.; Rao, Q.; Chen, S.; Song, L.; Hu, J.; Hou, Y.; Gao, X.; Lu, J.; Zhan, X.; Zhang, Q. Sea slug inspired smart marine antifouling coating with reversible chemical bonds: Controllable UV-responsive coumarin releasing and efficient UV-healing properties. *Chem. Eng. J.* **2022**, *429*, 132471. [CrossRef]
60. Carman, M.L.; Estes, T.G.; Feinberg, A.W.; Schumacher, J.F.; Wilkerson, W.; Wilson, L.; Callow, M.; Callow, J.; Brennan, A. Engineered antifouling microtopographies—Correlating wettability with cell attachment. *Biofouling* **2006**, *22*, 11–21. [CrossRef] [PubMed]
61. Zheng, J.; Song, W.; Huang, H.; Chen, H. Protein adsorption and cell adhesion on polyurethane/Pluronic® surface with lotus leaf-like topography. *Colloids Surf. B* **2010**, *77*, 234–239. [CrossRef] [PubMed]
62. He, J.; Du, Y.; Wang, B.; Wang, X.; Ye, Q.; Liu, S. Grafting embedded poly(ionic liquid) brushes on biomimetic sharklet resin surface for anti-biofouling applications. *Prog. Org. Coat.* **2021**, *157*, 106298. [CrossRef]
63. Prime, K.L.; Whitesides, G.M. Self-Assembled Organic Monolayers: Model Systems for Studying Adsorption of Proteins at Aurfaces. *Science* **1991**, *252*, 1164–1167. [CrossRef] [PubMed]
64. Ionov, L.; Synytska, A.; Kaul, E.; Diez, S. Protein-Resistant Polymer Coatings Based on Surface-Adsorbed Poly(aminoethyl methacrylate)/Poly(ethylene glycol) Copolymers. *Biomacromolecules* **2010**, *11*, 233–237. [CrossRef] [PubMed]
65. Imbesi, P.M.; Finlay, J.A.; Aldred, N.; Eller, M.; Felder, S.E.; Pollack, K.; Lonnecker, A.T.; Raymond, J.E.; Mackay, M.; Schweikert, E.; et al. Targeted surface nanocomplexity: Two-dimensional control over the composition, physical properties and anti-biofouling performance of hyperbranched fluoropolymer–poly(ethylene glycol) amphiphilic crosslinked networks. *Polym. Chem.* **2012**, *3*, 3121–3131. [CrossRef]
66. Chen, L.; Guo, Z.; Liu, W. Biomimetic Multi-Functional Superamphiphobic FOTS-TiO(2) Particles beyond Lotus Leaf. *ACS Appl. Mater. Interfaces* **2016**, *8*, 27188–27198. [CrossRef]
67. Qiu, N.; Wang, L.; Wu, S.; Likhachev, D.S. Research on cavitation erosion and wear resistance performance of coatings. *Eng. Fail. Anal.* **2015**, *55*, 208–223. [CrossRef]
68. Caccese, V.; Light, K.; Berube, K.A. Cavitation erosion resistance of various material systems. *Ships Offshore Struct.* **2006**, *1*, 309–322. [CrossRef]

69. Xiao, S.; Laux, K.A.; Wang, H.; Hu, F.; Sue, H. Physical correlation between abrasive wear performance and scratch resistance in model polyurethane elastomers. *Wear* **2019**, *418*, 281–289. [CrossRef]
70. Xie, F.; Zhang, T.; Bryant, P.; Kurusingal, V.; Colwell, J.M.; Laycock, B. Degradation and stabilization of polyurethane elastomers. *Prog. Polym. Sci.* **2019**, *90*, 211–268. [CrossRef]
71. Akindoyo, J.O.; Beg, M.D.H.; Ghazali, S.; Islam, M.R.; Jeyaratnam, N.; Yuvaraj, A.R. Polyurethane types, synthesis and applications—A review. *RSC Adv.* **2016**, *6*, 114453–114482. [CrossRef]
72. Yu, L.M. Research on Anti-Cavitation Corrosion Coating for Hydraulic Turbine Working Wheel and Ship Propeller. Master's Thesis, Ocean University of China, Qingdao, China, 1996.
73. Qiao, X.; Chen, R.; Zhang, H.; Liu, J.; Liu, Q.; Yu, J.; Liu, P.; Wang, J. Outstanding cavitation erosion resistance of hydrophobic polydimethylsiloxane-based polyurethane coatings. *J. Appl. Polym. Sci.* **2019**, *136*, 47668. [CrossRef]
74. Dai, Q.H. Study on Preparation and Property of Anti-cavitation Composite Coating. Master's Thesis, China Academy of Machinery Science and Technology Group, Wuhan, China, 2017.
75. Lee, J.H.; Kim, J.H.; Kim, Y.P.; Kim, S.J. Evaluation of Anti-Cavitation Performance of Polyurethane Coatings in Seawater using Ultrasonic Vibratory Method. *J. Weld. Join.* **2019**, *37*, 455–462. [CrossRef]
76. Rijensky, O.; Rittel, D. Polyurea coated aluminum plates under hydrodynamic loading: Does side matter? *Int. J. Impact Eng.* **2016**, *98*, 1–12. [CrossRef]
77. Amini, M.R.; Isaacs, J.B. Nemat-nasser S. Investigation of effect of polyurea on response of steel plates to impulsive loads in direct pressure-pulse experiments. *Mech. Mater.* **2010**, *42*, 628–639. [CrossRef]
78. Shi, F. Study on Preparation of Polyaspartic Polyurea and Its Property of Anti-Cavitation. Master's Thesis, China Academy of Machinery Science and Technology Group, Wuhan, China, 2011.
79. Marlin, P.; Chahine, G.L. Erosion and heating of polyurea under cavitating jets. *Wear* **2018**, *414*, 262–274. [CrossRef]
80. Ding, X.; Huang, Y.; Yuan, C.; Ding, Z. Deposition and cavitation erosion behavior of multimodal WC-10Co4Cr coatings sprayed by HVOF. *Surf. Coat. Technol.* **2020**, *392*, 125757. [CrossRef]
81. Gu, X.B.; Chen, D.M.; Yang, D.M. Research on preventing the corrosion of manganese-copper alloy propellers by thermal spraying layer. *Ship. Eng.* **1997**, *6*, 33–34,53–56.
82. Amarendra, H.J.; Prathap, M.; Karthik, S.; Darshan, B.M.; Devaraj; Girish, P.C.; Runa, V.T. Combined Slurry and Cavitation Erosion Resistance of Hvof Thermal Spray Coated Stainless Steel. *Mater. Today Proc.* **2017**, *4*, 465–470. [CrossRef]
83. Szala, M.; Walczak, M.; Łatka, L.; Gancarczyk, K.; Özkan, D. Cavitation Erosion and Sliding Wear of MCrAlY and NiCrMo Coatings Deposited by HVOF Thermal Spraying. *Adv. Mater. Sci.* **2020**, *20*, 26–38. [CrossRef]
84. Kwok, C.T.; Man, H.C.; Cheng, F.T.; Lo, K. Developments in laser-based surface engineering processes: With particular reference to protection against cavitation erosion. *Surf. Coat. Technol.* **2016**, *291*, 189–204. [CrossRef]
85. Lian, F.; Zhang, H.C.; Gao, Y.Z.; Pang, L. Influence of Surface Texture and Surface Film on Cavitation Erosion Characteristics of Ti6Al4V Alloy. *Rare. Metal. Mat. Eng.* **2011**, *40*, 793–796.
86. Quazi, M.; Fazal, M.A.; Haseeb, A.S.M.A.; Yusof, F.; Masjuki, H.H.; Arslan, A. Laser-based Surface Modifications of Aluminum and its Alloys. *Crit. Rev. Solid. State Mater. Sci.* **2015**, *41*, 1–26. [CrossRef]
87. Yi, J.C. Preparation and Abrasion Properties of LaserHigh-Entropy Alloying Coatings. Master's Thesis, Shenyang University of Technology, Shenyang, China, 2015.
88. Ding, Y.P.; Yao, J.H.; Liu, R.; Wang, L.; Zhang, Q.L.; Sheng, J.J.; Xue, C.G. Effects of Surface Treatment on the Cavitation Erosion–Corrosion Performance of 17-4PH Stainless Steel in Sodium Chloride Solution. *J. Mater. Eng. Perform.* **2020**, *29*, 2687–2696. [CrossRef]
89. Zhang, S.; Zhang, C.H.; Liu, C.S.; Man, H.C. Cavitation Erosion Behaviour on Aluminium Alloy by Using Laser Surface Cladding of NiCrBSi. *Rare Metal Mater. Eng.* **2002**, *2*, 99–102.
90. Ren, Y.H. Preparation of Laser Modification Layer on 17-4ph Stainless Steel and Study On its Cavitation erosion Performance. Master's Thesis, Shenyang University of Technology, Shenyang, China, 2013.
91. Huang, W.H.; Chen, K.C.; He, J.L. A study on the cavitation resistance of ion-nitrided steel. *Wear* **2002**, *252*, 459–466. [CrossRef]
92. Chong, S.O.; Kim, S.J. Characterization of Cavitation-Erosion Resistance of Plasma Ion Nitrided 316L Stainless Steel Under Shock Wave in Seawater. *J. Nanosci. Nanotechnol.* **2019**, *19*, 3943–3949. [CrossRef]
93. Szkodo, M.; Sitko, A.; Gazda, M. Cavitation Erosion Resistance of Austenitic Stainless Steel after Glow-Discharge Nitriding Process. *Solid. State Phenom.* **2011**, *183*, 201–206. [CrossRef]
94. Mitelea, I.; Dimian, E.M.; Bordeasu, I.; Crăciunescu, C. Cavitation Erosion of Laser-Nitrided Ti–6Al–4V Alloys with the Energy Controlled by the Pulse Duration. *Tribol. Lett.* **2015**, *59*, 1–9. [CrossRef]
95. Krella, A.; Czyżniewski, A. Cavitation erosion resistance of Cr–N coating deposited on stainless steel. *Wear* **2006**, *260*, 1324–1332. [CrossRef]
96. Yang, H.C. Preparation of Polyolefin Polyurethane Coatings and Their Antifouling and Anti-cavitation Erosion Performance. Ph.D. Thesis, Harbin Engineering University, Harbin, China, 2021.
97. Shi, X.Z. Research Progress of Coatings Used on Marine Propeller. *Corros. Sci. Prot. Technol.* **2017**, *29*, 199–203.
98. Cong, W.W.; Gui, T.J.; Zhang, K.; Wang, Z.; Lv, Z. Property Study of Marine Propeller Coating for Ship. *Mater. Rep.* **2021**, *35*, 367–371.

99. Yang, H.; Guo, X.; Chen, R.; Liu, Q.; Liu, J.; Yu, J.; Lin, C.; Wang, J.; Zhang, M. Enhanced anti-biofouling ability of polyurethane anti-cavitation coating with ZIF-8: A comparative study of various sizes of ZIF-8 on coating. *Eur. Polym. J.* **2021**, *144*, 110212. [CrossRef]
100. Yang, H.; Zhang, M.; Chen, R.; Liu, Q.; Liu, J.; Yu, J.; Zhang, H.; Liu, P.; Lin, C.; Wang, J. Polyurethane coating with heterogeneity structure induced by microphase separation: A new combination of antifouling and cavitation erosion resistance. *Prog. Org. Coat.* **2021**, *151*, 106032. [CrossRef]

Disclaimer/Publisher's Note: The statements, opinions and data contained in all publications are solely those of the individual author(s) and contributor(s) and not of MDPI and/or the editor(s). MDPI and/or the editor(s) disclaim responsibility for any injury to people or property resulting from any ideas, methods, instructions or products referred to in the content.

Review

Polymer Waveguide-Based Optical Sensors—Interest in Bio, Gas, Temperature, and Mechanical Sensing Applications

Svetlana N. Khonina [1,2], Grigory S. Voronkov [3], Elizaveta P. Grakhova [3], Nikolay L. Kazanskiy [1,2], Ruslan V. Kutluyarov [3] and Muhammad A. Butt [1,*]

1 Samara National Research University, 443086 Samara, Russia
2 IPSI RAS-Branch of the FSRC "Crystallography and Photonics" RAS, 443001 Samara, Russia
3 Ufa University of Science and Technology, 32, Z. Validi Street, 450076 Ufa, Russia
* Correspondence: butt.m@ssau.ru

Abstract: In the realization of photonic integrated devices, materials such as polymers are crucial. Polymers have shown compatibility with several patterning techniques, are generally affordable, and may be functionalized to obtain desired optical, electrical, or mechanical characteristics. Polymer waveguides are a viable platform for optical connectivity since they are easily adaptable to on-chip and on-board integration and promise low propagation losses <1 dB/cm. Furthermore, polymer waveguides can be made to be extremely flexible, able to withstand bending, twisting, and even stretching. Optical sensing is an interesting field of research that is gaining popularity in polymer photonics. Due to its huge potential for use in several industries, polymer waveguide-based sensors have attracted a lot of attention. Due to their resilience to electromagnetic fields, optical sensors operate better in difficult situations, such as those found in electrical power generating and conversion facilities. In this review, the most widely used polymer materials are discussed for integrated photonics. Moreover, four significant sensing applications of polymer-waveguide based sensors which include biosensing, gas sensing, temperature sensing and mechanical sensing have been debated.

Keywords: polymer; waveguide; optical sensor; temperature sensor; biosensor; gas sensor

1. Introduction

Polymers have the potential to be useful for many passive and active sub-components. One benefit of polymers over other types of materials is that their physical and optical characteristics may be greatly customized by adjusting the composition and level of polymerization. By including the proper molecular moieties in the polymer chain or as side pendants, functionality may be introduced [1]. Different techniques may be used to process polymers, such as solution and gas-phase deposition, and they can be made compatible with substrate chemistry by appropriate surface functionalization (including inorganic building blocks). Large-scale, inexpensive production of polymers is also a possibility [2]. The optical waveguide (WG) is one of the fundamental components of integrated photonics [3–6]. Polymer WGs can operate in either single-mode (with core diameters between 2 µm and 5 µm) or multimode (with core dimensions generally between 30 µm and 500 µm) regimes. They are both entirely consistent with the matching optical fiber type due to the similar mode field diameter. Numerous methods, including photolithography [7], flexographic and inkjet printing [8], nanoimprint lithography [9,10], femtosecond laser processing [11], and hot embossing [12], can be used to create these WGs. A stamp structure is transferred from a stamp onto a substrate through the hot embossing process, which is a replication process. The method is appropriate for replicating structures with a millimeter to nanoscale size. Due to its potential for mass production at low cost and integration with roll-to-roll processes, hot embossing is a desirable manufacturing method for optical applications [13]. By using sustainable and efficient hybrid lithography, a three-dimensional polymer WG with a taper structure was exhibited and created [14]. A polymer WG and a polymer taper structure were created using grayscale lithography and hybrid lithography, respectively. Gray-scale lithography was intended for laser ablation and shadow aluminum evaporation. The laser strength, the rate of ablation, and the thickness of the aluminum may all be adjusted to alter the length of the grayscale zone, which ranges from 20 to 400 μm [14].

Due to their flexible processability and integration over inorganic counterparts, polymer optical WG devices are crucial in several rapidly evolving broadband communications domains, including optical networking, metropolitan/access communications, and computer systems [15]. Owing to their many benefits, they are also a perfect integration platform for the insertion of foreign material systems like YIG (yttrium iron garnet) and lithium niobate in addition to semiconductor devices like lasers, detectors, amplifiers, and logic circuits into etched grooves in planar lightwave circuits to enable full amplifier modules or optical add/drop multiplexers on a single substrate. Additionally, optical polymers may be vertically combined to produce 3D and even all-polymer integrated optics because of their flexibility and durability combination [14].

Polymer WG-based optical sensors can be competitors to devices based on photonic integrated circuits (PICs). Such devices are mainly manufactured based on MPW (multi-project wafer) on SOI, SiN platforms, or based on Group III–V semiconductors, primarily InP or GaAs. Type III–V platforms can provide a wide range of active devices, but they are poorly applicable to passive elements due to high attenuation and low contrast [16–18]. Figure 1 shows the emission wavelength coverage of semiconductor lasers based on III–V platforms [19]. One can see that their emission region is lower than the Si transparency window. The transparency range of a polymer WGs depends on the specific polymer material being used. In general, the transparency range is in the near-infrared (NIR) region of the electromagnetic spectrum, usually from 700 nm to 1700 nm. Therefore, one can easily combine them with III–V group light sources in the sensing system's design. However, some polymers have higher transparency windows extending into the visible region or even beyond the short-wave infrared (SWIR) region [20,21]. It is also worth noting that the transparency range of polymer WGs can be affected by various factors such as processing conditions, doping, and absorption or scattering losses. SOI platforms, which are well-compatible with traditional CMOS electronics, have high contrast and small allowable bend radii [22], making it possible to design small-sized passive sensor devices but not active components. For example, a sensor based on a Mach–Zehnder interferometer (MZI) with a double-slot hybrid plasmonic WG [23] provides a high sensitivity of up to 1061 nm/RIU in liquid refractometry. The micro-ring resonator-based sensors' sensitivity on the SOI platform is less than 100 nm/RIU. However, the application of a subwavelength grating micro-ring makes it possible to achieve a sensitivity of 672.8 nm/RIU [24,25]. Hybrid integration significantly expands the capabilities of the SOI and III–V platforms, allowing the design of complex sensing systems, including active devices [26,27]. It is also possible to expand the capabilities of the SOI platform using IMOS (indium phosphide membrane on silicon) technology [28].

The paper is organized in the following manner. Section 2 provides a piece of information on the characteristics of polymer WGs. There are several polymer materials commercially available and being used in research for the development of photonic devices as discussed in Section 3. The extraordinary optical properties of these polymer materials include low optical losses at operating wavelengths, well-controlled and tunable refractive indices, resistance to temperature and chemicals, mechanical stability in a variety of environments, and environmentally friendly fabrication techniques. Afterward in Section 4, polymer WG-based sensors are discussed. We have dedicated our research to polymer WG-based biosensors (Section 4.1), gas sensors (Section 4.2), temperature sensors (Section 4.3), and mechanical sensors (Section 4.4), which are at present the main focal point of investigation. The paper finishes with a brief conclusion and outlook as presented in Section 5. The applications presented in this paper are shown in Figure 2.

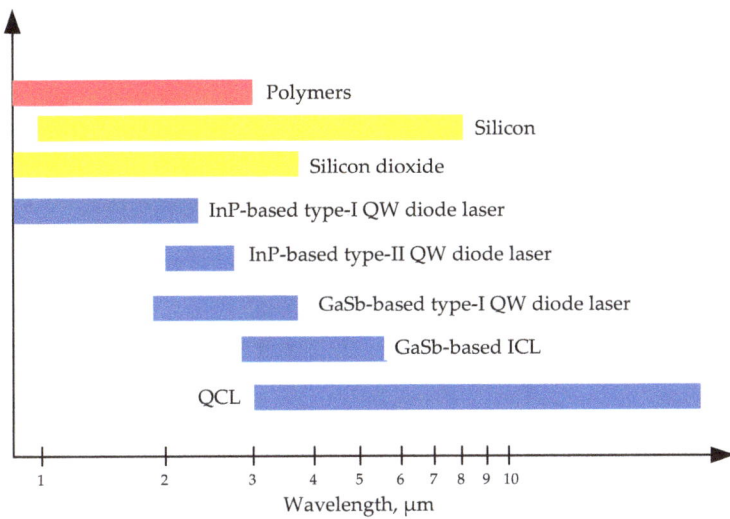

Figure 1. Transparent window of polymers, silicon, and silicon dioxide, and emission wavelength coverage of semiconductor lasers based on different III–V active regions. InP-based type-I, type-II, and GaSb-based type-I quantum well (QW) diode lasers, GaSb-based interband cascade lasers (ICLs), and quantum cascade lasers (QCLs) are included [19].

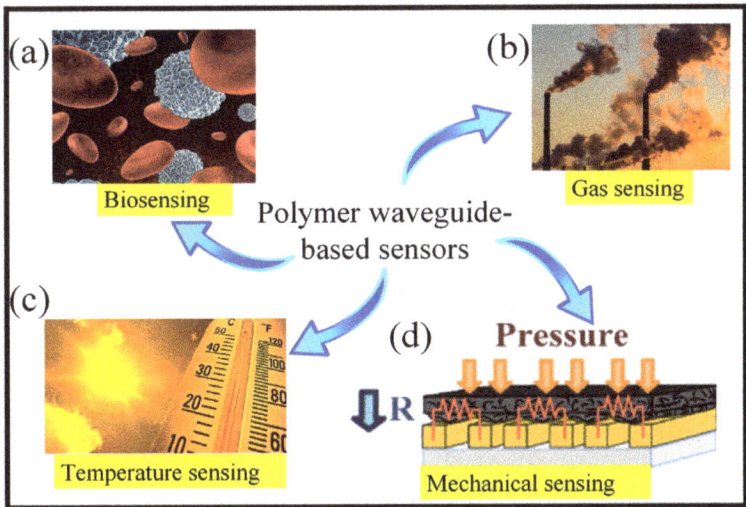

Figure 2. Polymer WG-based sensors employed in (**a**) biosensing [29], (**b**) gas sensing [30], (**c**) temperature sensing [31], and (**d**) mechanical sensing applications [32], are discussed in this review.

2. Characteristics of Polymer Waveguides

Using straightforward spin-coating processes, low-temperature processing, and compatibility with semiconductor electronics, optical polymers enable the flexible, large-area, and inexpensive production of WG devices [13]. These WGs are relatively cheap to produce compared to other materials such as glass or silicon. This makes them a popular choice for cost-sensitive applications. Glass, quartz, oxidized silicon, glass-filled epoxy printed circuit board substrate, and flexible polyimide film are just a few examples of the numerous rigid and flexible substrates that can be employed. Film thicknesses between 0.1 and 100 microns can be achieved by adjusting the polymer/solvent ratio and spin speed in the film coating process. Contrary to other optical material systems, polymers are created and designed by chemically altering their component molecules to have the desired properties, such as melt or solution processibility in the form of monomers or prepolymers, enhanced mechanical properties from photo- or thermo-crosslinking, and matched refractive indices between the core and cladding layers. Polymer materials have high flexibility, making them ideal for applications where flexible WGs are required, such as flexible displays or wearable devices [33]. These WGs are much lighter than traditional glass or silicon WGs, making them easier to integrate into portable or mobile devices.

Additionally, these characteristics are modifiable by formulation changes. In addition to the traditional photoresist patterning, there are other methods for forming polymer WGs, such as direct lithography, soft lithography, embossing, molding, and casting. This enables the quick and inexpensive structuring of components like active films, Faraday rotators, or half-wave plates for the creation of WGs as well as material removal for grating. This adaptability also makes polymers an excellent hybrid integration platform, allowing for the insertion of semiconductor devices like lasers, detectors, and logic circuits into an etched groove in a planar lightwave circuit to facilitate full amplifier modules or optical add/drop multiplexers on a single substrate. These foreign material systems include Yttrium iron garnet (YIG), lithium niobate, and foreign material systems [34].

The fact that polymeric materials' refractive indices change more quickly with temperature than more traditional optical materials like the glass is a key distinction between the two. When ambient temperature increases, the refractive index of the polymer material drops at a rate of $10^{-4}/°C$, which is significantly—faster than inorganic glasses. As opposed to the interferometric devices that are required in silica-on-silicon, for example, this large thermo-optic coefficient and poor thermal conductivity allow for the realization of thermo-optic switches with low power consumption, digital thermo-optic switches, and thermo-optic switches based on adiabatic WG transitions.

Polymers are more sensitive to humidity compared to inorganic materials. The impact of humidity on the refractive index may also be examined using the return-loss method [35]. If the core and cladding modifications were different, the refractive index change caused by humidity would have an impact on single-mode WG performance. It would also have an impact on the device's return loss if index matching were utilized as a return loss reduction technique. Devices like Bragg gratings and AWGs may function differently even though the core and cladding change is the same because the humidity may alter the WG's effective index.

Many optical systems rely on the exclusion of any wavelength-dependent optical effects outside those that were physically intended. Material dispersion should thus typically be avoided. The values for the polymers on the order of 10^{-6} nm^{-1} are substantially lower than those for semiconductors or doped glasses, although being equivalent to those for SiO_2 [7]. Because polymeric materials are susceptible to yellowing with thermal aging, the thermal stability of optical qualities is a crucial property for practical applications. Such aging is often caused by the production of partly conjugated chemical groups, which exhibit wide UV absorption bands and weaken over the visible spectrum. The chemical composition of the original polymer has a significant impact on this yellowing.

Polymer materials have a limited operating temperature range, which can make them unsuitable for high-temperature environments. These materials are more prone to mechanical degradation, such as cracking or breaking, compared to other materials such as glass or silicon. They are also sensitive to environmental factors such as moisture, heat, and UV light, which can affect the performance and longevity of the WG. Overall, polymer WGs have some advantages such as low cost and flexibility, but also have some limitations in terms of optical quality and mechanical stability. Whether or not to use polymer WGs will depend on the specific requirements and constraints of the application.

3. Polymer Materials for Integrated Optics and Fabrication Methods

In the past, optical crystals like lithium niobate [36], lithium tantalate [37], and rubidium titanyl phosphate [38,39] as well as semiconductors like silicon [4,5,40], silicon nitride [41,42], indium phosphide [43,44], III–V compound, and silica have been utilized to create optical WGs. In recent years, there has been much study on highly integrated optics and photonic devices made of polymers. Polymer materials, as compared to the materials mentioned above, have easier manufacturing procedures, which result in produced optical devices with much-reduced material and production costs [45]. As a result, new types of polymers for optical and photonics uses have been produced in several laboratories throughout the world in recent decades, and some of them are now accessible on the market. These materials include UV-curable epoxy polymers Su-8, EpoCore/EpoClad [46,47], siloxane LIGHTLINKTM XP-6701A core, LIGHTLINKTM XH-100145 clad, and benzocyclobutene (Dow Chemical, Midland, MI, USA) Polymers [48], ZPU resin, and polymers (ChemOptics Inc., Daejeon, Republic of Korea) [49], OrmoClear®FX (micro resist technology GmbH), and SUNCONNECT are examples of inorganic–organic hybrid polymers (Nissan Chemical Ltd., Tokyo, Japan) [20,50], Truemode Backplane Polymer (Dow Corning, Midland, MI, USA) [51], UV exposure optical elastomer OE-4140 core, and OE-4141 cladding (Exxelis, Ltd., Washington, DC, USA) [52], Sylgard 184, LS-6943, polydimethylsiloxane (PDMS) [53], etc.

Polymethyl Methacrylate (PMMA) is a widely used polymer material for optical WGs due to its transparency, high refractive index, and low cost. Polycarbonate (PC) is a thermoplastic polymer that is commonly used for optical WGs due to its high transparency, high mechanical strength, and good thermal stability. Polystyrene (PS) is a transparent polymer that is widely used for WGs due to its low cost, easy processability, and good transparency in the visible and near-infrared regions of the spectrum. Polyimide (PI) is a heat-resistant polymer that is widely used for high-temperature applications due to its excellent mechanical stability, high transparency, and high refractive index. Polyethylene (PE) is a flexible and lightweight polymer that is widely used for applications such as fiber-optic sensing and flexible displays due to its low cost and high transparency. Polyvinyl chloride (PVC) is a low-cost polymer that is widely used for optical WGs due to its good processability and high transparency in the visible and near-infrared regions of the spectrum. Cytop and cyclic olefin copolymer (COC) polymer materials are widely used for WGs in optical communication systems and other optical applications. These materials have a low-loss, high-refractive-index polymer material that offers a combination of transparency, high mechanical strength, and good thermal stability. They are some of the most used polymer materials for optical WGs, but many other polymers can also be used, depending on the specific requirements of the application.

These polymer materials have exceptional optical qualities, for instance, low optical losses at working wavelengths (including in the infrared spectrum), well-controlled and tunable refractive indices, resistance to temperature and chemicals, mechanical stability in a variety of environments, and environmentally friendly fabrication methods [54,55]. Polymers are used to create common types of fiber sensors: fiber Bragg gratings (FBG) [56,57], surface plasmon resonance (SPR) sensors [58,59], and intensity variation-based sensors [60,61]. To realize optical and photonic devices, optical planar WGs are essential building blocks. Several distinct manufacturing methods for polymer optics WG devices have been documented. These fabrication techniques include the use of mask photolithography in conjunction with wet etching [62], photo-resist patterning and reactive ion etching [63], two-photon polymerization [64], laser direct writing [47], electron beam writing [65], flexographic and inkjet printing [8], the hot embossing process [66–68], photo-bleaching [69], and others. These processes need numerous processing steps, which can result in lengthy fabrication periods and poor yield. For mass production, technologies like stamping processes [70] are being researched. The roll-to-roll (R2R) nanoimprint lithography technologies [71,72] and roll-to-plate nanoimprinting are two examples of these techniques. Polymers, which are excellent material candidates for applications needing low-cost mass manufacturing, can be used with these roller-based methods [72]. Flexible electronics are created using these methods [73]. Additionally, current research is looking into potential uses in optics and photonics. Additional information on thin-film coating processes can be found in [13]. We should notice that the soft-lithography fabrication process provides high geometrical accuracy [74], which guarantees sensors' fabrication reproducibility.

Several researchers have looked at the development of optical WGs using hot embossing in recent years [12,66,68,75]. Numerous stamp production methods, including LIGA technology, photolithography, micro-machining, and etching, were investigated. The investigation of several polymer materials also included thermoplastics and photoresists [76]. The manufactured WGs were utilized for optical communication and sensing applications and worked in the single-mode and multi-mode regimes. The WG transmission loss varies from several dB/cm to values under 1 dB/cm depending on the embossing stamp and the polymer materials. Despite its advantages, the hot embossing process also has some disadvantages such as requiring specialized equipment, which can be expensive and may not be readily available for all users. The hot embossing process requires high temperatures, which can cause thermal stress and degradation of the polymer material. It requires complex tooling, such as molds and stamps, which can be difficult to design and fabricate. Moreover, it is limited to certain types of polymer materials such as thermoplastics, which may not be suitable for all applications. The surface finish of the embossed component may not be as high as other fabrication methods, which can affect the optical properties of the component and can result in non-uniform embossing, which can then affect the performance of the component. Furthermore, it may not be suitable for high-volume production, as it is a relatively slow and labor-intensive process. Table 1 shows the most frequently used polymer materials and their physical characteristics.

Table 1. Most commonly used polymer materials and their characteristics.

Polymer	Chemical Formula	Young Modulus	Optical Loss	Transparency Region, nm	Refractive Index
NOA 73	N/A	11 MPa [77]	N/A	370–1250 [78]	1.559 [78]
PDMS	$(CH_3)_3SiO[Si(CH_3)_2O]nSi(CH_3)_3$ [79]	1.32–2.97 MPa [80]	0.027 dB/cm [81]	400–1600 [81]	$n(T) = 1.4176 - 4.5 \times 10^{-4}T$ [82]
PMMA [9,83]	$(C_5H_8O_2)n$ [84]	3–3.7 GPa [85]	2.5 dB/cm [83]	400–700 with C-H absorption peak at 630 nm [86]	1.48–1.505 [87]
NOA 68	N/A	N/A	N/A	450–1250 [88]	1.54 [88]
COC	cyclic olefin copolymer	29–237 MPa [89]	0.5 dB/cm (830 nm); 0.7 dB/cm (1550 nm) [90]	N/A	1.5–1.54 [90]
SU-8	epoxy polymer	4.54 GPa at 19 kHz to 5.24 GPa at 318 kHz [91]	1.36 and 2.01 dB/cm (TE_{00} and TM_{00}) [46]	N/A	1.67 at UV [92]
Ormocer	inorganic-organic hybrid polymers	1–17,000 MP [93]	0.64 dB/cm [94]	N/A	From 1.5382 to 1.59 at 633 nm [95]
ZIF-8	2-Methylimidazole zinc salt	3 GPa [96]	N/A	N/A	1.355 ± 0.004 at 589 nm [97]
PHMB	polihexanide	N/A	N/A	N/A	1.48–1.5 [98]
BCB	benzocyclobutene	9.58 GPa [99]	0.81 dB/cm at 1300 nm [100]	N/A	1.5589 at λ = 632 nm; 1.5489 at λ = 838 nm [101]
FSU-8	fluorinated epoxy resin	N/A	N/A	N/A	1.495 to 1.565 at 1550 nm [102]
NOA 63	N/A	N/A	N/A	350–1250	1.56 [103]
PEI	polyethylenimine	3.5–3.6 GPa [104]	N/A	400–800 [27,105]	1.66 at 546 nm [106]
PSS	polystyrene sulfonate	N/A	N/A	N/A	N/A
PAH	polyallylamine hydrochloride	100 MPa [107]	N/A	N/A	1.51 at 550 nm [108]
Ma-P 1205	n/a	N/A	N/A	N/A	1.644 at 633 nm [109]
PMATRIFE	poly(2,2,2) MethAcrylate of TRIFluoro-Ethyle	N/A	N/A	N/A	1.409 [110]
PC [33]	plastic polycarbonate	N/A	10 dB/m	750–850	1.586 [111]
CYTOP [33]	amorphous perfluorinated polymer	N/A	10 db/km	950–1100	1.34 at 587.6 nm
PDLLA	poly(D,L-lactide)	N/A	From 0.4 dB/cm on 500 nm to 0.12 dB/cm on 800 nm with C-H adsorbtion peak at 720–740 nm [45]	500–850 [86]	N/A

Rapid and extensive duplication of structures with dimensions ranging from the microscale to centimeter scale is possible using hot embossing [68]. Figure 3 provides a detailed illustration of the fabrication processes. The top and lower phases, respectively, were initially covered with a structured stamp and a PMMA sheet. Then, they were heated to 140 °C, which is both lower than the PMMA melting point and higher than the glass transition temperature. The PMMA sheet was in a rubber condition at this temperature, making it appropriate for the ensuing embossing of structural patterns. The structural stamp was applied with an embossing force to the PMMA sheet during the embossing process. A further cooling procedure was used to maintain the embossing force while chilling the stamp and PMMA sheet. A further cooling procedure was used to maintain the embossing force while chilling the stamp and PMMA sheet. The embossing force was withdrawn once the temperature reached the demolding temperature (50 °C). Manual separation of the PMMA sheet and stamp is possible [68].

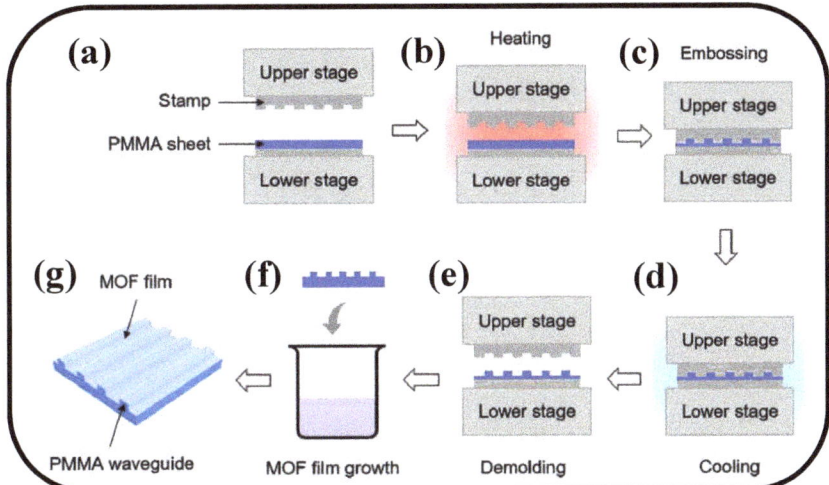

Figure 3. Manufacturing process of MOF-coated WGs by employing hot embossing, (**a–g**) Step by step fabrication process of PMMA based WG [68].

4. Polymer WG-Based Sensors

The characteristics and needs of a certain application will determine how a sensor is designed for that purpose. It is helpful to list the qualities that one would need in a perfect sensor for chemical and biological species. In a perfect world, the sensor would have sufficient sensitivity (in certain situations down to the level of a single molecule) and a wide dynamic range. High selectivity for the target species and resistance to sample-matrix interferences should also be present. A perfect sensor would also be suited for multicomponent measurements, have a quick reaction time, be reversible, and have high long-term stability. The ideal sensor should also be capable of self-calibration and be strong, dependable, easy to make, affordable, and have a compact size.

The most significant benefit of optical sensors over other types is their broad range of applications: optical sensors may detect analytes for which other (bio)chemical sensors are ineffective [112]. Additionally, optical sensors can be used for "indirect" analyte identification, which makes use of an auxiliary reagent, as well as "direct" analyte detection, in which the spectroscopic characteristics of the analyte are detected. Such a reagent experiences a modification in an optical property, such as elastic or inelastic scattering, optical path length, absorption, luminescence intensity, luminescence lifespan, or polarization state, when it interacts with the analyte species. This type of indirect detection is essential because it merges chemical selectivity with the spectroscopic measurement's capabilities and frequently overcomes interference issues that would otherwise be problematic.

There are several subcategories of optical sensors. There are "intrinsic" and "extrinsic" sensors, for example. A WG is simply utilized as a light link to connect external instruments to a sampling point or an optical sensing element in an extrinsic sensor. In biomedicine, environmental monitoring, process control, and safety, extrinsic sensors are already widely employed [113]. Due to several appealing qualities they possess, including simplicity, electrical passivity, inherent safety, chemical robustness, thermal tolerance, compatibility with telemetry, and use of a common technology to produce sensors intended for various (bio)chemical and physical measurements, they have found success in both commercial and practical purposes. Extrinsic sensors may also frequently be made simpler, work with in vivo applications, and be applied in situations that are contaminated by electromagnetic and microwave waves.

In an intrinsic sensor, the WG itself contributes to the conversion of (bio)chemical data into an analytical signal that is usable [114]. In addition to the common qualities of most extrinsic sensors, intrinsic sensors offer a variety of useful features [115,116]. By using evanescent-wave-based optical discrimination, intrinsic sensors may be employed to study interfacial processes and thin films. By adjusting the parameters of the waveguiding configuration and, in the case of indirect sensors, the parameters of immobilized reagents, they also provide flexibility in the selection of their sensing mode and their analytical properties. For instance, the architecture of the WG, its cladding, and any reagents nearby the WG can all affect how much an evanescent wave interacts [117,118]. The choice of the physical, chemical, or biological characteristics of the analyte species to be monitored is often the first stage in the construction of a good sensor. One could decide to analyze an analyte's absorption, fluorescence, or Raman spectra, for instance. This decision will determine the sensor's instrumental needs as well as its analytical performance. It is frequently helpful to take into consideration a combination of detection techniques since doing so may frequently result in a significant performance gain. A fluorescence-based sensor combined with time-resolved detection, for instance, can enhance a sensor's selectivity, sensitivity, and long-term stability.

Surface plasmon resonance (SPR) [119], grating, micro-ring [120–122], and MZI [123] structures are only a few of the basic structural types found in polymer optical WG sensors, as shown in Figure 4a–d [124–126]. A commercial chemical detection device based on prism-coupling technology is included in the SPR structure. Its limitations include susceptibility to temperature and test media composition, as well as optical loss due to the gold film. The grating structure has a very low grating period—only a few hundred nanometers—but it also has demanding production and spectral analysis constraints. Temperature and outside stress might affect longer grating times. The micro-ring structure's radius is many tens of micrometers, making it desirable for sensor downsizing. The equivalent contact length of the WG and test medium is greatly increased by the optical signal resonance phenomena to provide adequate sensitivity. Nevertheless, it is difficult to manage the coupling distance between the micro-ring and straight WG without using a high-precision construction procedure. The micro-ring may also result in more bending loss.

In the MZI structure, one WG branch acts as the sensing arm and the other branch as the reference arm. The evanescent field of the light wave in the core layer may be made to make contact with the test medium by detaching the upper cladding layer of the WG from the sensing arm [127]. The refractive index (RI) of the test medium varies, which affects how the optical fields between the sensing and reference arms are phased. The output optical power varies because of the phase shift. The MZI construction is affordable, simple to construct, and capable of simultaneous multi-channel identification without the need for spectral detection. An integrated MZI-based methane sensor that is covered in a styrene-acrylonitrile film that contains cryptophane-A is presented [126]. A supramolecular molecule called cryptophane-A that can selectively trap methane increases sensitivity by a factor of 17 when it is present in the cladding [126].

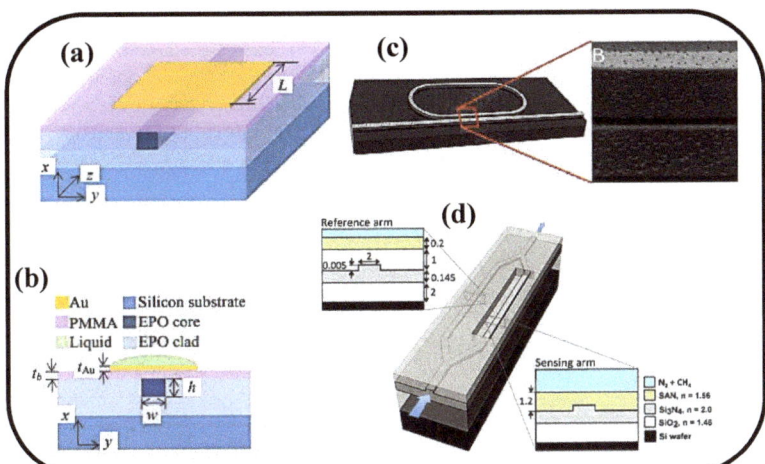

Figure 4. Different types of sensors based on polymer WGs, (**a**,**b**) SPR [124], (**c**) Micro-ring resonator [125], (**d**) MZI structure [126].

4.1. Polymer WG-Based Low-Cost Biosensors

Today, biosensors are widely used in biological diagnostics as well as a variety of other fields, including forensics, environmental monitoring, food control, drug development, and point-of-care monitoring of illness progression [128,129]. The creation of biosensors may make use of a broad variety of methodologies. As a result of their interaction with high-affinity biomolecules, a variety of analytes may be sensitively and specifically detected [130,131]. Every biosensor contains a particular set of static and dynamic properties [132,133]. The efficiency of the biosensor is enhanced by the optimization of these properties [134].

4.1.1. Selectivity

Perhaps the most crucial component of a biosensor is selectivity. A bioreceptor's selectivity refers to its capacity to identify a particular analyte in a sample that contains various admixtures and impurities. The interplay of an antigen and an antibody is the greatest illustration of selectivity. Antibodies often serve as receptors and are immobilized on the transducer's surface. The antigen is then introduced to a solution (often a buffer including salts), which is subjected to the transducer, where antibodies only bind with the antigens. Selectivity is the key factor to be taken into account while developing a biosensor [135].

4.1.2. Limit of Detection

The limit of detection (LOD) of a biosensor is the lowest conc. of analyte that it can detect [136]. A biosensor is necessary for several medical and environmental monitoring applications to confirm the existence of traces of analytes in a sample at analyte conc. as low as ng/mL or even fg/mL. For instance, prostate cancer is linked to blood levels of the prostate-specific antigen (PSA) of 4 ng/mL, for which doctors recommend biopsy procedures. As a result, LoD is thought to be a key characteristic of a biosensor.

4.1.3. Stability

The biosensing system's stability refers to how susceptible it is to environmental perturbations within and outside of it. A biosensor under study may experience a drift in its output signals because of these disruptions [137]. This may result in a conc. measurement inaccuracy and compromise the biosensor's quality and precision. In situations where a biosensor needs lengthy incubation periods or ongoing monitoring, stability is the most important component. The reaction of electronics and transducers may be temperature-sensitive, which might affect a biosensor's stability. To achieve a steady response from the sensor, proper tuning of the electronics is necessary. The degree to which the analyte attaches to the bioreceptor—the affinity of the bioreceptor—can also have an impact on stability. High-affinity receptors promote the analyte's covalent or strong electrostatic connection, which strengthens a biosensor's stability. The deterioration of the bioreceptor over time is another element that has an impact on a measurement's stability.

4.1.4. Repeatability

The biosensor's repeatability refers to its capacity to provide the same results under identical testing conditions. The transducer and electronics in a biosensor are precise and accurate, which defines repeatability. When a sample is tested more than once, accuracy refers to the sensor's capability to offer a mean value that is near to the real value while precision refers to the sensor's ability to produce identical findings every time. The inference built on a biosensor's response is very reliable and robust when the signals are reproducible [137].

The optical sensors based on evanescent wave monitor changes in the RI. These sensors make use of the confinement of the electromagnetic waves in a dielectric and/or metal structure to generate a propagating or localized electromagnetic mode. The evanescent wave is created when a portion of the confined light disperses into the surrounding medium as shown in Figure 5. Local changes in the optical properties of the excited electromagnetic mode, most significantly a change in the effective RI, are brought on by RI shifts in the surrounding medium through this evanescent wave [116]. When a receptor layer is affixed to the guiding structure's surface, the corresponding analyte is exposed to it, and the resulting (bio)chemical interactions between them change the RI locally. The interaction's amplitude may be measured by comparing it to the analyte's conc. and the interaction's affinity constant. The evanescent wave only reaches the exterior medium up to hundreds of nanometers and perishes exponentially; consequently, the background from the external medium will be little impacted. As a result, only variations near the sensor surface will be noticed.

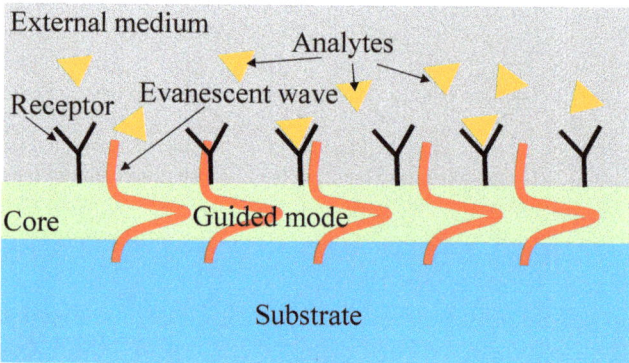

Figure 5. Evanescent field sensing mechanism.

The interferometric evanescent wave sensing technique is used by planar-integrated optical biosensors to provide highly sensitive label-free detection of biomolecules. Using injection molding and spin-coating, a novel polymer WG device design is proposed that enables the production of disposable sensor chips at a cheap cost [138]. To couple light into and out of the biosensor, surface grating couplers and lateral tapers were included. By adding a thin layer of inorganic high-index material to these polymer gratings, the coupling strength is improved, allowing for grating size reduction and effective lateral tapering into single-mode WGs [138].

For detecting effective medication doses of ginkgolide A for the suppression of pulmonary microvascular endothelial cell (PMVEC) apoptosis, a fluorinated cross-linked polymer Bragg WG grating-based optical biosensor is developed [139]. PMMA was created as the sensing window cladding and fluorinated photosensitive polymer SU-8 (FSU-8) as the sensing core layer. Pharmacological experiments were used to examine and examine the ginkgolide A medication conc. range that was most efficient for inhibiting PMVEC apoptosis (5–10 g/mL). The device's framework was built to be created and manufactured using direct UV writing technology. With varying refractive indices of various drug conc., the characteristics of the biosensor were simulated. The biosensor's sensitivity was determined to be 1606.2 nm/RIU. The limit of detection (LoD) and resolution were specified as 3×10^{-5} RIU and 0.05 nm, respectively [139].

A multilayer polymeric-inorganic composite WG arrangement was used to produce an evanescent field sensor [140]. The low RI polymer layers are covered by layers of a Ta_2O_5 thin film that was produced by sputtering, creating the composite WG structure. According to the results, the polymer-based sensor can detect molecules adhering to surfaces down to a limit of around 100 fg/mm^2 for molecular adsorption detection and an LoD of 3×10^{-7} RIU for RI sensing. The inorganic coating on the polymer layer was discovered to successfully limit water absorption into the WG, which led to stabilized sensor operation in addition to greatly increasing sensitivity. By examining antibody–antigen binding interactions, it was demonstrated that the created sensor can be used in precise molecular detection [140].

For the detection of biomolecules in the lower nano-molar (nM) range, a surface plasmon resonance (SPR) biosensor based on a planar-optical multi-mode polymer WG structure is presented in [141]. With a measuring resolution of 4.3×10^{-3} RIU, the fundamental sensor exhibits a sensitivity of 608 nm/RIU when subjected to variations in RI. C-reactive protein (CRP) was detected in a buffer solution with a response of 0.118 nm/nM by integrating the SPR sensor with an aptamer-functionalized, gold-nanoparticle (AuNP)-enhanced sandwich assay. The biosensor is highly suited for low-cost disposable lab-on-a-chip operations because of the multi-mode polymer WG structure and the straightforward concept. It may also be employed with very straightforward and affordable equipment. The sensor specifically offers the ability for quick and multiplexed identification of various biomarkers on a single integrated technology [141].

There is much interest in polymer-based materials used in photonic circuits, such as benzocyclobutene (BCB), SU8, and PMMA, for label-free, cost-effective biosensing and communications applications [142,143]. It is simple to embed optical components and electronics into polymers [144]. In comparison to low-contrast polymer-based WGs, high-index contrast materials like silicon and silicon nitride give substantial loss inside wall scattering [145,146]. This eliminates any manufacturing limitations and enables the construction of polymer-based WGs with a large footprint in a silicon wafer and minimal side wall scattering loss. These polymer materials may be used in a variety of devices, and by doping impurities, one can adjust the material's optical, thermal, and electrical properties. High sensitivity is provided by an RI-based biosensor using a unique horizontal slot WG structure composed of cost-effective polymer material in an MZI configuration as shown in Figure 6a [147]. The possibility of creating novel hybrid WGs for sensors using silicon wafer-based polymer material has arisen because of the recent need for low-cost point-of-care biosensors. The core of the sensor is made of SU8 material, the outside layer of the sensor is made of PMMA, and the inner layer is made of BCB.

It is suggested and quantitatively proved that a novel evanescent wave biosensor based on the modal interaction between the fundamental mode and the second-order mode is possible [148]. It is feasible to create a device where only the fundamental and second-order modes may propagate, without stimulation of the first-order mode, by considering the characteristics of their symmetry as shown in Figure 6b. Due to the significant contact between the evanescent field and the outer surface as a comparison to prior evanescent wave-based biosensor designs, it is feasible to obtain a high sensitivity response in the biosensor arrangement with this mode selection. The LoD of the device is ~7.34×10^{-7} RIU.

Although polymeric WGs have been suggested as a means of lowering production costs, it has not been feasible to create biosensors with sensitivity levels that are on par with those made available by silicon photonic technology due to the polymeric material's low RI. Therefore, a multimode interferometer has two special features that combine to improve the sensitivity of a modal biosensor interferometer: first, a novel modal interferometer with greater penetration depth by high order mode evanescent tail; and second, a high modal interaction by coupling mode theory optimization. This feature is essential for making lab-on-chip technologies accessible and widely manufactured in underdeveloped parts of the globe [148].

It is the first time that an all-optical plasmonic sensor platform based on integrated planar-optical WG structures in a polymer chip has been published as shown in Figure 6c [149]. The detection of 25-hydroxyvitamin D (25OHD) in human serum samples using an aptamer-based assay was performed to show the sensor system's usefulness for biosensing. The devised assay allowed for the achievement of 25OHD conc. between 0 and 100 nM with a sensitivity of 0.752 pixels/nM. It is possible to simultaneously detect several analytes, including biomarkers, because of the WG structure of the sensor's miniaturization and parallelization capabilities. It is possible to fabricate large-scale, economically priced sensors by integrating the entire optical setup onto a single polymer chip. The proposed concept is particularly appealing for its wider use in lab-on-chip solutions due to the widespread use and accessibility of smartphone electronics [149].

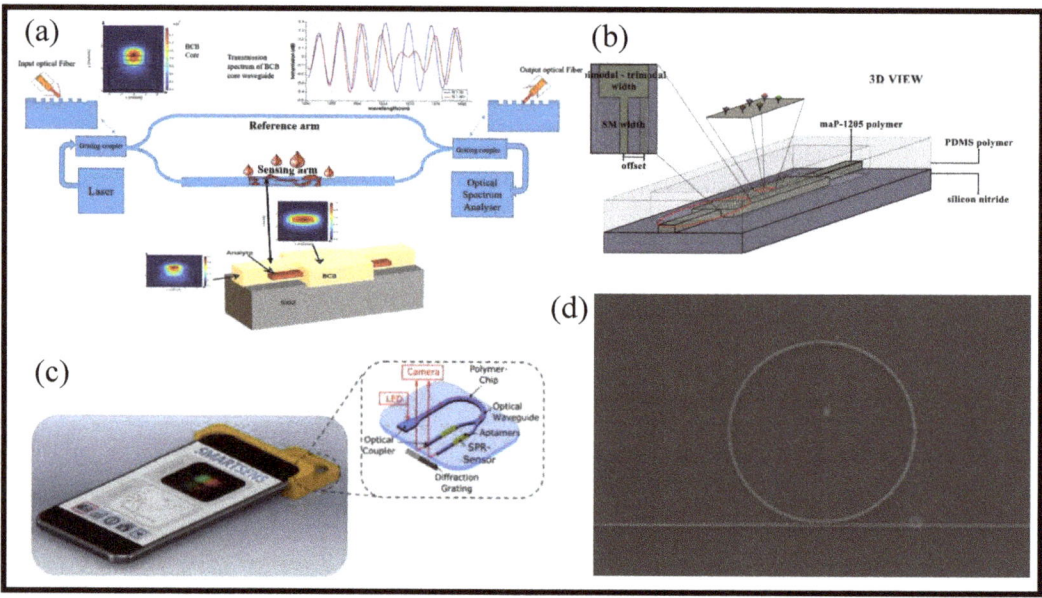

Figure 6. Polymer WG-based biosensors, (**a**) BCB and SU-8 photonic WG in MZI architecture for point-of-care device [147], (**b**) Multimode interferometer [148], (**c**,**d**) SEM image of the polymer WG resonator [150].

Numerous studies are being conducted on optical micro-ring resonators as a possible label-free biosensing technology for use in environmental monitoring and medical diagnostics. The surface mass loading or the change in RI brought on by the presence of analytes in the surrounding media is investigated in these micro-rings using the evanescent fields of the resonant light [3]. It is suggested to use the high-quality SU-8 micro-ring resonators made by NIL for biosensing applications [150]. The SEM image of the ring resonator is shown in Figure 6d. Due to its exceptional optical and mechanical qualities, strong corrosion resistance, and high thermal stability, SU-8 polymer has received extensive research in the fields of photonics and microfluidics. Its high level of cross-linking has a special ability to build sidewalls with straight profiles and large aspect ratios. A record-high intrinsic Q-factor of 8.0×10^5 is attained by UV imprinting the device with a transparent polymer mold that was copied from a smooth-sidewall silicon master mold.

4.2. Polymer WG-Based Gas Sensors

For detecting various harmful gases, several optical WG-based techniques have been put forth [151]. These methods may be roughly divided into two subgroups: RI sensing and optical absorption sensing [118,152]. The principle behind RI sensing is to track changes in the analyte's RI, which would affect the output light's frequency or phase. Since each gas has a distinct absorption wavelength, optical absorption spectroscopy-based WG sensors are more selective than RI sensing. The conc. of the gas can be estimated by analyzing the light attenuation that occurs when the light of a certain wavelength flows through the gas [153]. According to the Lambert–Beer Law, increasing the optical path—the distance over which light interacts with the analyte—will lower the LoD [154]. For optical WG sensors, the ambient gas interacts with the WG's evanescent field and serves as its cladding [155].

Beginning in the early 1980s, conducting polymers including polypyrrole (PPy), polyaniline (Pani), polythiophene (PTh), and their derivatives were utilized as the active layers of gas sensors [156]. The sensors built of conducting polymers offer numerous better properties in contrast to most of the widely viable sensors, which are often based on metal oxides and operated at high temperatures. These feathers in particular are guaranteed to be at room temperature and have high sensitivity and rapid reaction times. Conducting polymers are simple to make using chemical or electrochemical methods, and it is simple to change their molecular chain structure via copolymerization or structural derivations. Additionally, conducting polymers have outstanding mechanical qualities that make it simple to fabricate sensors.

The selective and sensitive sensing of harmful greenhouse gases is a significant challenge in the environmental and industrial domains due to the growing emphasis on environmental protection and monitoring [113,118]. Carbon dioxide (CO_2) is one of the principal greenhouse gases that is created in the environment. According to the operating environments and application areas, several specific kinds of CO_2 sensors have been produced during the last few decades. CO_2 sensors can be broadly categorized as electrochemical gas sensors, optical gas sensors, and acoustic gas sensors based on the detecting processes they use [118,157,158]. The choice and design of the sensing material used in a sensing device have a substantial impact on how well the sensor performs since a gas sensor relies on active interaction between the sensing layer and the target gas. Metal oxide, polymers, carbon compounds including carbon nanotubes (CNTs) and graphene, metal–organic frameworks (MOFs), and composites of these materials are the most often employed active materials in CO_2 sensing [159,160]. A flexible CO_2 gas sensor working at room temperature based on CNTs is developed on a low-cost polyimide substrate [159]. Resistive networks are utilized in gas detection to create very uniform CNT thin films using a trustworthy and repeatable transfer technique. When the ambient CO_2 gas conc. was 800 ppm, the flexible gas sensor had a high sensitivity of 2.23%.

MOF-based optical gas sensors, which focus primarily on light–gas interaction within a thin MOF layer, are preferable due to their properties of minimal drift, high gas selectivity to other gases—pertinent for the optical gas sensing part—and the high porosity, large surface area, and tailor-made pore sizes related for the optical gas sensing part. Metal–organic frameworks are porous crystalline solids that are put together by the coordination of inorganic building units by organic linker molecules. Because MOFs' pores may be replaced with different substances, they are appealing for use in a variety of applications, including gas storage, gas separation, catalysis, and sensing.

Air and hydrogen mixes may catch fire easily. Therefore, hydrogen sensors are crucial for quick leak identification during handling. Existing solutions, nevertheless, fall short of the high-performance standards established by stakeholders, and deactivation brought on by poisoning—such as that caused by carbon monoxide—remains a significant issue. In a plasmonic metal–polymer hybrid nanomaterial idea, deactivation resistance is supplied by a specially designed tandem polymer membrane, while the polymer coating lowers the apparent activation energy for hydrogen transport into and out of the plasmonic nanoparticles [161]. This provides subsecond sensor response times in conjunction with the optimum volume-to-surface ratio of the signal transducer given exclusively by nanoparticles. In addition, sensor LoD is improved, hydrogen sorption hysteresis is reduced, and sensor operation in challenging chemical conditions is made possible without long-term deactivation symptoms [161].

The development of affordable PCB-integrated optical WG sensors is shown using a unique platform [162]. The sensor design depends on the utilization of multimode polymer WGs that can be created directly on common PCBs and chemical dyes that are readily accessible in the market, allowing the assembly of all crucial sensing elements (electronic, photonic, and chemical) on low-cost substrates. Furthermore, it uses WG arrays functionalized with various chemical dyes to permit the detection of many analytes from a single device. The devices may be made using PCB manufacturing techniques that are standard practice, such as pick-and-place assembly and solder-reflow operations. An FR4 substrate is used to construct an ammonia gas sensor that is PCB integrated as a proof of concept. The sensor's functionality depends on the way ammonia molecules affect the optical transmission properties of chemically functionalized optical WGs. In addition to the basic modeling and characterization investigations, the manufacturing and assembly of the sensor unit are discussed. At normal temperatures, the device achieves a sensitivity of around 30 ppm and a linear response of up to 600 ppm. Finally, principal-component assessment is used to show how it is possible to identify numerous analytes from a single device [162].

For the detection and sensing of CO_2, a cheap gas sensor based on planar polymer optical WGs with an embedded zeolite imidazole framework-8 (ZIF-8) thin film is presented [68]. The PMMA planar optical WGs are formed by hot embossing, which makes them flexible and economical. A simple solution approach is used to evenly produce thin ZIF-8 films on the surface of WGs, which is essential for the envisioned mass manufacturing of metal–organic framework-based sensing devices. The microscope image of the WG without MOF film and with MOF film is shown in Figure 7a,b, respectively.

Figure 7c depicts the recording of optical signal transmission out of the MOF-coated WG with a gas switching time interval of $t = 1$ min. Several cycles were repeated. The alternate purging of N_2 and CO_2 caused the transmission signal to shift frequently and reproducibly. When CO_2 was released into the gas cell, the transmission rapidly fell and maintained a largely constant level. The repeated cycle tests provide additional evidence of the proposed MOF-based polymer WGs' resilience and effectiveness as sensors. A further experiment with $t = 30$ s was carried out to test the optical response of the planar WG sensor to CO_2 exposure during shorter gas changeover times. Figure 7d depicts the outcome. The optical signal leaving the WG here also changes regularly due to the alternate purging of N_2 and CO_2. Additionally, it can be shown that the short gas switching time results in a high-power level when N_2 is purged that lacks a clear steady state. The time from the response's beginning to 90% of its maximum in a steady state is referred to as the adsorption and desorption times. The MOF-coated PMMA WG used in this study exhibits CO_2 adsorption and desorption times of approximately 6 s and 16 s, respectively, as shown in Figure 7e. These findings show that the diffusion of gases into the MOF sensing layer from the surrounding medium occurs at a very high rate. According to experimental findings, the developed optical elements have good reversibility of the gas molecules' adsorption and desorption, a sensitivity of 2.5 µW/5 vol% toward CO_2 [68].

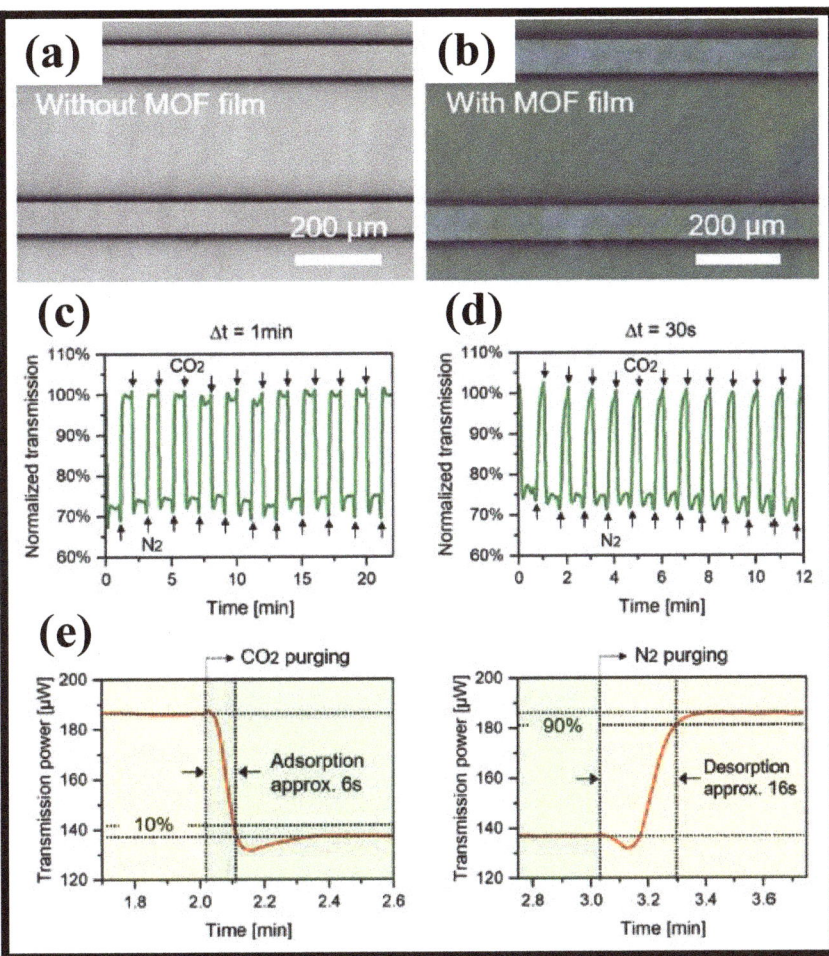

Figure 7. Microscope images of (**a**) uncoated [68], (**b**) ZIF-8 coated WG [68], (**c**) response of the MOF-coated WG to CO_2 with gas switching time intervals of 1 min [68], (**d**) response of the MOF-coated WG to CO_2 with gas switching time intervals of the 30 s [68], and (**e**) adsorption and desorption time of MOF-coated WG sensor [68].

4.3. Polymer WG-Based Temperature Sensors

In several industries and applications, including healthcare, consumer electronics, transportation, and aerospace, temperature sensors are crucial [163,164]. To meet the expanding demands for automation in production and monitoring, there is an expanding market for temperature sensors that are high-performing, trustworthy, and affordable [165]. Such sensors are increasingly being included in materials (such as composites) during the manufacturing process. Conventional electronic-based temperature sensors are unsuitable for several tasks because of their sensitivity to electromagnetic interference [166].

Optical WGs on a planar substrate are an intriguing solution to fibers for sensing since they may incorporate splitters, optoelectronic components, or even whole Bragg grating interrogation systems [167,168]. Comparatively to single fiber sensors, the use of planar foils makes it easier to position and align the sensors during integration. Due to the wide range of materials that are readily accessible, each with unique features and optimized for certain manufacturing methods, Bragg grating-based sensors in polymer WGs are becoming more and more popular in several applications [169–175]. Because of its high thermo-optic coefficient, superior thermal stability, and relatively low absorption loss from visible to telecom wavelengths, Ormocer, an inorganic-organic hybrid polymer, is an excellent option to be utilized in the implementation of a Bragg grating-based temperature sensor. Additionally, the material is economical, simple to utilize, and risk-free to handle without specific safeguards. However, due to its oxygen inhibition and liquid condition during UV exposure, traditional contact mask lithography makes it challenging to define nanometer-scale WG features. An equivalent has been imprinting-based technology; however, because of the rather thick residual layer, the reported structures made with this approach are inverted-rib WGs [176]. To create a highly sensitive Ormocer-based WG Bragg grating temperature sensor, a novel capillary filling-based duplication technique is suggested [177]. A series of polymer WGs imprinted with a wide area grating is patterned on top of an under-cladding layer to create the sensor, which has a sensitivity of −249 pm/°C at ambient temperature and a working wavelength of 1530 nm [177].

The use of switchable molecular compounds in a polymer WG-based temperature sensor system is described in [178]. A maskless lithographic optical system is used to manufacture the polymer WG cladding, and hot embossing equipment is used to copy it onto polymer material (for instance polymethyl methacrylate (PMMA)). The substance used to monitor changes in outside temperature is a molecular combination of iron, amino, and triazole. A mixture of core material (NOA68) is used to fill the WG's core for this purpose, and doctor blading and UV curing are used to solidify the molecular complex. In the low-spin state, two distinct absorption bands are present in the molecular complex's UV/VIS light spectrum. A spin-crossover transition takes place when the temperature gets close to room temperature, which causes the molecular complex to go from violet pink to white (or spectral characteristics). With a hysteresis width of around 12 °C and a sensitivity of 0.08 mW/°C, the measurement of optical power transmitted through the WG as a function of temperature displays a memory effect. In situations where electromagnetic interference might skew the results, this permits optical rather than electrical temperature detection [178].

A polymer WG incorporated in an optical fiber micro-cavity-based MZI was presented [179]. Femtosecond laser micromachining, fiber splicing, and single-mode fibers were used to create the micro-cavity with two symmetric apertures. The 70 μm long polymer WG was then fabricated using the two-photon polymerization manufacturing technique and incorporated into the micro-cavity. A complete interference spectrum and over 25 dB of fringe visibility were displayed by the MZI. The suggested MZI's temperature sensitivity exceeded 447 pm/°C because of the strong thermo-optical coefficient of the polymer material. Due to its flawless linearity (99.7%) and persistence, it may be utilized as a trustworthy temperature sensor [179].

It is suggested to use an asymmetric MZI with varying widths in the two interferometer arms as the foundation for an ultra-sensitive polymeric WG temperature sensor [180]. The device's sensitivity was improved by using a polymer with a higher thermo-optic coefficient (TOC). The effect of the two arms' distinct widths and the cladding materials' various TOCs on the sensor's sensitivity was investigated and experimentally proven. The devices were created by combining a straightforward all-wet etching method with conventional photolithography. The sensitivity of the WG temperature sensor was found to be 30.8 nm/°C when the cladding material Norland optical adhesive 73 (NOA 73) and the width difference of 6.5 μm were used. Additionally, the lowest temperature resolution was almost 0.97×10^{-3} °C. The sensor has the specific benefits of high sensitivity, high resolution, simple manufacture, low cost, and biological compatibility, making it potentially useful for temperature detection of organisms, molecular analysis, and biotechnology [180].

With the use of bottom metal printing technology, multimodal responsive optical WG sensors that make use of a stable cross-linking gel polymer electrolyte have been successfully created [181]. Figure 8a depicts the prototype optical WG multimodal sensor's schematic structure diagram. First, Si substrates with SiO_2 buffer layers are used to fabricate metal thermo-inducting electrode designs. Spin-coating and UV curing are used to cure a gel polymer electrolyte that was self-created as a sensing WG material on metal electrodes. By using the bottom metal printing process, the MMI WG sensing structure is directly defined through electrode areas. The lift-off technique is used to detach the polymer films off the thermo-inducting electrodes' pads on both sides. Next, a printed circuit board (PCB) is attached to the overall sensing WG chip using adhesive. On either side of the chip, two further PCBs are connected. Copper and aluminum metal electrode pads on two different PCBs with holes that are aligned to them might make a thermo-inducting contact. Figure 8b provides the packed chip's complete model diagram. Figure 8c,d show the measurement systems of the multimodal photonic chip for temperature and humidity sensing applications, respectively [181].

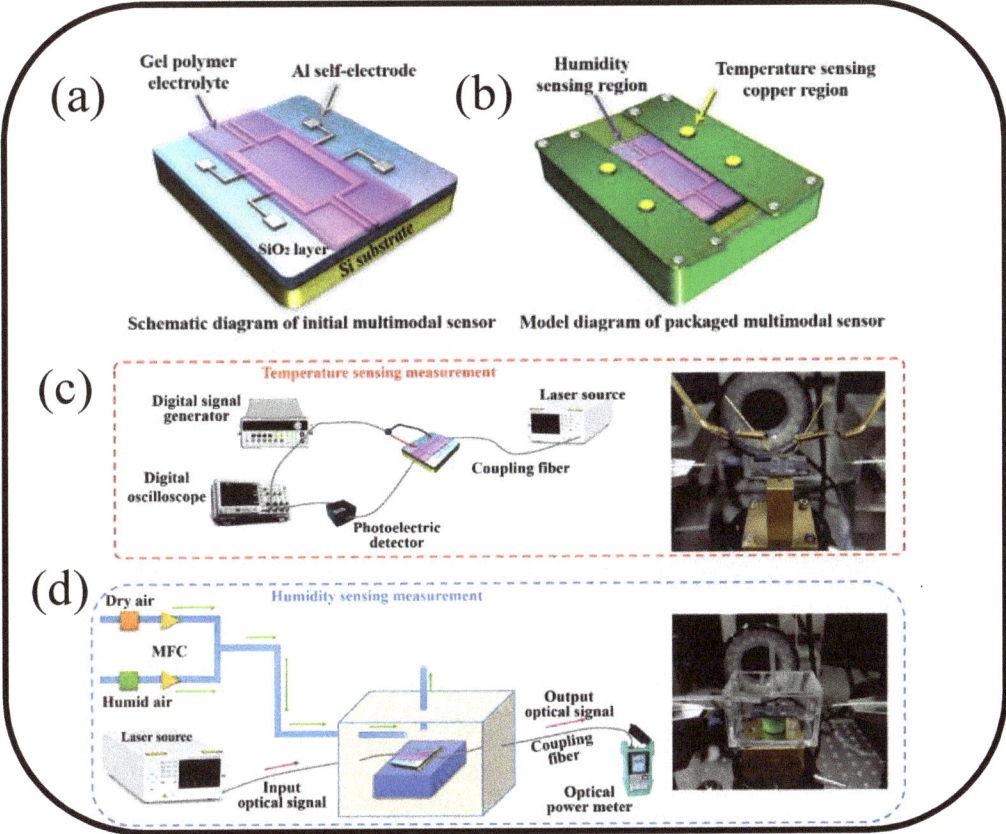

Figure 8. Graphical illustration of (**a**) the prototype of optical WG multimodal sensor [181], (**b**) packaged photonic multimodal sensor [181], (**c**) measuring systems of the sensor for temperature [181], and (**d**) measuring systems of the sensor for humidity [181].

It is simulated to characterize temperature and humidity sensing using a polymer electrolyte. Considering the study of ion relaxation dynamics, the multimodal responsive features of the photonic chip are established: optical attenuation and phase variables for temperature and humidity sensing, respectively. The device's temperature and humidity sensitivities were measured at 0.5 rad/°C and 1.14 dB/% RH, respectively, in the monitoring temperature (36.0–38.0 °C) and relative humidity (45%–65%) ranges. The multifunctional sensor's quick reaction times may be calculated to be 4.21 ms and 1.32 s, for temperature and humidity, respectively. This study offers a workable plan for the development and use of temperature and humidity sensors in possible medical procedures [181].

4.4. Polymer WG-Based Mechanical Sensors

Polymer WG-based mechanical sensors are devices that use the optical properties of a polymer WG to detect mechanical changes. When the WG is deformed due to mechanical stress or strain, the refractive index of the material changes, which alters the way that light propagates through it. This change in the optical properties of the WG can be measured and used to detect the presence and magnitude of the mechanical stress or strain. Polymer WG mechanical sensors have a number of potential applications, including in structural health monitoring, biomedical sensing, and environmental monitoring. There are different types of polymer WG-based mechanical sensors, including cantilever-based sensors, micro-ring resonator sensors, and Mach–Zehnder interferometer sensors. Each of these sensors has its own advantages and disadvantages, and the choice of sensor depends on the specific application requirements.

Optical sensors with low residual stress, straightforward manufacturing, and low cost have been created employing polymer-based WG systems. In general, the mechanical characteristics of polymers may be described in a manner quite similar to those of metals or other common crystalline materials. This is especially true for elastic moduli and other classes of strength measures like yield and tensile strengths. A wide range of biomedical applications in sensing, diagnostics, and phototherapy have demonstrated considerable potential for biocompatible polymeric optical WGs with soft and flexible mechanical features [182]. Today, transparent touch-sensitive panels for portable devices like smartphones and tablet computers are commonly used. These panels have become more functional and versatile, enabling multipoint touch interaction in addition to single point touch, which is a growing trend in touch sensing in electronic devices. Besides contact detection, pressure-based touchscreen interaction increases usability. For instance, a touch-sensitive keyboard performs better when pressing the keys. Recently, flexible or even stretchy touch sensors were developed so they can work with flexible screens [166].

A high sensitivity of 8.2 ppm/Pa has been established for an optomechanical pressure sensor employing polymer multimode interference (MMI) couplers [183]. It has been established that a polymer WG sensor with a symmetric multilayer design may be used to detect low humidity conc. [184]. When water molecules diffuse into the polymer WG, the sensor records the resultant optical phase shifts. It is possible to reach a sensitivity of several parts per million for humidity levels. Additionally, the sensor provides the absolute sign of the movement of the generated interference fringes, which makes it simple to identify trends in the index (increase or reduction) in the sensing layer. This work shows a very promising future for the development of a small, disposable optical sensor with a cheap cost for humidity sensing applications [184].

Stretchable polymeric optical WGs have also been researched for wearable body temperature readings in addition to mechanical sensing [185]. One of the most important physiological indicators that accurately identifies stages of health is body temperature. The reading of wearable temperature sensors should be resistant to body motions and independent of mechanical deformation for continuous and long-term temperature monitoring. Utilizing PDMS optical fibers that have integrated upconversion nanoparticles, a unique stretchable optical temperature sensor that can maintain its sensing capability despite massive strain deformations (up to 80%) has been produced [185].

The main criteria, such as thin-film design, localized force sensing, multiple-point identification, and bending resilience, as well as a quick response for a tactile sensor functioning on curved surfaces, are satisfied by a polymer WG-based transparent and flexible force sensor array [186]. A contact force with a location at 27 different spots is detected by the force sensor array separately. The sensor array is elastic, thin (total thickness: 150 µm), as well as very transparent (transmittance: up to 90%). The force sensor can detect contact forces at one or more points with a quick response (response delay: 10 ms), high replicability (Pearson correlation coefficient: as high as 0.994, hysteresis: as low as 6.7%), high sensitivity (as high as 16% N^{-1}), and high bendability (10.8% sensitivity degradation at a bending radius of 1.5 mm), all without utilizing any electronic components. Without noticeably degrading its function, the sensor can detect pressure on curved objects as well as soft ones like the human body. For detecting dynamic contact forces on different surfaces, the force sensor could be employed in a variety of applications [186]. The fabrication process of the sensor is shown in Figure 9a. The SEM image and an optical microscope image of the sensing area is shown in Figure 9b,c, respectively. Despite its multilayered configuration, the WG layer exhibits excellent optical transparency of 90% in the range of 550–1000 nm, allowing removal from the stiff substrate without mechanical damage as shown in Figure 9d. Due to the soft nature of the materials utilized for both the clad and core, the WG-based thin force sensor is mechanically resilient to bending, twisting, or folding and is capable of close contact with a forearm with great transparency and flexibility as shown in Figure 9e–g [186].

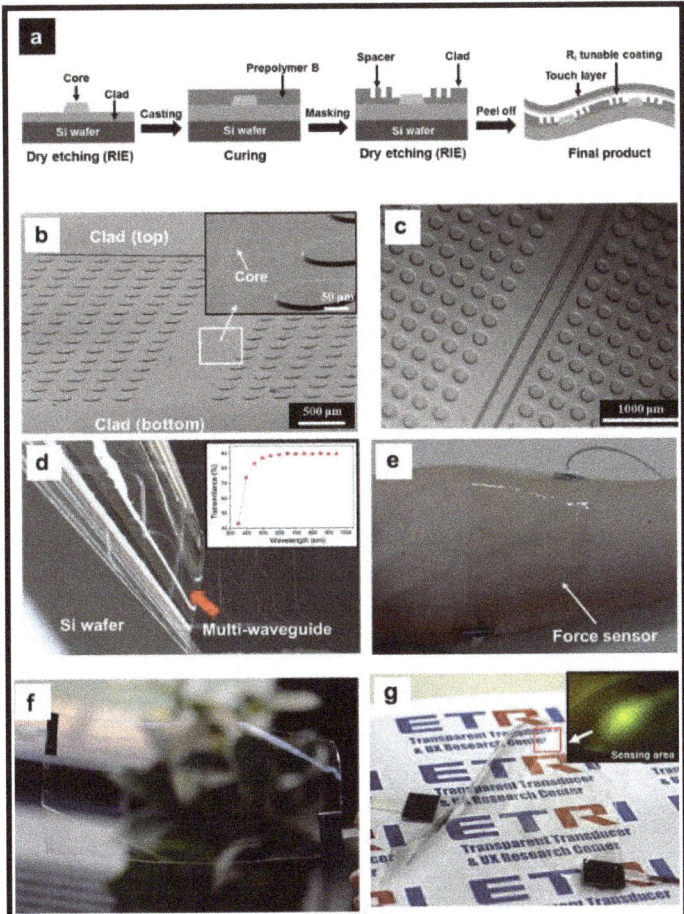

Figure 9. (**a**) Fabrication process of a WG-based sensor, (**b**) SEM image, and (**c**) optical microscope image of a WG [186], (**d**) an image showing the construction of the waveguide layer being peeled from a Si wafer [186], (**e**) photo demonstrating the close proximity of a thin-film sensor to a forearm [186], (**f**) the transparency of the sensor [186], and (**g**) a photograph and an optical microscope image (inset) of a force sensor deformed by mechanical bending [186].

A photopolymerizable resin solution that self-writes a strain sensor WG is described as an experimental demonstration [187]. The sensor is created between two multi-mode optical fibers using ultraviolet (UV) light waves, and it functions as a sensor by examining the power that is passed through the WG in the infrared (IR) wavelength range. Following sensor failure brought on by loading, the WG uses UV resin to re-bridge the gap between the two optical fibers. Measurements reveal similarities in the responses of the original and self-repaired sensors under strain [187].

Due to the weak mechanical strengths of most synthetic hydrogels, hydrogel optical WGs are frequently brittle and unstable when subjected to strain deformations. Due to the possibility of WG structural damage from body/tissue movement, this property limits their use in wearable or implanted sensors. For the creation of optical fibers, hybrid ionic/covalent hydrogels with high stretchability and toughness were used to produce great deformability and robustness. Hybrid alginate–polyacrylamide hydrogels have been molded and dip-coated to create robust, stretchable hydrogel optical fibers with an overall step-index profile [188]. Table 2 presents the recently proposed polymer WGs employed for biosensing, gas sensing, temperature sensing, and mechanical sensing.

Table 2. Recently proposed polymer WG-based biosensors, gas sensors, temperature sensors, and mechanical sensors.

Polymer	Sensor Design	Application	Temperature Range (°C)	Sensitivity	Fabrication Method	Numerical/Experimental	Ref.
NOA 73	MZI	Temperature	-	30.8 nm/°C	Wet etching	Experimental	[180]
Gel polymer	MZI	Temperature	36–38	0.5π rad/°C	Bottom metal-printing	Experimental	[181]
NOA 73	MZI	-Temperature	25–75	−431 pm/°C	CMOS	Experimental	[189]
PDMS	PhC	Temperature	10–90	0.109 nm/°C	-	Numerical	[82]
PMMA	BG	Temperature	−10–70	−48.6 pm/°C	Direct laser writing	Experimental	[190]
NOA68	Planar WG	Temperature	18–35	0.08 mW/°C	Hot embossing	Experimental	[178]
COC	BG	Temperature	30–160	−7.3 pm/K	Single writing step	Experimental	[191]
PDMS	Metasurface	Temperature	30–60	−0.18 nm/°C	-	Numerical	[192]
PDMS	FPI-AWG	Temperature	30–40	−0.854 dB/°C	-	Experimental	[193]
SU-8	Trimodal WG interferometer	Temperature	22–27	0.0586 dB/°C	Direct laser writing	Experimental	[194]
Ormocer	BG	Temperature	20–110	−150 pm/°C	Imprinting	Experimental	[195]
ZIF-8	Planar WG	CO_2	-	2.5 µW/5 vol%	Hot embossing	Experimental	[68]
PHMB	Modified BG	CO_2	-	226 pm/ppm	-	Numerical	[196]
-	Planar WG	CO	-	-	NIL	Experimental	[197]
PMMA	Photonic crystal nanocavity		-	287 ppb/$\sqrt{Hz}$	EBL	Experimental	[198]
PHMB	Plasmonic WG	CO_2	-	135.95 pm/ppm	-	Numerical	[199]
PHMB	Metasurface	CO_2	-	17.3 pm/ppm	-	Numerical	[98]
BCB and SU8	MZI	Biosensing	-	19,280 nm/RIU and 16,500 nm/RIU	N/A	Numerical	[147]
FSU-8 and PMMA	BG	Drug conc.	-	1606.2 nm/RIU	Direct UV writing	Experimental	[139]
PMMA and NOA 63	Planar WG structure	Vitamin D	-	0.752 pixel/nM	Hot embossing and doctor blading	Experimental	[149]
PMMA and NOA 63	Planar optical multi-mode WG	C-reactive protein	-	608.6 nm/RIU	Hot embossing and doctor blading	Experimental	[141]
PEI, PSS and PAH	Phase-shifted BG	NaCl	-	579.2 nm/RIU	CMOS	Experimental	[200]
Ormocore	Micro-ring	Biosensing	-	-	Soft UV NIL	Experimental	[201]

Table 2. Cont.

Polymer	Sensor Design	Application	Temperature Range (°C)	Sensitivity	Fabrication Method	Numerical/Experimental	Ref.
Ma-P 1205	Trimodal interferometer	Biosensing	-	2050 2π/RIU	-	Numerical	[109]
SU8/PMATRIFE	Ring resonator and MZI	Glucose	-	17,558 nm/RIU	CMOS	Experimental	[202]
Ormocore	MZI	Biosensing	-	10^4 nm/RIU	-	Numerical	[203]
PEI-B	MZI	STA-biotin	-	-	Inkjet printing	Experimental	[204]
PMMA	Polymer optical fiber	Breath and heartbeat	-	Error compared with reference: 1 cpm (breath) 4 bpm (heartbeat)	-	Experimental	[205]
PMMA	Polymer optical fiber	Smart textile: Bending, compression	-	-	Melt-spinning	Experimental	[206]
PDMS	Polymer optical fiber	Smart textile: Bending, compression	-	-	Moulding	Experimental	[206]
PMMA	Polymer optical fiber BG	Pressure	-	Up to 71.9 ± 0.3 μm/MPa	-	Experimental	[207]
-	Optical fiber-based polymer Fabry–Perot interferometer	Gas pressure	-	3.959 nm/MPa	3D-printed	Experimental	[208]

5. Conclusions and Outlook

The global sensor market is rapidly expanding due to the establishment and expansion of new applications, such as medical devices that use noninvasive optical sensors and environmental monitoring, which is gaining prominence due to an increase in the demand for both indoor and outdoor air quality measurements. Several planar WG systems have appeared in the last 10 years. They are characterized by the materials systems employed and their distinct qualities, which confer restrictions and benefits on each. For a very long period, most of these platforms were developed primarily to support applications from the telecommunications industry. Because of the wide variety of chemical structures and inherent features that polymers may take on, they have long been regarded in the engineering community as exceptional materials systems. However, due to their ease of manufacture in recent years, polymer optics have drawn increased interest.

Polymer WGs can be designed with small core diameters, which increases the sensitivity of the sensor to changes in the environment. These WGs are typically less expensive to manufacture than other types of optical fibers, making them an attractive option for cost-sensitive applications and can be bent and shaped into a variety of configurations, which allows for easy integration into various applications and environments. These WGs are typically lighter than other types of optical fibers, which makes them ideal for portable and mobile applications. They are typically more durable and resistant to damage than other types of optical fibers, which makes them suitable for use in harsh environments, and can be easily integrated with other optical components, such as lenses, detectors, and electronics, which makes them a flexible and versatile option for a variety of applications. Keeping in mind the potential of polymer WG sensors, four main sensing applications which include biosensing, gas sensing, temperature sensing, and mechanical sensing are reviewed.

Nevertheless, polymer sensors are not significantly inferior to sensors based on integrated photonics or classical FBG sensors. We have previously mentioned that the sensitivity of FBGs based on polymer fibers can exceed 1600 nm/RIU. At the same time, the sensitivity of refractive index sensors based on inorganic FBGs varies in different sources. Thus, in [209], a sensitivity of 1210.49 nm/RIU is shown. In [210], the sensitivity of a phase-shifting FBG is 463.7953 nm/RIU. At the same time, in [211], the sensitivity varies from 1008 dB/RIU for the high RI range to 8160 dB/RIU for the low RI range; however, a high sensitivity value is achieved by optimizing the cladding diameter, which is a technologically complex process. The sensitivity of sensors based on integrated photonics devices is often inferior to polymer sensors. In this case, the advantage of integrated photonics is the ability to implement the entire sensor system (including the interrogator) on a single chip [212,213].

A wide range of polymeric materials and methods of their use suggests expanding areas of application and production methods. For example, [214] shows the possibility of creating polymer fibers based on natural cellulose, transparent in the wavelength range from 500 nm to 4000 nm. The work [215] shows broad prospects for the creation of MWIR-range (Mid-Wavelength Infrared) devices, opening up with the use of organically modified chalcogenides (ORMOCHALC). The work shows the prospects for the creation of polymer optical fibers based on multi-materials (PMMA and Polycaprolactone (PCL)) for use in sensors. Finally, work [216] demonstrates the possibility of increasing the sensitivity of the SPR sensor when using meta-dielectric materials up to 1700 nm/RIU. Thus, the development and application of polymer-based sensors is a rapidly developing field of knowledge with a wide range of applications and prospects.

Author Contributions: Conceptualization, M.A.B. and G.S.V.; methodology, S.N.K., M.A.B. and N.L.K.; software, S.N.K., M.A.B. and N.L.K.; validation, S.N.K., M.A.B. and N.L.K.; formal analysis, G.S.V., S.N.K., M.A.B. and N.L.K.; investigation, S.N.K., E.P.G., M.A.B. and N.L.K.; resources, S.N.K., M.A.B. and N.L.K.; data curation, R.V.K., S.N.K., M.A.B. and N.L.K.; writing—original draft preparation, S.N.K., M.A.B. and N.L.K.; writing—review and editing, S.N.K., E.P.G., R.V.K., G.S.V., M.A.B. and N.L.K.; visualization, S.N.K., M.A.B. and N.L.K.; supervision, S.N.K., M.A.B. and N.L.K.; project administration, S.N.K., E.P.G., M.A.B. and N.L.K.; funding acquisition, S.N.K., M.A.B. and N.L.K. All authors have read and agreed to the published version of the manuscript.

Funding: The research was supported by the Ministry of Science and Higher Education of the Russian Federation in the financing of new laboratories under the guidance of young scientists within the framework of the national Project "Science and Universities" (project FSSS-2021-0016) in the part of sensors' applications and under the FSRC "Crystallography and Photonics" of the Russian Academy of Sciences (the state task No. 007-GZ/Ch3363/26) in the part of sensors' specifications, and by the Ministry of Science and Higher Education of the Russian Federation within the state assignment for the UUST (agreement No. 075-03-2021-014) and conducted in the research laboratory "Sensor systems based on integrated photonics devices" of the Eurasian Scientific and Educational Center in the part of photonics integrated-based sensors' analysis.

Acknowledgments: We acknowledge the equal contribution of all the authors in the completion of this work.

Conflicts of Interest: The authors declare no conflict of interest.

Abbreviations

Waveguide = WG; Limit of detection = LoD; Polymethyl methacrylate = PMMA; Polydimethylsiloxane = PDMS; Surface plasmon resonance = SPR; Mach-Zehnder interferometer = MZI; Polyethyleninmine = PEI; Polystyrene sulfonate (PSS); Polyallylamine hydrochloride = PAH; Polyvinylpyrrolidone = PVP; Cyclic olefin copolymer = COC; Polyhexamethylene biguanide = PHMB; Polypyrrole = PPy; Polyaniline = Pani; Polythiophene = PTh; Refractive index = RI.

References

1. Paquet, C.; Kumacheva, E. Nanostructured polymers for photonics. *Materialstoday* **2008**, *11*, 48–56. [CrossRef]
2. Khan, M.; Rahlves, M.; Lachmayer, R.; Roth, B. Low-cost fabrication of polymer based micro-optical devices for application in illumination, sensing, and optical interconnects. In Proceedings of the Conference on Lasers and Electro-Optics Europe and European Quantum Electronics Conference, Munich, Germany, 23–27 June 2019.
3. Butt, M.A.; Khonina, S.N.; Kazanskiy, N.L. A highly sensitive design of subwavelength grating double-slot waveguide microring resonator. *Laser Phys. Lett.* **2020**, *17*, 076201. [CrossRef]
4. Butt, M.; Khonina, S.; Kazanskiy, N. Optical elements based on silicon photonics. *Comput. Opt.* **2019**, *43*, 1079–1083. [CrossRef]

5. Butt, M.A.; Kozlova, E.S. Multiport optical power splitter design based on coupled-mode theory. *J. Phys. Conf. Ser.* **2019**, *1368*, 022006. [CrossRef]
6. Butt, M.A.; Kozlova, E.S.; Khonina, S.N. Conditions of a single-mode rib channel waveguide based on dielectric TiO$_2$/SiO$_2$. *Comput. Opt.* **2017**, *41*, 494–498. [CrossRef]
7. Ma, H.; Jen, A.-Y.; Dalton, L. Polymer-based optical waveguides: Materials, processing, and devices. *Adv. Mater.* **2002**, *14*, 1339–1365. [CrossRef]
8. Wolfer, T.; Bollgruen, P.; Mager, D.; Overmeyer, L.; Korvink, J. Flexographic and inkjet printing of polymer optical waveguides for fully integrated sensor systems. *Procedia Technol.* **2014**, *15*, 521–529. [CrossRef]
9. Han, T.; Madden, S.; Zhang, M.; Charters, R.; Luther-Davies, B. Low loss high index contrast nanoimprinted polysiloxane waveguides. *Opt. Express* **2009**, *17*, 2623–2630. [CrossRef]
10. Hiltunen, J.; Kokkonen, A.; Puustinen, J.; Hiltunen, M.; Lappalainen, J. UV-imprinted single-mode waveguides with low loss at visible wavelength. *IEEE Photonics Technol. Lett.* **2013**, *25*, 996–998. [CrossRef]
11. Wang, S.; Vaidyanathan, V.; Borden, B. Polymer optical channel waveguide components fabricated by using a laser direct writing system. *J. App. Sci. Eng. Technol.* **2009**, *3*, 47–52.
12. Rezem, M.; Gunther, A.; Rahlves, M.; Roth, B.; Reithmeier, E. Hot embossing of polymer optical waveguides for sensing applications. *Procedia Technol.* **2014**, *15*, 514–520. [CrossRef]
13. Butt, M. Thin-film coating methods: A successful marriage of high-quality and cost-effectiveness—A brief exploration. *Coatings* **2022**, *12*, 1115. [CrossRef]
14. Wang, H.; Liu, Y.; Jiang, M.; Chen, C.; Wang, X.; Wang, F.; Zhang, D.; Yi, Y. Three dimensional polymer waveguide using hybrid lithography. *Appl. Opt.* **2015**, *54*, 8412–8416. [CrossRef] [PubMed]
15. Dangel, R.; Hofrichter, J.; Horst, F.; Jubin, D.; Porta, A.; Meier, N.; Soganci, I.; Weiss, J.; Offrein, B. Polymer waveguides for electro-optical integration in data centers and high-performance computers. *Opt. Express* **2015**, *23*, 4736–4750. [CrossRef] [PubMed]
16. Marin, Y.; Nannipieri, T.; Oton, C.; Pasquale, F. Current status and future trends of photonic-integrated FBG interrogators. *J. Light. Technol.* **2018**, *36*, 946–953. [CrossRef]
17. Lin, H.; Luo, Z.; Gu, T.; Kimerling, L.; Wada, K.; Agarwal, A.; Hu, J. Mid-infrared integrated photonics on silicon:A perspective. *Nanophotonics* **2017**, *7*, 393–420. [CrossRef]
18. Sharma, T.; Wang, J.; Kaushik, B.; Cheng, Z.; Kumar, R.; Wei, Z.; Li, X. Review of Recent Progress on Silicon Nitride-Based Photonic Integrated Circuits. *IEEE Access* **2020**, *8*, 195436–195446. [CrossRef]
19. Wang, R.; Vasiliev, A.; Muneeb, M.; Malik, A.; Sprengel, S.; Boehm, G.; Amann, M.-C.; Simonyte, I.; Vizbaras, A.; Vizbaras, K. III-V-on-silicon photonic integrated circuits for spectroscopic sensing in the 2–4 mm wavelength range. *Sensors* **2017**, *17*, 1788. [CrossRef]
20. Prajzler, V.; Jasek, P.; Nekvindova, P. Inorganic–organic hybrid polymer optical planar waveguides for micro-opto-electro-mechanical systems (MOEMS). *Microsyst. Technol.* **2019**, *25*, 2249–2258. [CrossRef]
21. Ibrahim, N.; Riduwan, M.; Rusli, A. Synthesis of Siloxane-polyimide Copolymer with Low Birefringence and Low Loss for Optical Waveguide. *J. Phys. Sci.* **2019**, *30*, 103–113. [CrossRef]
22. Bahadori, M.; Nikdast, M.; Cheng, Q.; Bergman, K. Universal design of waveguide bends in silicon-on-insulator photonics platform. *J. Light. Technol.* **2019**, *37*, 3044–3054. [CrossRef]
23. Sun, X.; Dai, D.; Thylen, L.; Wosinski, L. High-Sensitivity Liquid Refractive-Index Sensor Based on a Mach-Zehnder Interferometer with a Double-Slot Hybrid Plasmonic Waveguide. *Opt. Express* **2015**, *23*, 25688. [CrossRef] [PubMed]
24. Huang, W.; Luo, Y.; Zhang, W.; Li, C.; Li, L.; Yang, Z.; Xu, P. High-sensitivity refractive index sensor based on Ge-Sb-Se chalcogenide microring resonator. *Infrared Phys. Technol.* **2021**, *116*, 103792. [CrossRef]
25. Wu, N.; Xia, L. Side-Mode Suppressed Filter Based on Anangular Grating-Subwavelength Grating Microring Resonator with High Flexibility in Wavelength Design. *Appl. Opt.* **2019**, *58*, 7174. [CrossRef] [PubMed]
26. Xie, W.; Komljenovic, T.; Huang, J.; Tran, M.; Davenport, M.; Torres, A.; Pintus, P.; Bowers, J. Heterogeneous silicon photonics sensing for autonomous cars. *Opt. Express* **2019**, *27*, 3642. [CrossRef]
27. Ramirez, J.; Malhouitre, S.; Gradkowski, K.; Morrissey, P.; O'Brien, P.; Caillaud, C.; Vaissiere, N. III-V-on-Silicon Integration: From Hybrid Devices to Heterogeneous Photonic Integrated Circuits. *IEEE J. Sel. Top. Quantum Electron.* **2020**, *26*, 6100213. [CrossRef]
28. Van der Tol, J.; Jiao, Y.; Van Engelen, J.; Pogoretskiy, V.; Kashi, A.; Williams, K. InP Membrane on Silicon (IMOS) Photonics. *IEEE J. Quantum Electron.* **2020**, *56*, 6300107. [CrossRef]
29. Available online: https://www.diabetes.co.uk/body/white-blood-cells.html (accessed on 10 January 2023).
30. Available online: https://www.gasalarmsystems.co.uk/latest-news/3-little-known-facts-about-flammable-gas-that-could-save-your-life/ (accessed on 10 January 2023).
31. Available online: https://www.lihealthcollab.org/news-and-blog/heat-alert--heart-health-at-risk-when-temperatures-rise (accessed on 10 January 2023).
32. Huang, K.-H.; Tan, F.; Wang, T.-D.; Yang, Y.-J. A highly sensitive pressure-sensing array for blood pressure estimation assisted by machine-learning techniques. *Sensors* **2019**, *19*, 848. [CrossRef]
33. Kazanskiy, N.; Butt, M.; Khonina, S. Recent advances in wearable optical sensor automation powered by battery versus skin-like battery-free devices for personal healthcare-A review. *Nanomaterials* **2022**, *12*, 334. [CrossRef]

34. Zhang, M.; Zhang, W.; Wang, F.; Zhao, D. High-gain polymer optical waveguide amplifiers based on core-shell NaYF$_4$/NaLuF$_4$: Yb^{3+}, Er^{3+} NPs-PMMA covalent-linking nanocomposites. *Sci. Rep.* **2016**, *6*, 36729. [CrossRef]
35. Park, Y.-J.; Lee, S.; Kim, B.; Kim, J.-H.; So, J.-H.; Koo, H.-J. Impedance study on humidity dependent conductivity of polymer composites with conductive nanofillers. *Compos. Part B Eng.* **2020**, *202*, 108412. [CrossRef]
36. Geiss, R.; Saravi, S.; Sergeyev, A.; Diziain, S.; Setzpfandt, F.; Schrempel, F.; Grange, R.; Kley, E.-B.; Tunnermann, A.; Pertsch, T. Fabrication of nanoscale lithium niobate waveguides for second-harmonic generation. *Opt. Lett.* **2015**, *40*, 2715–2718. [CrossRef] [PubMed]
37. Eknoyan, O.; Yoon, D.; Taylor, H. Low-loss optical waveguides in lithium tantalate by vapor diffusion. *Appl. Phys. Lett.* **1987**, *51*, 384. [CrossRef]
38. Butt, M.; Kozlova, E.S.; Khonina, S.N.; Skidanov, R.V. Optical planar waveguide sensor based on (Yb,Nb): RTP/RTP(001) system for the estimation of metal coated cells. *CEUR Workshop Proc.* **2016**, *1638*, 16–23.
39. Butt, M.; Sole, R.; Pujol, M.; Rodenas, A.; Lifante, G.; Choudhary, A.; Murugan, G.; Shepherd, D.; Wilkinson, J.; Aguiló, M.; et al. Fabrication of Y-Splitters and Mach-Zehnder Structures on (Yb,Nb):RbTiOPO$_4$/RbTiOPO$_4$ Epitaxial Layers by Reactive Ion Etching. *J. Light. Technol.* **2015**, *33*, 1863–1871. [CrossRef]
40. Zhang, L.; Lin, Q.; Yue, Y.; Yan, Y.; Beausoleil, R.; Willner, A. Silicon waveguide with four zero-dispersion wavelengths and its application in on-chip octave-spanning supercontinuum generation. *Opt. Express* **2012**, *20*, 1685–1690. [CrossRef]
41. Senichev, A.; Peana, S.; Martin, Z.; Yesilyurt, O.; Sychev, D.; Lagutchev, A.; Boltasseva, A.; Shalaev, V. Silicon nitride waveguides with intrinsic single-photon emitters for integrated Quantum photonics. *ACS Photonics* **2022**, *9*, 3357–3365. [CrossRef]
42. Sacher, W.; Luo, X.; Yang, Y.; Chen, F.-D.; Lordello, T.; Mak, J.; Liu, X.; Hu, T.; Xue, T.; Lo, P.-Q.; et al. Visible-light silicon nitride waveguide devices and implantable neurophotonic probes on thinned 200 mm silicon wafers. *Opt. Express* **2019**, *27*, 37400–37418. [CrossRef]
43. Takenaka, M.; Nakano, Y. InP photonic wire waveguide using InAlAs oxide cladding layer. *Opt. Express* **2007**, *15*, 8422–8427. [CrossRef]
44. Soleimannezhad, F.; Nikoufard, M.; Mahdian, M. Low-loss indium phosphide-based hybrid plasmonic waveguide. *Microw. Opt. Technol. Lett.* **2021**, *63*, 2242–2251. [CrossRef]
45. Rahlves, M.; Rezem, M.; Boroz, K.; Schlangen, S.; Reithmeier, E.; Roth, B. Flexible, fast, and low-cost production process for polymer based diffractive optics. *Opt. Express* **2015**, *23*, 3614–3622. [CrossRef] [PubMed]
46. Beche, B.; Pelletier, N.; Gaviot, E.; Zyss, J. Single-mode TE00-TM00 optical waveguides on SU-8 polymer. *Opt. Commun.* **2004**, *230*, 91–94. [CrossRef]
47. Elmogi, A.; Bosman, E.; Missinne, J.; Steenberge, V. Comparison of epoxy- and siloxane-based single-mode optical waveguides defined by direct-write lithography. *Opt. Mater.* **2016**, *52*, 26–31. [CrossRef]
48. Ibrahim, M.; Kassim, N.; Mohammad, A.; Lee, S.-Y.; Chin, M.-K. Single mode optical waveguides based on photodefinable benzocyclobutene (BCB 4024-40) polymer. *Microw. Opt. Technol. Lett.* **2006**, *49*, 479–481. [CrossRef]
49. Available online: http://www.chemoptics.co.kr/eng/main/main.php (accessed on 10 January 2023).
50. Buestrich, R.; Kahlenberg, F.; Popall, M.; Dannberg, P.; Muller-Fiedler, R.; Rosch, O. ORMOCER (R) s for optical interconnection technology. *J. Sol-Gel Sci. Technol.* **2001**, *20*, 181–186. [CrossRef]
51. Bamiedakis, N.; Beals, J.; Penty, R.; White, I.; DeGroot, J.; Clapp, T. Cost-effective multimode polymer waveguides for high-speed on-board optical interconnects. *IEEE J. Quantum Electron.* **2009**, *45*, 415–424. [CrossRef]
52. Bosman, E.; Van Steenberge, G.; Christiaens, W.; Hendrickx, N.; Vanfleteren, J.; Van Daele, P. Active optical links embedded in flexible substrates. In Proceedings of the 58th Electronic Components and Technology Conference, Lake Buena Vista, FL, USA, 27–30 May 2008; pp. 1150–2013.
53. Cai, Z.; Qiu, W.; Shao, G.; Wang, W. A new fabrication method for all-PDMS waveguides. *Sens. Actuators A Phys.* **2013**, *204*, 44–47. [CrossRef]
54. Hirose, C.; Fukuda, N.; Sassa, T.; Ishibashi, K.; Ochiai, T.; Furukawa, R. Fabrication of a fluorophore-doped cylindrical waveguide structure using elastomers for visual detection of stress. *Fibers* **2019**, *7*, 37. [CrossRef]
55. Kalathimekkad, S.; Missinne, J.; Schaubroeck, D.; Mandamparambil, R. Alcohol vapor sensor based on fluorescent dye-doped optical waveguides. *IEEE Sens. J.* **2015**, *15*, 76–81. [CrossRef]
56. Marques, C.; Pospori, A.; Demirci, G.; Çetinkaya, O.; Gawdzik, B.; Antunes, P.; Bang, O.; Mergo, P.; André, P.; Webb, D. Fast Bragg Grating Inscription in PMMA Polymer Optical Fibres: Impact of Thermal Pre-Treatment of Preforms. *Sensors* **2017**, *17*, 891. [CrossRef]
57. Woyessa, G.; Theodosiou, A.; Markos, C.; Kalli, K.; Bang, O. Single Peak Fiber Bragg Grating Sensors in Tapered Multimode Polymer Optical Fibers. *J. Light. Technol.* **2021**, *39*, 6934–6941. [CrossRef]
58. Pasquardini, L.; Cennamo, N.; Malleo, G.; Vanzetti, L.; Zeni, L.; Bonamini, D.; Salvia, R.; Bassi, C.; Bossi, A. A Surface Plasmon Resonance Plastic Optical Fiber Biosensor for the Detection of Pancreatic Amylase in Surgically-Placed Drain Effluent. *Sensors* **2021**, *21*, 3443. [CrossRef] [PubMed]
59. Kadhim, R.A.; Abdul, K.K.; Yuan, L. Advances in Surface Plasmon Resonance-Based Plastic Optical Fiber Sensors. *IETE Tech. Rev.* **2022**, *39*, 442–459. [CrossRef]
60. Cennamo, N.; Pesavento, M.; Zeni, L. A review on simple and highly sensitive plastic optical fiber probes for bio-chemical sensing. *Sens. Actuators B Chem.* **2021**, *331*, 129393. [CrossRef]

61. Rajamani, A.S.; Divagar, M.; Sai, V. Plastic fiber optic sensor for continuous liquid level monitoring. *Sens. Actuators A Phys.* **2019**, *296*, 192–199. [CrossRef]
62. Prajzler, V.; Neruda, M.; Jasek, P.; Nekvindova, P. The properties of free-standing epoxy polymer multi-mode optical waveguides. *Microsyst. Technol.* **2019**, *25*, 257–264. [CrossRef]
63. Kagami, M.; Kawasaki, A.; Ito, H. A polymer optical waveguide with out-of-plane branching mirrors for surface-normal optical interconnections. *J. Light. Technol.* **2001**, *19*, 1949–1955. [CrossRef]
64. Klein, S.; Barsella, A.; Leblond, H.; Bulou, H.; Fort, A.; Andraud, C.; Lemercier, G.; Mulatier, J.; Dorkenoo, K. One-step waveguide and optical circuit writing in photopolymerizable materials processed by two-photon absorption. *Appl. Phys. Lett.* **2005**, *86*, 211118. [CrossRef]
65. Wong, W.; Zhou, J.; Pun, E. Low-loss polymeric optical waveguides using electron-beam direct writing. *Appl. Phys. Lett.* **2001**, *78*, 2110–2112. [CrossRef]
66. Choi, C.-G.; Kim, J.-T.; Jeong, M.-Y. Fabrication of optical waveguides in thermosetting polymers using hot embossing. In Proceedings of the Integrated Photonics Research, Washington, DC, USA, 15–20 June 2003.
67. Choi, C.-G.; Han, S.-P.; Kim, B.-C.; Ahn, S.-H.; Jeong, M.-Y. Fabrication of large-core 1 × 16 optical power splitters in polymers using hot-embossing process. *IEEE Photon. Technol. Lett.* **2003**, *15*, 825–827. [CrossRef]
68. Zheng, L.; Keppler, N.; Zhang, H.; Behrens, P.; Roth, B. Planar polymer optical waveguide with metal-organic framework coating for carbon dioxide sensing. *Adv. Mater. Technol.* **2022**, *7*, 2200395. [CrossRef]
69. Fan, R.; Hooker, R. Tapered polymer single-mode waveguides for mode transformation. *J. Light. Technol.* **1999**, *17*, 466–474. [CrossRef]
70. Kobayashi, J.; Yagi, S.; Hatakeyama, Y.; Kawakami, N. Low loss polymer optical waveguide replicated from flexible film stamp made of polymeric material. *Jpn. J. Appl. Phys.* **2013**, *52*, 072501. [CrossRef]
71. Bruck, R.; Muellner, P.; Kataeva, N.; Koeck, A.; Trassl, S.; Rinnerbauer, V.; Schmidegg, K.; Hainberger, R. Flexible thin-film polymer waveguides fabricated in an industrial roll-to-roll process. *Appl. Opt.* **2013**, *52*, 4510–4514. [CrossRef]
72. Shneidman, A.; Becker, K.; Lukas, M.; Torgerson, N.; Wang, C.; Reshef, O.; Burek, M.; Paul, K.; McLellan, J.; Loncar, M. All-polymer integrated optical resonators by roll-to-roll nanoimprint lithography. *ACS Photonics* **2018**, *5*, 1839–1845. [CrossRef]
73. Shao, J.; Chen, X.L.; Li, X.; Tian, H.; Wang, C.; Lu, B. Nanoimprint lithography for the manufacturing of flexible electronics. *Sci. China Technol. Sci.* **2019**, *62*, 175–198. [CrossRef]
74. Šakalys, R.; Kho, K.; Keyes, T. A reproducible, low cost microfluidic microcavity array SERS platform prepared by soft lithography from a 2 photon 3D printed template. *Sens. Actuators B Chem.* **2021**, *340*, 129970. [CrossRef]
75. Azadegan, R.; Nagarajan, P.; Yao, D.; Ellis, T. Polymer micro hot embossing for the fabrication of three-dimensional millimeter-wave components. In Proceedings of the IEEE Antennas and Propagation Society International Symposium, Charleston, SC, USA, 1–5 June 2009; pp. 1–4.
76. Aldada, L.; Shacklette, L. Advances in polymer integrated optics. *IEEE J. Sel. Top. Quant.* **2000**, *6*, 54–68. [CrossRef]
77. Andolfi, L.; Greco, S.L.M.; Tierno, D.; Chignola, R.; Martinelli, M.; Giolo, E.; Luppi, S.; Delfino, I.; Zanetti, M.; Battistella, A.; et al. Planar AFM macro-probes to study the biomechanical properties of large cells and 3D cell spheroids. *Acta Biomater.* **2019**, *94*, 505–513. [CrossRef]
78. Norland Optical Adhesive 73. Available online: https://www.amstechnologies-webshop.com/media/pdf/09/bb/1a/NOA-73-TDS-Datasheet.pdf (accessed on 13 February 2023).
79. Santiago-Alvarado, A.; Cruz-Felix, A.; Iturbide, F.; Licona-Morán, B. Physical-chemical properties of PDMS samples used in tunable lenses. *Int. J. Eng. Sci. Innov. Technol.* **2014**, *3*, 563–571.
80. Johnston, D.; McCluskey, D.K.; Tan, C.K.L.; Tracey, M.C. Mechanical characterization of bulk Sylgard 184 for microfluidics and microengineering. *J. Micromech. Microeng.* **2014**, *24*, 035017. [CrossRef]
81. Cai, D.K.; Neyer, A.; Kuckuk, R.; Heise, H.M. Optical absorption in transparent PDMS materials applied for multimode waveguides fabrication. *Opt. Mater.* **2008**, *30*, 1157–1161. [CrossRef]
82. Khan, Y.; Butt, M.; Khonina, S.; Kazanskiy, N. Thermal sensor based on polydimethylsiloxane polymer deposited on low-index-contrast dielectric photonic crystal structure. *Photonics* **2022**, *9*, 770. [CrossRef]
83. Demir, M.M.; Koynov, K.; Akbey, Ü.; Bubeck, C.; Park, I.; Lieberwirth, I.; Wegner, G. Optical Properties of Composites of PMMA and Surface-Modified Zincite Nanoparticles. *Macromolecules* **2007**, *40*, 1089–1100. [CrossRef]
84. Abdelrazek, E.M.; Hezma, A.M.; El-khodary, A.; Elzayat, A.M. Spectroscopic studies and thermal properties of PCL/PMMA biopolymer blend. *Egypt. J. Basic Appl. Sci.* **2016**, *3*, 10–15. [CrossRef]
85. Ishiyama, C.; Higo, Y. Effects of humidity on Young's modulus in poly(methyl methacrylate). *J. Polym. Sci. Part B Polym. Phys.* **2002**, *40*, 460–465. [CrossRef]
86. Min, R.; Hu, X.; Pereira, L.; Soares, M.S.; Silva, L.C.B.; Wang, G.; Martins, L.; Qu, H.; Antunes, P.; Marques, C.; et al. Polymer optical fiber for monitoring human physiological and body function: A comprehensive review on mechanisms, materials, and applications. *Opt. Laser Technol.* **2022**, *147*, 107626. [CrossRef]
87. Beadie, G.; Brindza, M.; Flynn, R.A.; Rosenberg, A.; Shirk, J.S. Refractive index measurements of poly(methyl methacrylate) (PMMA) from 0.4–1.6 μm. *Appl. Opt.* **2015**, *54*, F139–F143. [CrossRef]
88. NOA68. Available online: https://www.norlandprod.com/adhesives/noa%2068.html (accessed on 13 February 2023).

89. Sabzekar, M.; Chenar, M.P.; Maghsoud, Z.; Mostaghisi, O.; García-Payo, M.C.; Khayet, M. Cyclic olefin polymer as a novel membrane material for membrane distillation applications. *J. Membr. Sci.* **2021**, *621*, 118845. [CrossRef]
90. Khanarian, G. Optical properties of cyclic olefin copolymers. *Opt. Eng.* **2001**, *40*, 1024–1029. [CrossRef]
91. Schmid, S. *Electrostatically Actuated All-Polymer Microbeam Resonators—Characterization and Application*; ETH Zurich: Zurich, Switzerland, 2009.
92. Lee, B.; Choi, K.-H.; Yoo, K. Innovative SU-8 Lithography Techniques and Their Applications. *Micromachines* **2015**, *6*, 1–18. [CrossRef]
93. Wolter, H.; Storch, W.; Ott, H. New Inorganic/Organic Copolymers (Ormocer®s) for Dental Applications. *MRS Online Proc. Libr.* **1994**, *346*, 143. [CrossRef]
94. Schmidt, V.; Kuna, L.; Satzinger, V.; Houbertz, R.; Jakopic, G.; Leising, G. Application of two-photon 3D lithography for the fabrication of embedded ORMOCER waveguides. In Proceedings of the Optoelectronic Integrated Circuits IX, San Jose, CA, USA, 22–24 January 2007.
95. Declerck, P.; Houbertz, R.; Jakopic, G.; Passinger, S.; Chichkov, B. High Refractive Index Inorganic-Organic Hybrid Materials for Photonic Applications. *MRS Online Proc. Libr.* **2007**, *1007*, 1007. [CrossRef]
96. Mahdi, E.M.; Chaudhuri, A.K.; Tan, J.-C. Capture and immobilisation of iodine (I 2) utilising polymer-based ZIF-8 nanocomposite membranes. *Mol. Syst. Des. Eng.* **2016**, *1*, 122–131. [CrossRef]
97. Keppler, N.C.; Hindricks, K.D.J.; Behrens, P. Large refractive index changes in ZIF-8 thin films of optical quality. *RSC Adv.* **2022**, *12*, 5807–5815. [CrossRef]
98. Kazanskiy, N.L.; Butt, M.A.; Khonina, S.N. Carbon Dioxide Gas Sensor Based on Polyhexamethylene Biguanide Polymer Deposited on Silicon Nano-Cylinders Metasurface. *Sensors* **2021**, *21*, 378. [CrossRef]
99. Wang, J.; Luo, Y.; Jin, K.; Yuan, C.; Sun, J.; He, F.; Fang, Q. A novel one-pot synthesized organosiloxane: Synthesis and conversion to directly thermo-crosslinked polysiloxanes with low dielectric constants and excellent thermostability. *Polym. Chem.* **2015**, *6*, 5984–5988. [CrossRef]
100. Kane, C.F.; Krchnavek, R.R. Benzocyclobutene optical waveguides. *IEEE Photonics Technol. Lett.* **1995**, *7*, 535–537. [CrossRef]
101. Guo, S.; Gustafsson, G.; Hagel, O.J.; Arwin, H. Determination of refractive index and thickness of thick transparent films by variable-angle spectroscopic ellipsometry: Application to benzocyclobutene films. *Appl. Opt.* **1996**, *35*, 1693–1699. [CrossRef]
102. Wang, J.; Chen, C.; Wang, C.; Wang, X.-B.; Yi, Y.; Sun, X.; Wang, F.; Zhang, D. Metal-Printing Defined Thermo Optic Tunable Sampled Apodized Waveguide Grating Wavelength Filter Based on Low Loss Fluorinated Polymer Material. *Appl. Sci.* **2019**, *10*, 167. [CrossRef]
103. NOA63. Available online: https://www.norlandprod.com/adhesives/noa%2063.html (accessed on 13 February 2023).
104. Bucknall, C.B.; Gilbert, A.H. Toughening tetrafunctional epoxy resins using polyetherimide. *Polymer* **1989**, *30*, 213–217. [CrossRef]
105. Wang, X.; Wang, Y.; Bi, S.; Wang, Y.; Chen, X.; Qiu, L.; Sun, J. Optically Transparent Antibacterial Films Capable of Healing Multiple Scratches. *Adv. Funct. Mater.* **2014**, *24*, 403–411. [CrossRef]
106. Jiang, F.; Cakmak, M. Real time optical and mechano-optical studies during drying and uniaxial stretching of Polyetherimide films from solution. *Polymer* **2015**, *68*, 168–175. [CrossRef]
107. Vinogradova, I.; Andrienko, D.; Lulevich, V.V.; Nordschild, S.; Sukhorukov, G.B. Young's Modulus of Polyelectrolyte Multilayers from Microcapsule Swelling. *Macromolecules* **2004**, *37*, 1113–1117. [CrossRef]
108. Kim, J.-H.; Fujita, S.; Shiratori, S. Design of a thin film for optical applications, consisting of high and low refractive index multilayers, fabricated by a layer-by-layer self-assembly method. *Colloids Surf. A Physicochem. Eng. Asp.* **2006**, *284–285*, 290–294. [CrossRef]
109. Liang, Y.; Zhao, M.; Wu, Z.; Morthier, G. Investigation of Grating-Assisted Trimodal Interferometer Biosensors Based on a Polymer Platform. *Sensors* **2018**, *18*, 1502. [CrossRef] [PubMed]
110. Bosc, D.; Maalouf, A.; Messaad, K.; Mahé, H.; Bodiou, L. Advanced analysis of optical loss factors in polymers for integrated optics circuits. *Opt. Mater.* **2013**, *35*, 1207–1212. [CrossRef]
111. Kaino, T. Plastic optical fibers. In *Polymers in Optics: Physics, Chemistry, and Applications: A Critical Review*; Society of Photo Optical Loaction: Bellingham, WA, USA, 1996.
112. Butt, A.M.; Kazanskiy, N.L.; Khonina, S.N. Highly integrated plasmonic sensor design for the simultaneous detection of multiple analytes. *Curr. Appl. Phys.* **2020**, *20*, 1274–1280. [CrossRef]
113. Butt, A.M.; Voronkov, G.; Grakhova, E.; Kutluyarov, R.; Kazanskiy, N.; Khonina, S. Environmental Monitoring: A Comprehensive Review on optical waveguide and fiber-based sensors. *Biosensors* **2022**, *12*, 1038. [CrossRef]
114. Butt, A.M. Numerical investigation of a small footprint plasmonic Bragg grating structure with a high extinction ratio. *Photonics Lett. Pol.* **2020**, *12*, 82–84. [CrossRef]
115. Butt, M.A.; Khonina, S.N.; Kazanskiy, N.L. A plasmonic colour filter and refractive index sensor applications based on metal-insulator-metal square micro-ring cavities. *Laser Phys.* **2020**, *30*, 016205. [CrossRef]
116. Butt, M.A.; Khonina, S.N.; Kazanskiy, N.L. Hybrid plasmonic waveguide-assisted metal-insulator-metal ring resonator for refractive index sensing. *J. Mod. Opt.* **2018**, *65*, 1135–1140. [CrossRef]
117. Butt, M.A.; Khonina, S.N.; Kazanskiy, N.L. Sensitivity enhancement of silicon strip waveguide ring resonator by incorporating a thin metal film. *IEEE Sens. J.* **2020**, *3*, 1355–1362. [CrossRef]

118. Butt, M.A.; Degtyarev, S.A.; Khonina, S.N.; Kazanskiy, N.L. An evanescent field absorption gas sensor at mid-IR 3.39 um wavelength. *J. Mod. Opt.* **2017**, *64*, 1892–1897. [CrossRef]
119. Sonawane, H.; Singh, A.; Gupta, A.; Tiwari, M.; Gaur, S. Polymer based surface plasmon resonance sensors: Theoretical study of sensing characteristics. *Mater. Proc.* **2022**, *67*, 726–731. [CrossRef]
120. Sahraeibelverdi, T.; Guo, L.; Veladi, H.; Malekshahi, M. Polymer ring resonator with a partially tapered waveguide for biomedical sensing: Computational study. *Sensors* **2021**, *21*, 5017. [CrossRef]
121. Morarescu, R.; Pal, P.; Beneitez, N.; Missinne, J.; Steenberge, G.; Bienstman, P.; Morthier, G. Fabrication and characterization of high-optical-quality factor hybrid polymer microring resonators operating at very near infrared wavelengths. *IEEE Photonics J.* **2016**, *8*, 6600409. [CrossRef]
122. Ma, X.; Zhao, Z.; Yao, H.; Deng, J.; Wu, J.; Hu, Z.; Chen, K. Compact and highly sensitive refractive index sensor based on embedded double-ring resonator using Vernier effect. *IEEE Photonics J.* **2022**, *15*, 6800109. [CrossRef]
123. Xiao, Y.; Hofmann, M.; Wang, Z.; Sherman, S.; Zappe, H. Design of all-polymer asymmetric Mach-Zehnder interferometer sensors. *Appl. Opt.* **2016**, *55*, 3566–3573. [CrossRef]
124. Ji, L.; Yang, S.; Shi, R.; Fu, Y.; Su, J.; Wu, C. Polymer waveguide coupled surface plasmon refractive index sensor: A theoretical study. *Photonic Sens.* **2020**, *10*, 353–363. [CrossRef]
125. Mancuso, M.; Goddard, J.; Erickson, D. Nanoporous polymer ring resonators for biosensing. *Opt. Express* **2011**, *20*, 245–255. [CrossRef]
126. Dullo, F.; Lindecrantz, S.; Jagerska, J.; Hansen, J.; Engqvist, M.; Solbo, S.; Helleso, O. Sensitive on-chip methane detection with a cryptophane- A cladded Mach-Zehnder interferometer. *Opt. Express* **2015**, *23*, 31564–31573. [CrossRef] [PubMed]
127. Lin, B.; Yi, Y.; Cao, Y.; Lv, J.; Yang, Y.; Wang, F.; Sun, X.; Zhang, D. A polymer asymmetric Mach-Zehnder interferometer sensor model based on electrode thermal writing waveguide technology. *Micromachines* **2019**, *10*, 628. [CrossRef] [PubMed]
128. Chen, S.; Lin, C. Sensitivity comparison of graphene based surface plasmon resonance biosensor with Au, Ag and Cu in the visible region. *Mater. Res. Express* **2019**, *6*, 056503. [CrossRef]
129. Ceylan, K.H.; Kulah, H.; Ozgen, C. Thin film biosensors. In *Thin Films and Coatings in Biology. Biological and Medical Physics, Biomedical Engineering*; Springer: Dordrecht, The Netherlands, 2013.
130. Bahabady, A.; Olyaee, S. Two-curve-shaped biosensor for detecting glucose concentration and salinity of seawater based on photonic crystal nano-ring resonator. *Sens. Lett.* **2015**, *13*, 774–777. [CrossRef]
131. Xu, D.-X.; Vachon, M.; Densmore, A.; Ma, R.; Janz, S.; Delâge, A.; Lapointe, J.; Cheben, P.; Schmid, J.H.; Post, E.; et al. Real-time cancellation of temperature induced resonance shifts in SOI wire waveguide ring resonator label-free biosensor arrays. *Opt. Express* **2010**, *18*, 22867–22879. [CrossRef]
132. Guo, J.; Yang, C.; Dai, Q.; Kong, L. Wearable and biomedical applications. *Sensors* **2019**, *19*, 3771. [CrossRef]
133. Ahmed, Z.; Reddy, J.; Malekoshoaraie, M.; Hassanzade, V.; Kimukin, I.; Jain, V.; Chamanzar, M. Flexible optoelectric neural interfaces. *Curr. Opin. Biotechnol.* **2021**, *72*, 121–130. [CrossRef]
134. Ang, L.; Por, L.; Yam, M. Study on different molecular weights of chitosan as an immobilization matrix for a glucose biosensor. *PLoS ONE* **2013**, *8*, e70597. [CrossRef]
135. Bucur, B.; Purcarea, C.; Andreescu, S.; Vasilescu, A. Addressing the selectivity of enzyme biosensors: Solutions and percpectives. *Sensors* **2021**, *21*, 3038. [CrossRef]
136. Lavin, A.; Vicente, J.; Holgado, M.; Laguna, M.; Casquel, R.; Santamaria, B.; Maigler, M.; Hernandez, A.; Ramirez, Y. On the determination of uncertainty and limit of detection in label-free biosensors. *Sensors* **2018**, *18*, 2038. [CrossRef] [PubMed]
137. Chen, L.-C.; Wang, E.; Tai, C.-S.; Chiu, Y.-C.; Li, C.-W.; Lin, Y.-R.; Lee, T.-H.; Huang, C.-W.; Chen, J.-C.; Chen, W. Improving the reproducibility, accuracy, and stability of an electrochemical biosensor platform for point-of-care use. *Biosens. Bioelectron.* **2020**, *155*, 112111. [CrossRef] [PubMed]
138. Bruck, R.; Hainberger, R. Polymer waveguide based biosensor. In *Photonics, Devices, and Systems IV*; SPIE: Bellingham, WA, USA, 2008.
139. Wang, C.; Yi, P.; Li, J.; Dong, H.; Chen, C.; Zhang, D.; Shen, H.; Fu, B. Polymer optical waveguide grating-based biosensor to detect effective drug concentrations of Ginkgolide A for Inhibition of PMVEC apoptosis. *Biosensors* **2021**, *11*, 264. [CrossRef] [PubMed]
140. Wang, M.; Hiltunen, J.; Liedert, C.; Pearce, S.; Charlton, M.; Hakalahti, L.; Karioja, P.; Myllyla, R. Highly sensitive biosensor based on UV-imprinted layered polymeric-inorganic composite waveguides. *Opt. Express* **2012**, *20*, 20309–20317. [CrossRef]
141. Walter, J.-G.; Eilers, A.; Alwis, L.; Roth, B.; Bremer, K. SPR Biosensor Based on Polymer Multi-Mode Optical Waveguide and Nanoparticle Signal Enhancement. *Sensors* **2020**, *20*, 2889. [CrossRef]
142. Han, X.-Y.; Wu, Z.-L.; Yang, S.-C.; Shen, F.-F.; Liang, Y.-X.; Wang, L.-H.; Wang, J.-Y.; Ren, J.; Jia, L.-Y.; Zhang, H.; et al. Recent progress of imprinted polymer photonic waveguide devices and applications. *Polymers* **2018**, *10*, 603. [CrossRef]
143. Nguyen, H.D.; Hollenbach, U.; Pfirrmann, S.; Ostrizinski, U.; Pfeiffer, K.; Hengsbach, S.; Mohr, J. Photo-structurable polymer for interlayer single-mode waveguide fabrication by femtosecond laser writing. *Opt. Mater.* **2017**, *66*, 110–116. [CrossRef]
144. Yi, L.; Changyuan, Y. Highly strechable hybrid silica/polymer optical fiber sensors for large-strain and high-temperature application. *Opt. Express* **2019**, *27*, 20107–20116. [CrossRef]
145. Wang, T.; Hu, S.; Chamlagain, B.; Hong, T.; Zhou, Z.; Weiss, S.; Xu, Y.-Q. Visualizing light scattering in silicon waveguides with black phosphorus photodetectors. *Adv. Mater.* **2016**, *28*, 7162–7166. [CrossRef]

146. Mirnaziry, S.; Wolff, C.; Steel, M.; Eggleton, B.; Poulton, C. Stimulated Brillouin scattering in silicon/chalcogenide slot waveguides. *Opt. Express* **2016**, *24*, 4786–4800. [CrossRef]
147. Kumaar, S.; Sivasubramanian, A. Analysis of BCB and SU 8 photonic waveguide in MZI architecture for point-of-care devices. *Sens. Int.* **2023**, *4*, 100207. [CrossRef]
148. Ramirez, J.; Lechuga, L.; Gabrielli, L.; Hernandez-Figueroa, H. Study of a low-cost trimodal polymer waveguide for interferometric optical biosensors. *Opt. Express* **2015**, *23*, 11985. [CrossRef] [PubMed]
149. Walter, J.-G.; Alwis, L.; Roth, B.; Bremer, K. All-Optical Planar Polymer Waveguide-Based Biosensor Chip Designed for Smartphone-Assisted Detection of Vitamin D. *Sensors* **2020**, *20*, 6771. [CrossRef] [PubMed]
150. Tu, X.; Chen, S.-L.; Song, C.; Huang, T.; Guo, L. Ultrahigh Q polymer microring resonators for biosensing applications. *IEEE Photonics J.* **2019**, *11*, 4200110. [CrossRef]
151. Girschikofsky, M.; Rosenberger, M.; Belle, S.; Brutschy, M.; Waldvogel, S.; Hellmann, R. Optical planar Bragg grating sensor for real-time detection of benzene, toluene and xylene in solvent vapour. *Sens. Actuators B Chem.* **2012**, *171–172*, 338–342. [CrossRef]
152. Zaky, Z.; Ahmed, A.; Shalaby, A.; Aly, A. Refractive index gas sensor based on the Tamm state in a one-dimensional photonic crystal: Theoretical optimisation. *Sci. Rep.* **2020**, *10*, 9736. [CrossRef]
153. Butt, M.A.; Khonina, S.N.; Kazanskiy, N.L. Modelling of rib channel waveguides based on silicon-on-sapphire at 4.67 um wavelength for evanescent field gas absorption sensor. *Optik* **2018**, *168*, 692–697. [CrossRef]
154. Kazanskiy, N.; Khonina, S.; Butt, M. Polarization-insensitive hybrid plasmonic waveguide design for evanescent field absorption gas sensor. *Photonic Sens.* **2021**, *11*, 279–290. [CrossRef]
155. Wanguemert-Perez, J.; Cheben, P.; Ortega-Monux, A.; Alonso-Ramos, C.; Perez-Galacho, D.; Halir, R.; Molina-Fernandez, I.; Xu, D.-X.; Schmid, J. Evanescent field waveguide sensing with subwavelength grating structures in silicon-on-insulator. *Opt. Lett.* **2014**, *39*, 4442–4445. [CrossRef]
156. Nylabder, C.; Armgrath, M.; Lundstrom, I. An ammonia detector based on a conducting polymer. In Proceedings of the International Meeting on Chemical Sensors, Fukuoka, Japan, 19–22 September 1983; pp. 203–207.
157. Hopper, R.; Popa, D.; Udrea, F.; Ali, S.; Stanley-Marbell, P. Miniaturized thermal acoustic gas sensor based on a CMOS microhotplate and MEMS microphone. *Sci. Rep.* **2022**, *12*, 1690. [CrossRef]
158. Sadaoka, Y.; Sakai, Y.; Manabe, T. Detection of CO_2 using a solid-state electrochemical sensor based on sodium ionic conductors. *Sens. Actuators B Chem.* **1993**, *15*, 166–170. [CrossRef]
159. Lin, Z.-D.; Young, S.-J.; Chang, S.-J. CO_2 gas sensors based on carbon nanotube thin films using a simple transfer method on flexible substrate. *IEEE Sens. J.* **2015**, *15*, 7017–7020. [CrossRef]
160. Gheorghe, A.; Lugier, O.; Ye, B.; Tanase, S. Metal-organic framework based systems for CO_2 sensing. *J. Mater. Chem. C* **2021**, *9*, 16132–16142. [CrossRef]
161. Nugroho, F.; Darmadi, I.; Cusinato, L.; Susarrey-Arce, A.; Schreuders, H.; Bannenberg, L.; Fanta, A.; Kadkhodazadeh, S.; Wagner, J.; Antosiewicz, T.; et al. Metal-polymer hybrid nanomaterials for plasmonic ultrafast hydrogen detection. *Nat. Mater.* **2019**, *18*, 489–495. [CrossRef] [PubMed]
162. Bamiedakis, N.; Hutter, T.; Penty, R.; Elliott, S. PCB-integrated optical waveguide sensors: An ammonia gas sensor. *J. Light. Technol.* **2013**, *31*, 1628. [CrossRef]
163. Batumalay, M.; Johari, A.; Khudus, M.I.M.A.; Bin Jali, M.H.; Al Noman, A.; Harun, S.W. Microbottle resonator for temperature sensing. *J. Phys. Conf. Ser.* **2019**, *1371*, 012006. [CrossRef]
164. Kong, Y.; Wei, Q.; Liu, C.; Wang, S. Nanoscale temperature sensor based on Fano resonance in metal-insulator-metal waveguide. *Opt. Commun.* **2017**, *384*, 85–88. [CrossRef]
165. Butt, M.; Khonina, S.; Kazanskiy, N. A compact design of a modified Bragg grating filter based on a metal-insulator-metal waveguide for filtering and temperature sensing applications. *Optik* **2022**, *251*, 168466. [CrossRef]
166. Butt, M.; Kazanskiy, N.; Khonina, S. Revolution in flexible wearable electronics for temperature and pressure monitoring—A review. *Electronics* **2022**, *11*, 716. [CrossRef]
167. Roth, G.-L.; Hessler, S.; Kefer, S.; Girchikofsky, M.; Esen, C.; Hellmann, R. Femtosecond laser inscription of waveguides and Bragg gratings in transparent cyclic olefin copolymers. *Opt. Express* **2020**, *28*, 18077–18084. [CrossRef]
168. Missinne, J.; Beneitez, N.; Lamberti, A.; Chiesura, G.; Luyckx, G.; Mattelin, M.-A.; Paepegem, W.; Steenberge, G. Thin and flexible polymer photonic sensor foils for monitoring composite structures. *Adv. Eng. Mater.* **2018**, *20*, 1701127. [CrossRef]
169. Butt, M.; Kazanskiy, N.; Khonina, S. Advances in Waveguide Bragg Grating Structures, Platforms, and Applications: An Up-to-Date Appraisal. *Biosensors* **2022**, *12*, 497. [CrossRef]
170. Eldada, L.; Blomquist, R.; Maxfield, M.; Pant, D.; Boudoughian, G.; Poga, C.; Norwood, R. Thermooptic planar polymer Bragg grating OADMs with broad tuning range. *IEEE Photon. Technol. Lett.* **1999**, *11*, 448–450. [CrossRef]
171. Yun, B.; Hu, G.; Cui, Y. Third-order polymer waveguide Bragg grating array by using conventional contact lithography. *Opt. Commun.* **2014**, *330*, 113–116. [CrossRef]
172. Binfeng, Y.; Guohua, H.; Yiping, C. Polymer waveguide Bragg grating Fabry-Perot filter using a nanoimprinting technique. *J. Opt.* **2014**, *16*, 105501. [CrossRef]
173. Song, F.; Xiao, J.; Xie, A.; Seo, S.-W. A polymer waveguide grating sensor integrated with a thin-film photodetector. *J. Opt.* **2014**, *16*, 015503. [CrossRef]

174. Lin, H.; Xing, Y.; Chen, X.; Zhang, S.; Forsberg, E.; He, S. Polymer-based planar waveguide chirped Bragg grating for high-resolution tactile sensing. *Opt. Express* **2022**, *30*, 20871–20882. [CrossRef]
175. Rosenberger, M.; Eisenbeil, W.; Schmauss, B.; Hellmann, R. Simultaneous 2D strain sensing using polymer planar Bragg gratings. *Sensors* **2015**, *15*, 4264–4272. [CrossRef]
176. Wang, M.; Hiltunen, J.; Uusitalo, S.; Puustinen, J.; Lappalainen, J.; Karioja, P.; Myllylä, R. Fabrication of optical inverted-rib waveguides using UV-imprinting. *Microelectron. Eng.* **2011**, *88*, 175–178. [CrossRef]
177. Beneitez, N.; Missinne, J.; Shi, Y.; Chiesura, G.; Luyckx, G.; Degrieck, J.; Steenberge, G. Highly sensitive waveguide Bragg grating temperature sensor using hybrid polymers. *IEEE Photonics Technol. Lett.* **2016**, *28*, 1150. [CrossRef]
178. Khan, M.; Farooq, H.; Wittmund, C.; Klimke, S.; Lachmayer, R.; Renz, F.; Roth, B. Polymer optical waveguide sensor based on Fe-Amino-Triazole complex molecular switches. *Polymers* **2021**, *13*, 195. [CrossRef]
179. Liu, Y.; Li, M.; Zhao, P.; Wang, X.; Qu, S. High sensitive temperature sensor based on a polymer waveguide integrated in an optical fibre micro-cavity. *J. Opt.* **2018**, *20*, 015801. [CrossRef]
180. Niu, D.; Wang, L.; Xu, Q.; Jiang, M.; Wang, X.; Sun, X.; Wang, F.; Zhang, D. Ultra-sensitive polymeric waveguide temperature sensor based on asymmetric Mach-Zehnder interferometer. *Appl. Opt.* **2019**, *58*, 1276–1280. [CrossRef]
181. Wang, C.; Zhang, D.; Yue, J.; Zhang, X.; Wu, Z.; Zhang, T.; Chen, C.; Fei, T. Optical waveguide sensors for measuring human temperature and humidity with gel polymer electrolytes. *ACS Appl. Mater. Interfaces* **2021**, *13*, 60384–60392. [CrossRef]
182. Nazempour, R.; Zhang, Q.; Fu, R.; Sheng, X. Biocompatible and implantable optical fibers and waveguides for biomedicine. *Materials* **2018**, *11*, 1283. [CrossRef]
183. Hah, D.; Yoon, E.; Hong, S. An optomechanical pressure sensor using multimode interference couplers with polymer waveguides on a thin p+-Si membrane. *Sens. Actuators* **2000**, *79*, 204–210. [CrossRef]
184. Ren, Y.; Mormile, P.; Petti, L.; Cross, G. Optical waveguide humidity sensor with symmetric multilayer configuration. *Sens. Actuators B Chem.* **2001**, *75*, 76–82. [CrossRef]
185. Guo, J.; Zhou, B.; Yang, C.; Dai, Q.; Kong, L. Stretchable and temperature-sensitive polymer optical fibers for wearable health monitoring. *Adv. Funct. Mater.* **2019**, *29*, 1902898. [CrossRef]
186. Yun, S.; Park, S.; Park, B.; Kim, Y.; Park, S.; Nam, S.; Kyung, K. Polymer-waveguide-based flexible tactile sensor array for dynamic response. *Adv. Mater.* **2014**, *26*, 4474–4480. [CrossRef]
187. Song, Y.; Peters, K. A self-repairing polymer waveguide sensor. *Smart Mater. Struct.* **2011**, *20*, 065005. [CrossRef]
188. Guo, J.; Liu, X.; Jiang, N.; Yetisen, A.; Yuk, H.; Yang, C.; Khademhosseini, A.; Zhao, X.; Yun, S. Highly stretchable, strain sensing hydrogel optical fibers. *Adv. Mater.* **2016**, *28*, 10244–10249. [CrossRef]
189. Niu, D.; Wang, X.; Sun, S.; Jiang, M.; Xu, Q.; Wang, F.; Wu, Y.; Zhang, D. Polymer/silica hybrid waveguide temperature sensor based on asymmetric Mach-Zehnder interferometer. *J. Opt.* **2018**, *20*, 045803. [CrossRef]
190. Mares, D.; Prajzler, V.; Martan, T.; Jerabek, V. Hybrid polymer–glass planar Bragg grating as a temperature and humidity sensor. *Opt. Quantum Electron.* **2022**, *54*, 590. [CrossRef]
191. Kefer, S.; Rosenberger, M.; Hessler, S.; Girschikofsky, M.; Belle, S.; Roth, G.-L.; Schmauss, B. Fabrication and applications of polymer planar Bragg grating sensors based on cyclic olefin copolymers. In Proceedings of the Photonics & Electromagnetics Research Symposium-Fall (PIERS-Fall), Xiamen, China, 17–20 December 2019; pp. 647–655.
192. Butt, M.; Khonina, S.; Kazanskiy, N.; Piramidowicz, R. Hybrid metasurface perfect absorbers for temperature and biosensing applications. *Opt. Mater.* **2022**, *123*, 111906. [CrossRef]
193. Li, J.; Zhao, C.; Dong, Q.; Kang, J.; Shen, C.; Wang, D. Multipoint temperature sensor based on PDMS-filled Fabry Perot interferometer with an array-waveguide grating. In Proceedings of the 18th International Conference on Optical Communications and Networks (ICOCN), Huangshan, China, 5–8 August 2019; pp. 1–3.
194. Ramirez, J.; Gabrielli, L.; Lechuga, L.; Hernandez-Figueroa, H. Trimodal Waveguide Demonstration and Its Implementation as a High Order Mode Interferometer for Sensing Application. *Sensors* **2019**, *19*, 2821. [CrossRef]
195. Missinne, J.; Beneitez, N.; Mattelin, M.-A.; Lamberti, A.; Luyckx, G.; Paepegem, W.; Steenberge, G. Bragg-Grating-Based Photonic Strain and Temperature Sensor Foils Realized Using Imprinting and Operating at Very Near Infrared Wavelengths. *Sensors* **2018**, *18*, 2717. [CrossRef]
196. Butt, M.; Kazanskiy, N.; Khonina, S. On-chip symmetrically and asymmetrically transformed plasmonic Bragg grating formation loaded with a functional polymer for filtering and CO_2 gas sensing applications. *Measurement* **2022**, *201*, 111694. [CrossRef]
197. Lim, J.; Kim, S.; Kim, J.-S.; Kim, J.; Kim, Y.; Lim, J.; Im, Y.; Park, J.; Hann, S. Polymer waveguide sensor with tin oxide thin film integrated onto optical-electrical printed circuit board. In *Optical Micro-and Nanometrology V*; SPIE: Brussels, Belgium, 2014; Volume 9132.
198. Clevenson, H.; Desjardins, P.; Gan, X.; Englund, D. High sensitivity gas sensor based on high-Q suspended polymer photonic crystal nanocavity. *Appl. Phys. Lett.* **2014**, *104*, 241108. [CrossRef]
199. Khonina, S.; Kazanskiy, N.; Butt, M.; Kazmierczak, A.; Piramidowicz, R. Plasmonic sensor based on metal-insulator-metal waveguide square ring cavity filled with functional material for the detection of CO_2 gas. *Opt. Express* **2021**, *29*, 16584–16594. [CrossRef]
200. Luan, E.; Yun, H.; Ma, M.; Ratner, D.; Cheung, K.; Chrostowski, L. Label-free biosensing with a multi-box sub-wavelength phase-shifted Bragg grating waveguide. *Biomed. Opt. Express* **2019**, *10*, 4825–4838. [CrossRef]

201. Liang, Y.; Liu, Q.; Wu, Z.; Morthier, G.; Zhao, M. Cascaded-Microrings Biosensors Fabricated on a Polymer Platform. *Sensors* **2019**, *19*, 181. [CrossRef]
202. Azuelos, P.; Girault, P.; Lorrain, N.; Poffo, L.; Guendouz, M.; Thual, M.; Lemaitre, J.; Pirasteh, P.; Hardy, I.; Charrier, J. High sensitivity optical biosensor based on polymer materials and using the Vernier effect. *Opt. Express* **2017**, *25*, 30799–30806. [CrossRef]
203. Han, X.; Han, X.; Shao, Y.; Wu, Z.; Liang, Y.; Teng, J.; Bo, S.; Morthier, G.; Zhao, M. Polymer integrated waveguide optical biosensor by using spectral splitting effect. *Photonic Sens.* **2017**, *7*, 131–139. [CrossRef]
204. Melnik, E.; Strasser, F.; Muellner, P.; Heer, R.; Mutinati, G.; Koppitsch, G.; Lieberzeit, P.; Laemmerhofer, M.; Hainberger, R. Surface modification of integrated optical MZI sensor arrays using inkjet printing technology. *Procedia Eng.* **2016**, *168*, 337–340. [CrossRef]
205. Leal-Junior, A.; Macedo, L.; Frizera, A.; Pontes, M.J. Polymer Optical Fiber Multimaterial: Flexible and Customizable Approach in Sensors Development. *IEEE Photonics Technol. Lett.* **2022**, *34*, 611–614. [CrossRef]
206. Bahin, L.; Tourlonias, M.; Bueno, M.-A.; Sharma, K.; Rossi, R.M. Smart textiles with polymer optical fibre implementation for in-situ measurements of compression and bending. *Sens. Actuators A Phys.* **2023**, *350*, 114117. [CrossRef]
207. Woyessa, G.; Pedersen, J.K.; Nielsen, K.; Bang, O. Enhanced pressure and thermal sensitivity of polymer optical fiber Bragg grating sensors. *Opt. Laser Technol.* **2020**, *130*, 106357. [CrossRef]
208. Wei, H.; Chen, M.; Krishnaswamy, S. Three-dimensional-printed Fabry–Perot interferometer on an optical fiber tip for a gas pressure sensor. *Appl. Opt.* **2020**, *59*, 2173–2178. [CrossRef]
209. Liu, Y.; Liu, X.; Zhang, T.; Zhang, W. Integrated FPI-FBG Composite All-Fiber Sensor for Simultaneous Measurement of Liquid Refractive Index and Temperature. *Opt. Lasers Eng.* **2018**, *111*, 167–171. [CrossRef]
210. Chen, M.; He, T.; Zhao, Y.; Yang, G. Ultra-Short Phase-Shifted Fiber Bragg Grating in a Microprobe for Refractive Index Sensor with Temperature Compensation. *Opt. Laser Technol.* **2023**, *157*, 108672. [CrossRef]
211. Korganbayev, S.; Sypabekova, M.; Amantayeva, A.; González-Vila, Á.; Caucheteur, C.; Saccomandi, P.; Tosi, D. Optimization of Cladding Diameter for Refractive Index Sensing in Tilted Fiber Bragg Gratings. *Sensors* **2022**, *22*, 2259. [CrossRef] [PubMed]
212. Voronkov, G.; Zakoyan, A.; Ivanov, V.; Iraev, D.; Stepanov, I.; Yuldashev, R.; Grakhova, E.; Lyubopytov, V.; Morozov, O.; Kutluyarov, R. Design and Modeling of a Fully Integrated Microring-Based Photonic Sensing System for Liquid Refractometry. *Sensors* **2022**, *22*, 9553. [CrossRef] [PubMed]
213. Voronkov, G.; Aleksakina, Y.; Ivanov, V.; Zakoyan, A.; Stepanov, I.; Grakhova, E.; Butt, M.; Kutluyarov, R. Enhancing the Performance of the Photonic Integrated Sensing System by Applying Frequency Interrogation. *Nanomaterials* **2023**, *13*, 193. [CrossRef] [PubMed]
214. Han, X.; Ding, L.; Tian, Z.; Song, Y.; Xiong, R.; Zhang, C.; Han, J.; Jiang, S. Potential New Material for Optical Fiber: Preparation and Characterization of Transparent Fiber Based on Natural Cellulosic Fiber and Epoxy. *Int. J. Biol. Macromol.* **2023**, *224*, 1236–1243. [CrossRef]
215. Islam, M.D.; Liu, S.; Boyd, D.A.; Zhong, Y.; Nahid, M.M.; Henry, R.; Taussig, L.; Ko, Y.; Nguyen, V.Q.; Myers, J.D.; et al. Enhanced mid-wavelength infrared refractive index of organically modified chalcogenide (ORMOCHALC) polymer nanocomposites with thermomechanical stability. *Opt. Mater.* **2020**, *108*, 110197. [CrossRef]
216. Afroozeh, A.; Zeinali, B. Improving the Sensitivity of New Passive Optical Fiber Ring Sensor Based on Meta-Dielectric Materials. *Opt. Fiber Technol.* **2022**, *68*, 102797. [CrossRef]

Disclaimer/Publisher's Note: The statements, opinions and data contained in all publications are solely those of the individual author(s) and contributor(s) and not of MDPI and/or the editor(s). MDPI and/or the editor(s) disclaim responsibility for any injury to people or property resulting from any ideas, methods, instructions or products referred to in the content.

Article

How do Graphene Composite Surfaces Affect the Development and Structure of Marine Cyanobacterial Biofilms?

Maria J. Romeu [1,2], Luciana C. Gomes [1,2], Francisca Sousa-Cardoso [1,2], João Morais [3], Vítor Vasconcelos [3,4], Kathryn A. Whitehead [5], Manuel F. R. Pereira [2,6], Olívia S. G. P. Soares [2,6] and Filipe J. Mergulhão [1,2,*]

[1] LEPABE—Laboratory for Process Engineering, Environment, Biotechnology and Energy, Faculty of Engineering, University of Porto, Rua Dr. Roberto Frias, 4200-465 Porto, Portugal
[2] ALiCE—Associate Laboratory in Chemical Engineering, Faculty of Engineering, University of Porto, Rua Dr. Roberto Frias, 4200-465 Porto, Portugal
[3] CIIMAR—Interdisciplinary Centre of Marine and Environmental Research, University of Porto, Terminal de Cruzeiros do Porto de Leixões, Av. General Norton de Matos s/n, 4450-208 Matosinhos, Portugal
[4] Department of Biology, Faculty of Sciences, University of Porto, Rua do Campo Alegre, 4169-007 Porto, Portugal
[5] Microbiology at Interfaces, Manchester Metropolitan University, Chester Street, Manchester M1 5GD, UK
[6] LSRE–LCM—Laboratory of Separation and Reaction Engineering–Laboratory of Catalysis and Materials, Faculty of Engineering, University of Porto, Rua Dr. Roberto Frias, 4200-465 Porto, Portugal
* Correspondence: filipem@fe.up.pt; Tel.: +351-225-081-668

Abstract: The progress of nanotechnology has prompted the development of novel marine antifouling coatings. In this study, the influence of a pristine graphene nanoplatelet (GNP)-modified surface in cyanobacterial biofilm formation was evaluated over a long-term assay using an in vitro platform which mimics the hydrodynamic conditions that prevail in real marine environments. Surface characterization by Optical Profilometry and Scanning Electron Microscopy has shown that the main difference between GNP incorporated into a commercially used epoxy resin (GNP composite) and both control surfaces (glass and epoxy resin) was related to roughness and topography, where the GNP composite had a roughness value about 1000 times higher than control surfaces. The results showed that, after 7 weeks, the GNP composite reduced the biofilm wet weight (by 44%), biofilm thickness (by 54%), biovolume (by 82%), and surface coverage (by 64%) of cyanobacterial biofilms compared to the epoxy resin. Likewise, the GNP-modified surface delayed cyanobacterial biofilm development, modulated biofilm structure to a less porous arrangement over time, and showed a higher antifouling effect at the biofilm maturation stage. Overall, this nanocomposite seems to have the potential to be used as a long-term antifouling material in marine applications. Moreover, this multifactorial study was crucial to understanding the interactions between surface properties and cyanobacterial biofilm development and architecture over time.

Keywords: antifouling surface; biofilm architecture; cyanobacterial biofilm; graphene; marine biofouling

1. Introduction

The oil, gas, and maritime industry are significantly impacted by biofouling, either through the costs related to the increased fuel consumption, hull cleaning, material deterioration, repainting, and corrosion [1], or even by the management intervention time and incorrect measurements in the submerged and moored sensors [2]. Ecologically, fouling events in marine environments promote species invasion and the establishment of exotic biofouling species in ports [3]. Moreover, biofouling can be a major concern in health-related problems since contamination of aquaculture facilities, such as fish cages, can occur by toxin accumulation, and air pollution may be increased through greenhouse gas emissions [4].

The non-toxic marine antifouling approaches available in the market are often expensive and not as effective as conventional biocides, which can accumulate in the marine

environment and affect non-target aquatic organisms. Inorganic ingredient-based coatings have been used for biofouling prevention, including silver nanoparticles [5], carbon nanotubes (CNTs) [6], graphene [7], and metal oxides semiconductors such as zinc oxide (ZnO) [8] and titanium dioxide (TiO_2) [9]. Overall, antifouling marine paints containing nanomaterials have been reported to offer superhydrophobicity, microbial resistance, high durability, water repellency, anti-sticking, and anti-corrosive properties [10], being novel solutions for the sustainable development of the maritime industry. Advancements in materials technology have introduced carbon nanomaterials such as CNTs and graphene as a powerful approach for various applications in the marine and shipping industries [11]. Graphene consists of a single layer of carbon atoms arranged in a sp^2-bonded hexagonal pattern. It is considered one of the strongest and thinnest materials available, which shows high specific surface area, electrical conductivity, and thermal stability, making it appealing for different applications [12,13]. Moreover, due to its high strength level [14], this material is a breakthrough alternative in the naval industry. All these features make the application of graphene in technical processes of maritime industries attractive, such as in water management systems, desalination, removing toxic pollutants and filtering gasses, and as a coating material [15–17]. As a coating on marine structures, besides the anti-corrosive properties, graphene can also be used in de-icing surfaces for ship operations in extremely low-temperature regions, such as the Arctic and Antarctica, due to its electrical conductivity [15].

Nanotechnology-based technologies can be of great interest in creating novel low-toxic antifouling coatings [18]. However, analysis of the literature indicates that there is little information about the use of pristine graphene. Indeed, most in situ studies on graphene-based surfaces were performed with functionalized graphene and graphene oxide (GO) coatings [19–23]. Diatom adhesion was completely inhibited after 10 days by surfaces containing 0.36 wt% GO [24]. In turn, GO-silver nanoparticle coatings improved antibacterial and anti-algal properties [25], and showed more than 80% *Halomonas pacifica* biofilm inhibition [26]. Since the current trend is to study the potential of modified and functionalized graphene, the antibiofilm performance of graphene alone is poorly understood. Moreover, most in vitro studies are usually performed for short periods and under hydrodynamic conditions that do not mimic the real marine environment [27]. In fact, some of the in vitro studies have been performed until 24 h [6,23,25,28,29], most of them between days and weeks [7,19,20,24,30–33], but only a study performed by Fazli-Shokouhi et al. [34] extended the assay period for 3 months to evaluate the antifouling potential of graphene-based coatings. Moreover, these studies focus on organisms other than cyanobacteria, namely diatoms, algae, and macrofoulers. Therefore, the main goal of this work was to evaluate the potential of a graphene composite surface to prevent and control the development of biofilms by marine microfoulers over a long-term assay and using an in vitro platform that mimics the hydrodynamic conditions found in real marine scenarios. Cyanobacterial biofilm architecture was evaluated by three different imaging techniques: Optical Coherence Tomography (OCT), Confocal Laser Scanning Microscopy (CLSM), and Scanning Electron Microscopy (SEM). Since the epoxy resin is a commercially available coating generally used to coat the hulls of small recreational vessels [35,36] due to its exceptional physical, chemical, and mechanical properties, no safety issues, and low cost [37], pristine graphene nanoplatelets (GNP) were incorporated into this polymer matrix. Furthermore, epoxy composites exhibited high durability and resistance to fatigue and UV irradiation [38]. Surface characterization was also performed by water contact angle measurements, Optical Profilometry, and SEM.

2. Materials and Methods

Figure 1 presents the flowchart of the experimental work fully described in the upcoming sections.

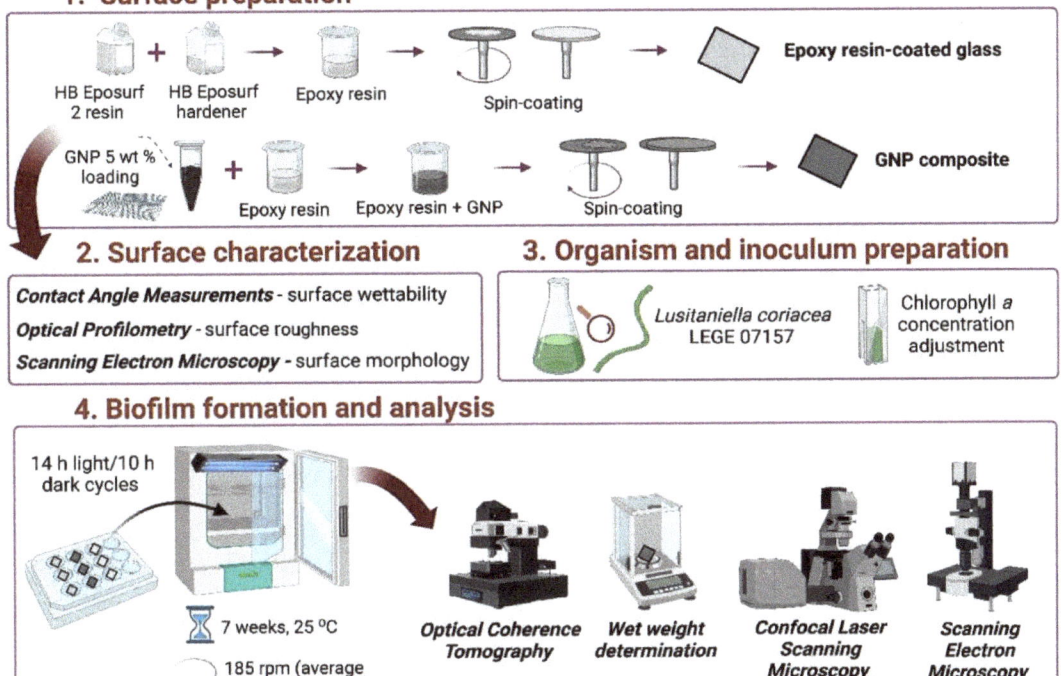

Figure 1. Scheme of the experimental steps of the present work.

2.1. Surface Preparation

Two control surfaces (glass, a commonly submerged artificial surface found on diverse equipment in aquatic environments, and epoxy resin, a commercially available marine coating) and a GNP composite were tested to determine their performance against cyanobacterial biofilm development. The epoxy resin-coated glass was prepared following the protocol described by Faria et al. [39]. Briefly, HB Eposurf 2 resin and HB Eposurf hardener, both from HB Química (Matosinhos, Portugal), were mixed in a ratio of 10:3 (v/v). To produce the epoxy resin-coated surfaces, 70 µL of the above mixture was placed on top of 1 cm × 1 cm glass coupons (Vidraria Lousada, Lda, Lousada, Portugal) using a spin coater (Spin150 PolosTM, Paralab, Porto, Portugal) at 6000 rpm for 40 s. After 12 h at room temperature, the surfaces were dried for 3 h at 60 °C.

The methodology for the preparation of the GNP composite was adapted from Oliveira et al. [40], in which the polydimethylsiloxane (PDMS) matrix was replaced by the epoxy resin. Briefly, after mixing the two epoxy resin components, 5 wt% of GNPs aggregates (Alfa Aesar, Thermo Fisher Scientific, Erlenbachweg, Germany) were incorporated into the epoxy resin mixture. The specific surface area (S_{BET}), external surface area (S_{meso}), micropore volume (V_{micro}), and total pore volume (V_p) of these GNPs are 464 $m^2\,g^{-1}$, 363 $m^2\,g^{-1}$, 0.045 $cm^3\,g^{-1}$, and 0.535 $cm^3\,g^{-1}$, respectively [40]. In the same study [40], 5 wt% was demonstrated to be an effective load in reducing single- and mixed-species biofilms of *Pseudomonas aeruginosa* and *Staphylococcus aureus* formation after 24 h. After the incorporation step, 70 µL of the composite (5 wt% GNP/epoxy resin) was deposited on glass squares by spin coating.

2.2. Surface Characterization

2.2.1. Wettability

The measurement of water contact angles (θ_w) was performed through the sessile drop method using an SL200C optical contact angle meter (Solon Information Technology Co., Ltd., Shanghai, China) as previously described [41,42]. At least 25 determinations for each material at room temperature were performed.

2.2.2. Optical Profilometry

Optical profilometry was used to determine the average roughness (S_a) of the glass, epoxy resin, and GNP composite surfaces, as previously performed by Whitehead et al. [43,44]. A MicroXAM surface mapping microscope (ADE Corporation, XYZ model 4400 mL system, Tucson, AZ, USA), with an objective of 50× and connected to an AD phase shift controller (Omniscan, Wrexham, UK), was used to image five areas (167 µm × 167 µm) of at least four different coupons. The MapView AE 2.17 software (Omniscan, Wrexham, UK) was chosen to determine the (S_a) values through extended range vertical scanning interferometry, and the 3D images were extracted by SPIP™ 6.7.9 software (Image Metrology A/S, Hørsholm, Denmark).

2.2.3. Scanning Electron Microscopy (SEM)

The surface morphology of the three surfaces used in this study was investigated using SEM. The surfaces were placed on SEM stubs (Agar Scientific, Stansted, UK) and sputter-coated with gold for 30 s in an SEM coating system (Polaron, London, UK). The secondary electron detector of a Supra 40VP scanning electron microscope (Carl Zeiss Ltd., Cambridge, UK) was used to obtain images of at least three coupons for each surface at an accelerating voltage of 2 kV.

2.3. Organism and Inoculum Preparation

A filamentous cyanobacterial strain *Lusitaniella coriacea* LEGE 07157 was obtained from the Interdisciplinary Centre of Marine and Environmental Research (CIIMAR), Matosinhos, Portugal [45]. *Lusitaniella coriacea* LEGE 07157 was isolated by rock surface scraping from a zone tide pool at Lavadores beach, Porto, Portugal (41.12919 N 8.668578 W). It was grown in Z8 medium [46] enhanced with 25 g L^{-1} of synthetic sea salts (Tropic Marin) and vitamin B$_{12}$ (Sigma Aldrich, Merck, Saint Louis, MO, USA), and at 25 °C under 14 h light (10–30 µmol photons m^{-2} s^{-1}, λ = 380–700 nm)/10 h dark cycles, as recommended by Ramos et al. [45].

2.4. Biofilm Formation

Cyanobacterial suspensions were adjusted to a chlorophyll *a* concentration of 1.22 ± 0.09 µg mL^{-1} since this pigment is unique and predominant in all groups of cyanobacteria and its quantification is a standard methodology to estimate the biomass in marine environments [47,48]. Briefly, cyanobacterial cells were collected by centrifugation (3202× *g* for 5 min) and a volume of 2 mL of 99.8% methanol (Methanol ACS Basic, Scharlau Basic, Barcelona, Spain) was added. After 24 h of dark incubation at 4 °C, cyanobacterial suspensions were centrifuged and absorbance measurements of the supernatant were performed at 750 nm (turbidity), 665 nm (chlorophyll *a*), and 652 nm (chlorophyll *b*) (V-1200 spectrophotometer, VWR International China Co., Ltd., Shanghai, China). The values obtained were used to calculate chlorophyll *a* concentration (µg·mL^{-1}) through Equation (1) [49]:

$$Chl\ a\ \left(\mu g \cdot mL^{-1}\right) = 16.29 \times A^{665} - 8.54 \times A^{652} \quad (1)$$

Biofilm formation was tested on agitated 12-well plates (VWR International, Carnaxide, Portugal) under previously optimized conditions for cyanobacterial biofilm development [47]. All coupons and plates were subjected to UV sterilization, after which the coupons were fixed on double-sided adhesive tape [40,47]. A volume of 3 mL of the adjusted cyanobacterial suspension was added to each well. To mimic the hydrodynamic conditions found in marine environments, microplates were then incubated at 25 °C in a shaker with a 25 mm orbital diameter (Agitorb 200ICP, Norconcessus, Ermesinde, Portugal) at 185 rpm, resulting in an average shear rate of 40 s^{-1} [47]. Biofilm formation in this platform includes the shear rate valued for a ship in a harbor, 50 s^{-1} [50], and it was shown to predict the biofouling behavior observed upon immersion in the sea for prolonged periods [51]. Moreover, to simulate marine biofilm formation on submerged surfaces, microtiter plates were kept under 14 h light (8–10 µmol photons m^{-2} s^{-1})/10 h dark cycles [47,52,53]. The light intensity was decreased from 10–30 µmol photons m^{-2} s^{-1} to 8–10 µmol photons m^{-2} s^{-1} because biofouling organisms have reduced access to light when in immersion (either by the effect of the marine equipment/device/ship to which they are attached, or by the influence of the biofilm structure in which the accumulation of different organisms occurs, and some of them are located in the inners layers of biofilm). Since a 2-month interval for maintenance is the minimum time for economically viable underwater monitoring systems [47], biofilm development was followed for 7 weeks (49 days), and during this incubation time, the medium was changed twice a week in the sample and control wells.

2.5. Biofilm Analysis

Two coupons of each surface were analyzed every 7 days. For that, the culture medium was removed and the wells were filled with 3 mL of sodium chloride solution (8.5 g L^{-1}) [47]. The solution was carefully removed and the wells were filled again with 3 mL of sodium chloride to assess the cyanobacterial biofilm structure through Optical Coherence Tomography (OCT). The determination of biofilm wet weight was also performed over 7 weeks, and at the end of the experiment (49 days), cyanobacterial biofilm architecture and morphology were analyzed by Confocal Laser Scanning Microscopy (CLSM) and SEM, respectively.

2.5.1. Optical Coherence Tomography (OCT)

Images from cyanobacterial biofilms were captured as reported by Romeu et al. [47] through OCT (Thorlabs Ganymede Spectral Domain Optical Coherence Tomography system with a central wavelength of 930 nm, Thorlabs GmbH, Dachau, Germany). Briefly, for each coupon, 2D imaging was performed (with a minimum of 2 fields of view) and evaluated through a routine developed in the Image Processing Toolbox from MATLAB 8.0 and Statistics Toolbox 8.1 (The MathWorks, Inc., Natick, MA, USA) [54]. The mean of biofilm thickness was calculated according to Equation (2):

$$\overline{L}_F = \frac{1}{N}\sum_{i=1}^{N} L_{F,i} \qquad (2)$$

where $L_{F,i}$ is a local biofilm thickness measurement at location i, N equals the number of thickness measurements, and $\overline{L}_F$ is the mean biofilm thickness. The percentage of empty spaces in the biofilm structure, as well as their average size, was also assessed [55].

2.5.2. Wet Weight Determination

The determination of the biofilm wet weight was performed as previously reported [56] and considered as the difference from the initial coupon weight determined before inoculation.

2.5.3. Confocal Laser Scanning Microscopy (CLSM)

Biofilms with 49 days were stained with 6 µM SYTO®61 (Thermo Fisher Scientific, Waltham, MA, USA), mounted on a microscopic slide, and observed in a Leica TCS SP5 II Confocal Laser Scanning Microscope (Leica Microsystems, Wetzlar, Germany) with a 40× water objective lens (Leica HCX PL APO CS 40.0x/1.10WATER UV) and 633-nm helium-neon laser. Image stacking was acquired with a z-step of 1 µm for each sample at a minimum of five random fields of 387.5 µm × 387.5 µm (equivalent to 512 pixels × 512 pixels. The "Easy 3D" tool of the IMARIS 9.1 software (Bitplane, Zurich, Switzerland) was used to create 3D projections of the biofilms. Additionally, biofilm architectural parameters such as biovolume ($\mu m^3\ \mu m^{-2}$) and surface coverage (%) were determined using the COMSTAT image-analysis software [57].

2.5.4. SEM

After 49 days of incubation, *Lusitaniella coriacea* LEGE 07157 biofilms grown on the three tested surfaces were observed by SEM. Samples were taken from the microplates, dehydrated with increasing ethanol concentrations (10, 25, 40, 50, 70, 80, 90, and 100% (v/v)), and left in a desiccator until microscopic analysis [58]. Then, they were sputter-coated using the equipment and conditions described before and imaged in the Supra 40VP scanning electron microscope.

2.6. Statistical Analysis

Two biological experiments with two technical replicates each were analyzed. Quantitative parameters obtained from biofilm wet weight, OCT, and CLSM were compared using one-way ANOVA with Tukey's multiple comparisons test (GraphPad Prism® version 6.01, GraphPad Software, Inc., San Diego, CA, USA). Statistically significant differences between the different surfaces for the same sampling day were considered for p values < 0.05 (corresponding to a confidence level greater than 95%). The error bars represent the standard deviation (SD) of the mean.

3. Results and Discussion

3.1. Surface Analysis

It is known that surface properties, including wettability, roughness, and morphology, are important factors governing initial bacterial adhesion to surfaces and consequently affect the development of mature biofilms [59]. Therefore, the three tested surfaces (glass, epoxy resin, and GNP composite) were evaluated regarding (i) wettability by water contact angle measurements, (ii) topography and roughness by Optical Profilometry, and (iii) morphology by SEM.

Results obtained from water contact angle measurements are shown in Figure 2a. Contact angle determination provides information about the wettability of the surface, i.e., the tendency of the fluid to spread on a surface, and the hydrophobicity of the surface, which describes the tendency of non-polar molecular aggregation and, consequently, water molecule repulsion [60]. Surfaces can be classified as superhydrophilic, hydrophilic, hydrophobic, or superhydrophobic, if the contact angle of water with the surfaces is lower than 10°, between 10° and 90°, between 90° and 150°, or over 150°, respectively [61]. While glass is considered hydrophilic (θ_w = 40.9° ± 7.4°), the GNP composite (θ_w = 68.6° ± 2.4°) and epoxy-coated surfaces (θ_w = 76.3° ± 2.5°) are slightly more hydrophobic than glass. The wettability of the epoxy resin was weakly affected by the incorporation of graphene on the coating; the GNP composite is slightly more hydrophilic than the resin. Similar results were observed in previous studies [40,62]. In fact, in a study performed by Rafiee et al. [62], the water contact angle of a copper substrate coated with graphene was up to 90.6°, while that of the pure copper substrate was about 85.9°. Oliveira and her coworkers [40] also verified that 5 wt% GNP/PDMS and PDMS surfaces presented similar water contact angles, 121.8° and 110.2°, respectively. Since the wetting properties at the water–graphene interface had little effect on the water–substrate interaction, this peculiar wettability of graphene has been described by the term "wetting transparency of graphene" [62]. In the present study, as dispersed GNPs were used instead of graphene layers, it is conceivable to assume that the wettability of the substrate may be even less affected by the presence of graphene.

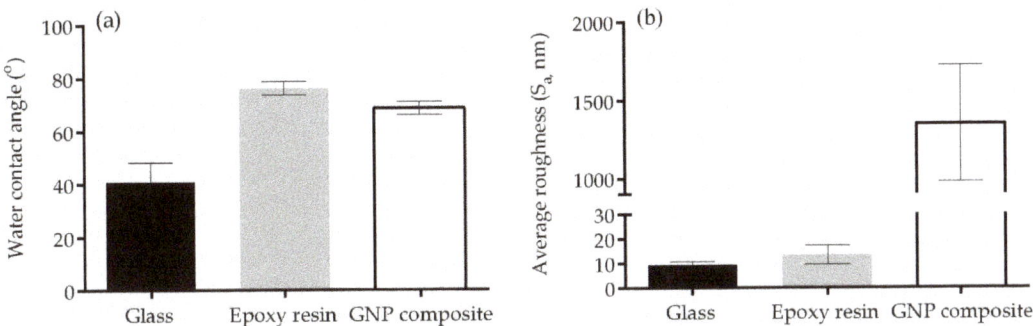

Figure 2. Water contact angles (**a**) and average roughness (S_a) (**b**) of the tested surfaces. The means ± SD are shown.

Results obtained from roughness analysis are shown in Figures 2b and 3. The GNP composite was the roughest surface (S_a = 1351 nm), followed by the epoxy resin (S_a = 13 nm) and glass (S_a = 10 nm). Therefore, the average roughness values of both control surfaces are on the nanometric scale, being about 1000 times lower than the roughness value determined for the GNP composite (Figure 2b). The difference in roughness and topography between the graphene composite and both control surfaces (glass and epoxy resin) is particularly evident in the 3D images of the surfaces given by profilometry (Figure 3). While glass and resin without incorporated carbon material were homogeneous and smooth surfaces (Figure 3a,b, respectively), the nanocomposite presented some irregularities distributed along the analyzed surface area (Figure 3c). Looking at the SEM images (Figure 4c,d), it was possible to confirm that these surface elevations correspond to agglomerated graphene nanoplatelets in the polymer matrix. Some of the larger agglomerates had sizes ranging from 7–30 μm.

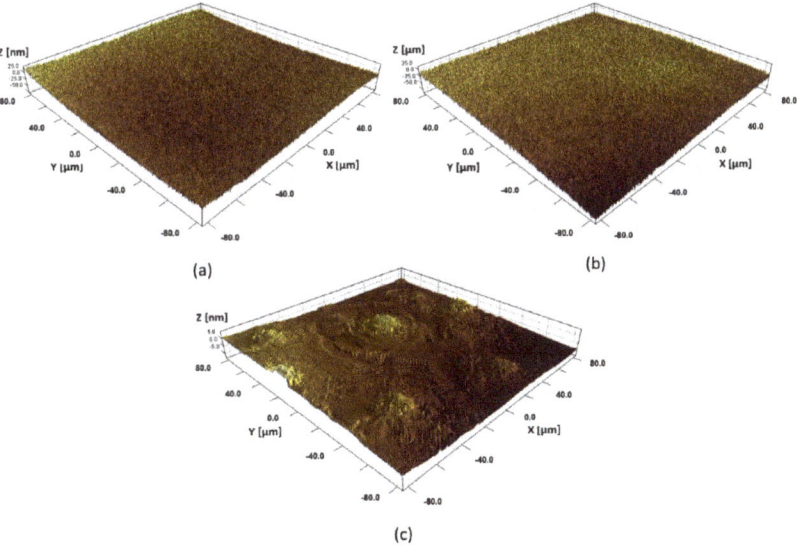

Figure 3. 3D profilometer images of glass (**a**), epoxy resin (**b**), and GNP composite (**c**).

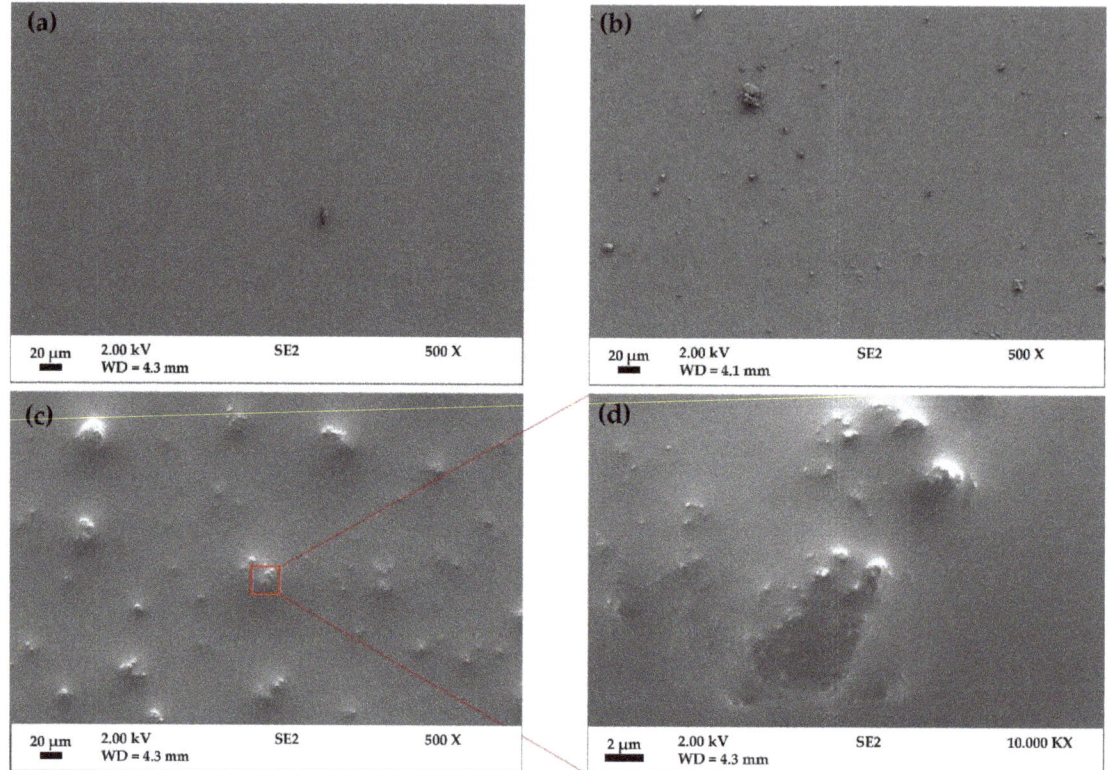

Figure 4. SEM images of glass (**a**), epoxy resin (**b**), GNP composite (**c**), and magnified GNP composite (**d**). The micrographs have a magnification of 500×, except for (**d**), which corresponds to a higher magnification image (10,000×) of the surface area marked in red in (**c**).

It is a challenge to disperse GNPs, especially compared to other carbon materials such as CNTs, due to the van der Waals forces and a strong π–π interaction between the individual GNP sheets, which are responsible for their layer-stacked compact structure [40,63]. Their incorporation in the polymeric resin may facilitate graphene dispersion, thus reducing the tendency to aggregate [64]. In the higher magnification image of the GNP composite (Figure 4d), it was interesting to detect some areas on top of the embedded GNP clusters where this carbon material was more exposed. It is described that one of the factors directing the antimicrobial activity of graphene is the agglomeration tendency, which causes a reduced surface area and shape alteration [65,66]. However, the presence of more exposed GNPs in the epoxy resin may facilitate their interaction with the microorganisms that come in contact with the surface, triggering the predominant antimicrobial mechanisms of this carbon nanomaterial—oxidative stress, mechanical damage of cell membrane/wall by nanoknives, and wrapping/trapping [66,67].

In vitro assays showed that no detectable leaching occured under the conditions used in this work as assessed by UV-Vis spectroscopy (data not shown).

3.2. Biofilm Development

As there are several mechanisms affecting the microbial response and the disruption of biofilm architecture, the long-term antifouling properties of a given surface for marine applications should be extensively studied. Although the Minimal Inhibitory Concentration (MIC) of the studied GNPs could not be determined for *Lusitaniella coriacea* LEGE 07167 (it is a filamentous cyanobacterium and therefore cell numbers cannot be accurately determined), it was found to be higher than 5% (w/v, in a VNSS suspension assay) for a model marine bacterium (*Cobetia marina*). Cyanobacterial biofilm development was monitored over 7 weeks (49 days) and the quantitative results obtained from wet weight determination and 2D OCT analysis are shown in Figure 5. The values regarding the biofilm thickness, the percentage of empty spaces, and the average size of empty spaces on biofilm structure (Figure 5b–d) are indicated from day 14 since the biofilm thickness was below the OCT range on the first sampling day. Although a progressive increase in biofilm development was observed from the results of wet weight (Figure 5a) and biofilm thickness (Figure 5b), this evolution over the 7 weeks was more noticeable on glass and epoxy resin than on the GNP composite. Indeed, for biofilm wet weight (Figure 5a), from day 7 to day 49, an increase of 84 %, 77 %, and 67 % was observed for glass, epoxy resin, and GNP composite, respectively, while for biofilm thickness (Figure 5b), from day 14 to day 49, an increase of 85 %, 86 %, and 58 % was observed for the same surfaces. This behavior was also seen regarding biofilm architecture (Figure 5c) given that an increase of 92 %, 98 %, and just 28 % for the percentage of empty spaces was observed on glass, epoxy resin, and GNP composite. This suggests that graphene-modified surfaces may delay biofilm development (Figure 5a,b) and change its structure to a less porous arrangement (Figure 5c). Additionally, on days 35, 42, and 49, a reduction of biofilm wet weight and thickness was observed on the GNP composite surface when compared to the epoxy resin itself. In fact, on days 35, 42 and 49, the wet weight values obtained for the GNP composite were 46 %, 47 %, and 44 % lower than those obtained on the epoxy resin surface, respectively (Figure 5a). For biofilm thickness on the same days, the values obtained on the GNP composite were 46 %, 42 %, and 54 % lower than those obtained on the epoxy resin surface, respectively (Figure 5b). The delay effect caused by this carbon nanomaterial on cyanobacterial biofilm development, as well as the greater antifouling effect on the maturation stage of biofilm development, was also observed in previous work where CNT-modified surfaces were used as potential antifouling surfaces [42]. In addition, in the maturation stage of biofilm development (days 42 and 49), a lower percentage of empty spaces (Figure 5c) and an average size of empty spaces on biofilm structure (Figure 5d) were detected for biofilms formed on the GNP composite when compared with control surfaces (glass and epoxy resin).

Figure 6 shows representative 2D cross-sectional images of *Lusitaniella coriacea* LEGE 07157 biofilm development on glass, epoxy resin, and GNP composite after 49 days. These images corroborate the quantitative data obtained from biofilm biomass (Figure 5a,b) and empty spaces on biofilm architecture (Figure 5c,d) since a lower amount of biofilm mass and percentage of empty spaces and its average size were observed on the GNP composite compared to the control epoxy resin. Although the reduced biofilm development on the GNP-based coating may be beneficial for the performance of marine devices and/or equipment, the efficacy of chemical methods to eradicate biofilms formed on this surface material may be hampered. Given the lower percentage of empty spaces and their smaller average size shown in biofilms formed on the GNP composite, the diffusion of chemical compounds typically used for biofouling control through the inner layers of the biofilm may be hindered [68]. The biofilm structure could also be compared between the three surfaces. While a flatter and homogeneous biofilm was observed on glass (Figure 6a), biofilms formed on the epoxy resin (Figure 6b) and GNP composite (Figure 6c) had heterogeneous shapes. Additionally, more streamers could be observed on top of the biofilm developed on the epoxy resin surface (Figure 6b), which probably contributed to the higher thickness of biofilms developed on this surface compared to the GNP composite (Figure 5b). Opposing results were obtained in a previous study focused on CNT-based surfaces [42], in which a flatter and homogeneous cyanobacterial biofilm was observed on the CNT composite. However, this divergent finding may be related to the use of a different cyanobacterial strain and carbon-based surface.

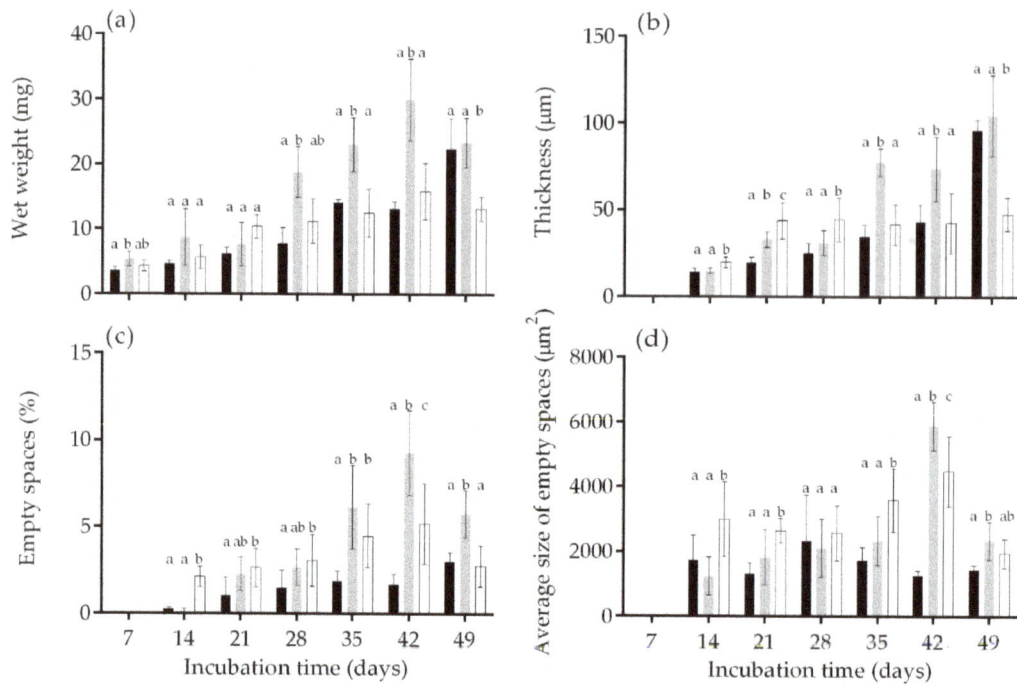

Figure 5. *Lusitaniella coriacea* LEGE 07157 biofilm development on different surfaces: glass (black), epoxy resin (grey), GNP composite (white). The parameters analyzed refer to biofilm wet weight (**a**), biofilm thickness (**b**), average percentage of empty spaces (**c**), and their respective average size (**d**) on biofilm structure. Mean values and SD from two biological assays with two technical replicates each are represented. For each sampling day, different lowercase letters (a, b, and c) indicate significant differences between surfaces ($p < 0.05$).

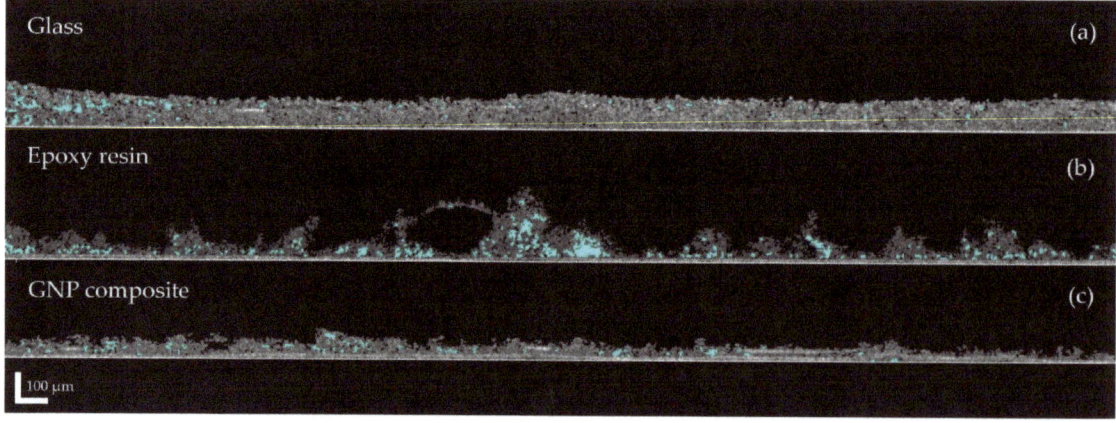

Figure 6. Representative 2D cross-sectional OCT images obtained for *Lusitaniella coriacea* LEGE 07157 biofilms on glass (**a**), epoxy resin (**b**), and GNP composite (**c**) after 49 days. The empty spaces on the biofilm structure are indicated in blue (scale bar = 100 µm).

Comparing the biofilm structure over the 49 days (Figure 5b,c), it was possible to observe different patterns between biofilms formed on the epoxy resin without and with embedded GNPs. In the early stages of biofilm formation (days 14 to 28), biofilms formed on the GNP composite were thicker and presented a higher percentage of empty spaces when compared to the epoxy resin surface, while in the maturation phase (days 35 to 49) these parameters became lower. These findings suggest that the higher roughness of the GNP composite may initially promote the development of a porous and thicker biofilm, and at the maturation stage, a biofilm structural arrangement may occur. In fact, 2D cross-sectional OCT images obtained for *Lusitaniella coriacea* LEGE 07157 biofilms after 28 days (Figure S1 in Supplementary Material) revealed that biofilms formed on the GNP composite presented a higher number of streamers than the older biofilm (Figure 6c). Initially, to avoid the direct contact with GNP present on epoxy resin, cyanobacterial cells may have adapted their adhesion for vertical biofilm growth rather than a homogeneous development along the surface, contributing to the higher values of thickness between days 14 and 28. Although some previous studies revealed that the first dead cells could prevent the direct contact of live cells with graphene oxide (GO) surfaces, serving as a nutrient for live cells and consequently enhancing biofilm formation [69], the differences in architectural adaptation observed over time may provide an additional explanation for the antibiofilm mechanisms of the GNP composite. Indeed, values obtained from the biofilm roughness coefficient (Figure S2 in Supplementary Material) showed that all biofilms formed on day 28 presented a higher roughness compared with biofilms developed on day 49 ($p < 0.05$), including biofilms formed on the GNP composite.

Moreover, it was also hypothesized that the inability of *Lusitaniella coriacea* LEGE 07157 to form a dense and robust biofilm on the GNP composite was related to the damaging effect of graphene in the first cells adhered to the surface. These bacterial cells could be affected by the higher roughness of the graphene-based nanocomposite since the surface peaks corresponding to GNP clusters may directly affect bacterial viability by increasing the surface contact area and, consequently, the piercing action of exposed graphene particles. These initial adhered cells would become more fragile, probably leading to the improper adherence of the following layers of cells and hindering long-term biofilm formation. In previous work, Oliveira et al. [70] showed that 5 wt% GNP/PDMS surfaces reduced the number of total (57%), viable (69%), culturable (55%), and VBNC cells (85%) of 24-h *S. aureus* biofilms compared to PDMS [40]. Besides cell death, the increased permeability of cell membranes caused by nanoparticles may affect extracellular polymeric substances (EPS) production, resulting in a disrupted 3D structure of biofilm.

The CLSM analysis was performed to analyze structural differences between the biofilms formed on the three tested surface materials (glass, epoxy resin, and GNP composite; Figure 7). While the 49-day-old biofilm developed on glass was dense, preventing the observation of individual filaments on the cyanobacterial cells in the top view (Figure 7a), the biofilms formed on epoxy resin surfaces without and with GNPs (Figure 7b,c, respectively) displayed randomly distributed filaments. Furthermore, a drastic reduction in total biofilm amount was observed on the GNP composite when compared to both glass and epoxy resin surfaces. This was corroborated by the quantitative data extracted from the confocal stacks (Figure 7d,e), in which a decrease in biovolume (on average 86%, $p < 0.05$; Figure 7d) and surface coverage (on average 65%, $p < 0.05$; Figure 7e) was observed on the graphene-based surface compared to control surfaces.

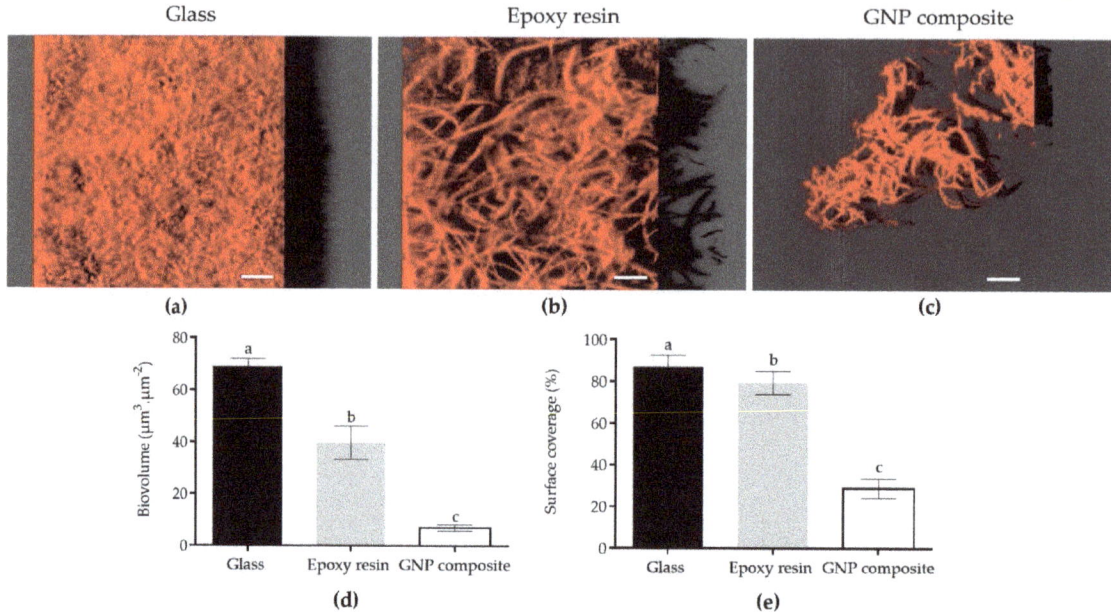

Figure 7. Three-dimensional confocal reconstructions of *Lusitaniella coriacea* LEGE 07157 biofilms grown on (**a**) glass, (**b**) epoxy resin, and (**c**) GNP composite after 49 days of incubation (white scale bars = 50 µm). These images show the biofilm aerial view, with the virtual shadow projection on the right (representative of biofilm thickness). Biovolume (**d**) and surface coverage (**e**) values of cyanobacterial biofilms were extracted from confocal files with the COMSTAT program. The means ± SD are presented. Different letters in the graphs (a, b, and c) indicate a significant difference among surfaces ($p < 0.05$).

The morphology of 49-day-old *Lusitaniella coriacea* LEGE 07157 biofilms on glass, epoxy resin, and GNP composite surfaces was also studied by SEM (Figure 8). The appearance of the biofilm developed on each surface was different, particularly when comparing the biofilm formed on glass (Figure 8a) with that formed on the epoxy resin without and with GNPs (Figure 8b,c, respectively). Indeed, the biofilm formed on the glass surface was very dense, and filamentous cyanobacteria were completely embedded in a complex matrix, which is a crucial feature of biofilm formation, making it impossible to distinguish the morphology of the individual *Lusitaniella coriacea* cells. Contrariwise, the electron micrograph of the epoxy resin surface (Figure 8b) displayed mesh-like structures around the *Lusitaniella* cells (possibly resulting from exopolysaccharide secretion), and filaments made of chains of cells that covered the entire surface and which were arranged in random directions. The cyanobacterial filaments were even more visible in the biofilm formed on the GNP composite, where the amount of extracellular material seemed to be lower and where uncovered areas of surface areas could be seen (Figure 8c). Therefore, SEM observations corroborated the results of the biofilm wet weight (Figure 5a), and those obtained by OCT (Figure 6) and CLSM (Figure 7), indicating that the graphene composite surface had a lower biofilm amount than the control surfaces after 49 days. Likewise, SEM micrographs also supported the finding that GNP-modified surfaces may decrease EPS production, resulting in a disrupted 3D structure of biofilm at the maturation stage. Similar results were obtained on CNT-based surfaces compared to the control surfaces (glass and epoxy resin) since SEM images of cyanobacterial biofilm grown on the CNT composite presented lower-density cell aggregates [42].

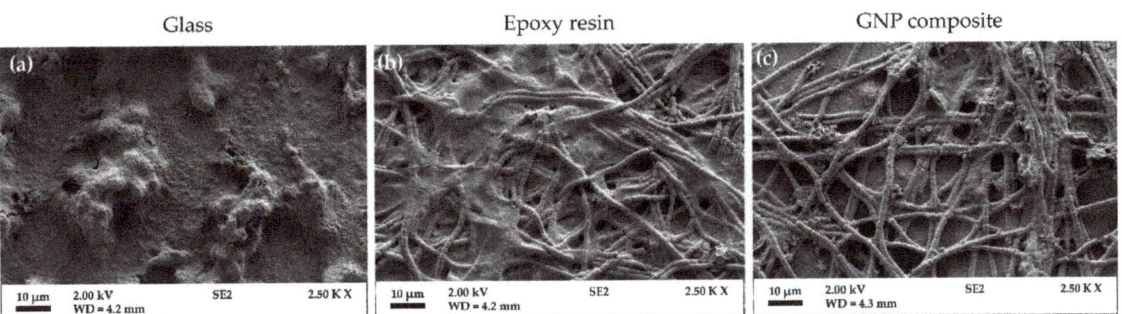

Figure 8. SEM micrographs of *Lusitaniella coriacea* LEGE 07157 biofilms grown on glass (**a**), epoxy resin (**b**), and GNP composite (**c**) after 49 days of incubation. Magnification = 2500×; scale bar = 10 μm.

This work may be limited by the use of cyanobacterial monocultures, the lack of field tests, and the absence of surface tribological characterization for long periods of immersion. Taking into account that the most studied forms of graphene within the marine context are GO and GO functionalized with metal nanoparticles [27], further experimental assays with other graphene-based surfaces for the marine environment should be performed. These assays can comprise different physical and chemical properties (e.g., dimension, number of layers, and functionalization), but also focus on factors related to the production of the surface (e.g., graphene loading, nanoparticle dispersion, and aggregation) [71–73]. Future in situ studies should also include other foulers, as well as being performed for a longer period since, in the present study, a distinct effect was observed over time. In fact, few studies on graphene-based surfaces were performed over one [19,20] or more months [34]. Although different optical techniques have been used in this study, further work should assess the metabolic state and the viability of biofilm cyanobacterial cells [74,75]. Omics approaches such as transcriptomic analysis to consider the expression of virulence factors and adhesion genes related to biofilm development [30], and proteomic analysis to elucidate cyanobacterial defense mechanisms to graphene toxicity, should also be considered [52,54,56]. Moreover, computational simulation approaches have been widely used for the characterization of novel materials [76]. Strong coupling between experimental data and numerical simulation is required for predicting macroscopic behavior with a fine description of local fields. By combining modeling approaches and corresponding experiments, the studies take advantage of the simplicity, efficiency, and mechanistic insight gained from the models, and the physicality, meaningful reproducibility, and reality check provided by the experiments [77]. Thus, future studies should include computational simulations to evaluate the potential use of GNP composites as a marine coatings.

4. Conclusions

Biofilms developed on the GNP composite had reduced wet weight, thickness, biovolume, and surface coverage in the maturation stage when compared to the control surfaces (glass and epoxy resin). Moreover, the GNP composite delayed cyanobacterial biofilm development and promoted the development of a less porous biofilm. As the graphene coating should serve as a long-term antifouling material for marine applications, the analysis of cyanobacterial biofilm behavior over time performed in this study is particularly relevant. The analysis of these GNP-based surfaces in vitro to assess their performance against mixed-species biofilms, and in situ to assess their effect on biofilm formation by additional microfoulers or to evaluate the potentially damaging effect of harsh marine environments on these nanostructured surfaces, will be performed in future studies. Therefore, upcoming tests should focus on tribological parameters such as the friction coefficient, wear, temperature, and durability related to the application, and surface characterization for long-term assays.

Supplementary Materials: The following supporting information can be downloaded at: https://www.mdpi.com/article/10.3390/coatings12111775/s1, Figure S1: Representative 2D cross-sectional OCT images obtained for *Lusitaniella coriacea* LEGE 07157 biofilms on glass (a), epoxy resin (b), and GNP composite (c) after 28 days. The empty spaces on the biofilm structure are indicated in blue (scale bar = 100 µm).; Figure S2: Roughness coefficient[1] values of *Lusitaniella coriacea* LEGE 07157 biofilms on different surfaces: glass (black), epoxy resin (grey), and GNP composite (white). Mean values and SD from two biological assays with two technical replicates each are represented. For each surface, different lowercase letters indicate significant differences between the sampling day (day 28 vs. day 49; $p < 0.05$; unpaired t-test). The p value obtained for glass and epoxy resin was < 0.0001, and for GNP composite was 0.0104.

Author Contributions: Conceptualization, M.J.R. and F.J.M.; methodology, M.J.R., L.C.G. and F.S.-C.; formal analysis, M.J.R. and L.C.G.; investigation, M.J.R. and L.C.G.; resources, M.F.R.P., V.V., K.A.W. and F.J.M.; funding acquisition, M.F.R.P., V.V. and F.J.M.; data curation, M.J.R. and L.C.G.; writing—original draft preparation, M.J.R. and L.C.G.; writing—review and editing, J.M., M.F.R.P., O.S.G.P.S., K.A.W. and F.J.M.; supervision, V.V. and F.J.M. All authors have read and agreed to the published version of the manuscript.

Funding: This research was funded by: LA/P/0045/2020 (ALiCE), UIDB/00511/2020 and UIDP/00511/2020 (LEPABE) and project PTDC/CTM-COM/4844/2020, funded by national funds through FCT/MCTES (PIDDAC); project HealthyWaters (NORTE-01-0145-FEDER-000069), supported by Norte Portugal Regional Operational Programme (NORTE 2020), under the PORTUGAL 2020 Partnership Agreement, through the European Regional Development Fund (ERDF); Strategic Funding UIDB/04423/2020 and UIDP/04423/2020 through national funds provided by the Foundation for Science and Technology (FCT); ATLANTIDA (NORTE-01-0145-FEDER-000040), financed by the FEDER through the NORTE 2020 Program and the European Regional Development Fund (ERDF) in the framework of the program PT2020; project SurfSAFE supported by the European Union's Horizon 2020 Research and Innovation Programme under grant agreement no. 952471. M.J.R, L.C.G., and O.S.G.P.S. thank FCT for the financial support of a Ph.D. grant (SFRH/BD/140080/2018), and work contracts through the Scientific Employment Stimulus—Individual Call—[CEECIND/01700/2017] and the Scientific Employment Stimulus—Institutional Call—CEECINST/00049/2018, respectively. Support from the EURO-MIC COST Action (CA20130) is also acknowledged.

Institutional Review Board Statement: Not applicable.

Informed Consent Statement: Not applicable.

Data Availability Statement: The data presented in this study are available from the corresponding author upon request.

Conflicts of Interest: The authors declare no conflict of interest.

References

1. Schultz, M.P.; Bendick, J.A.; Holm, E.R.; Hertel, W.M. Economic impact of biofouling on a naval surface ship. *Biofouling* **2011**, *27*, 87–98. [CrossRef]
2. Delauney, L.; Compère, C.; Lehaitre, M. Biofouling protection for marine environmental sensors. *Ocean Sci.* **2010**, *6*, 503–511. [CrossRef]
3. Giakoumi, S.; Katsanevakis, S.; Albano, P.G.; Azzurro, E.; Cardoso, A.C.; Cebrian, E.; Deidun, A.; Edelist, D.; Francour, P.; Jimenez, C.; et al. Management priorities for marine invasive species. *Sci. Total Environ.* **2019**, *688*, 976–982. [CrossRef] [PubMed]
4. Bannister, J.; Sievers, M.; Bush, F.; Bloecher, N. Biofouling in marine aquaculture: A review of recent research and developments. *Biofouling* **2019**, *35*, 631–648. [CrossRef] [PubMed]
5. Kumar, S.; Bhattacharya, W.; Singh, M.; Halder, D.; Mitra, A. Plant latex capped colloidal silver nanoparticles: A potent anti-biofilm and fungicidal formulation. *J. Mol. Liq.* **2017**, *230*, 705–713. [CrossRef]
6. Zhang, D.; Liu, Z.; Wu, G.; Yang, Z.; Cui, Y.; Li, H.; Zhang, Y. Fluorinated carbon nanotube superamphiphobic coating for high-efficiency and long-lasting underwater antibiofouling surfaces. *ACS Appl. Bio Mater.* **2021**, *4*, 6351–6360. [CrossRef] [PubMed]
7. Jin, H.; Zhang, T.; Bing, W.; Dong, S.; Tian, L. Antifouling performance and mechanism of elastic graphene-silicone rubber composite membranes. *J. Mater. Chem. B* **2018**, *7*, 488–497. [CrossRef]

8. Kumar, S.; Boro, J.C.; Ray, D.; Mukherjee, A.; Dutta, J. Bionanocomposite films of agar incorporated with ZnO nanoparticles as an active packaging material for shelf life extension of green grape. *Heliyon* **2019**, *5*, e01867. [CrossRef]
9. Kim, J.; Baek, Y.; Hong, S.P.; Yoon, H.; Kim, S.; Kim, C.; Kim, J.; Yoon, J. Evaluation of thin-film nanocomposite RO membranes using TiO_2 nanotubes and TiO_2 nanoparticles: A comparative study. *Desalin. Water Treat.* **2016**, *57*, 24674–24681. [CrossRef]
10. Callow, J.A.; Callow, M.E. Trends in the development of environmentally friendly fouling-resistant marine coatings. *Nat. Commun.* **2011**, *2*, 244. [CrossRef]
11. Al-Jumaili, A.; Alancherry, S.; Bazaka, K.; Jacob, M.V. Review on the antimicrobial properties of carbon nanostructures. *Materials* **2017**, *10*, 1066. [CrossRef]
12. Xu, H.; Ma, L.; Jin, Z. Nitrogen-doped graphene: Synthesis, characterizations and energy applications. *J. Energy Chem.* **2018**, *27*, 146–160. [CrossRef]
13. Parra, C.; Montero-Silva, F.; Henríquez, R.; Flores, M.; Garín, C.; Ramírez, C.; Moreno, M.; Correa, J.; Seeger, M.; Häberle, P. Suppressing bacterial interaction with copper surfaces through graphene and hexagonal-boron nitride coatings. *ACS Appl. Mater. Interfaces* **2015**, *7*, 6430–6437. [CrossRef]
14. Naddaf, A.; Heris, S.Z.; Pouladi, B. An experimental study on heat transfer performance and pressure drop of nanofluids using graphene and multi-walled carbon nanotubes based on diesel oil. *Powder Technol.* **2019**, *352*, 369–380. [CrossRef]
15. Smaradhana, D.F.; Prabowo, A.R.; Ganda, A.N.F. Exploring the potential of graphene materials in marine and shipping industries—A technical review for prospective application on ship operation and material-structure aspects. *J. Ocean Eng. Sci.* **2021**, *6*, 299–316. [CrossRef]
16. Pan, L.; Liu, S.; Oderinde, O.; Li, K.; Yao, F.; Fu, G. Facile fabrication of graphene-based aerogel with rare earth metal oxide for water purification. *Appl. Surf. Sci.* **2018**, *427*, 779–786. [CrossRef]
17. Hu, M.; Cui, Z.; Li, J.; Zhang, L.; Mo, Y.; Dlamini, D.S.; Wang, H.; He, B.; Li, J.; Matsuyama, H. Ultra-low graphene oxide loading for water permeability, antifouling and antibacterial improvement of polyethersulfone/sulfonated polysulfone ultrafiltration membranes. *J. Colloid Interface Sci.* **2019**, *552*, 319–331. [CrossRef] [PubMed]
18. Balaure, P.C. Special Issue: Advances in engineered nanostructured antibacterial surfaces and coatings. *Coatings* **2022**, *12*, 1041. [CrossRef]
19. Selim, M.S.; El-Safty, S.A.; Fatthallah, N.A.; Shenashen, M.A. Silicone/graphene oxide sheet-alumina nanorod ternary composite for superhydrophobic antifouling coating. *Prog. Org. Coatings* **2018**, *121*, 160–172. [CrossRef]
20. Selim, M.S.; Fatthallah, N.A.; Higazy, S.A.; Hao, Z.; Mo, P.J. A comparative study between two novel silicone/graphene-based nanostructured surfaces for maritime antifouling. *J. Colloid Interface Sci.* **2022**, *606*, 367–383. [CrossRef]
21. Gu, J.; Li, L.; Huang, D.; Jiang, L.; Liu, L.; Li, F.; Pang, A.; Guo, X.; Tao, B. In situ synthesis of graphene@cuprous oxide nanocomposite incorporated marine antifouling coating with elevated antifouling performance. *Open J. Org. Polym. Mater.* **2019**, *9*, 47–62. [CrossRef]
22. Li, Y.; Huang, Y.; Wang, F.; Liang, W.; Yang, H.; Wu, D. Fabrication of acrylic acid modified graphene oxide (AGO)/acrylate composites and their synergistic mechanisms of anticorrosion and antifouling properties. *Prog. Org. Coatings* **2022**, *168*, 106910. [CrossRef]
23. Zhang, Z.; Chen, R.; Song, D.; Yu, J.; Sun, G.; Liu, Q.; Han, S.; Liu, J.; Zhang, H.; Wang, J. Guanidine-functionalized graphene to improve the antifouling performance of boron acrylate polymer. *Prog. Org. Coatings* **2021**, *159*, 106396. [CrossRef]
24. Jin, H.; Bing, W.; Tian, L.; Wang, P.; Zhao, J. Combined effects of color and elastic modulus on antifouling performance: A study of graphene oxide/silicone rubber composite membranes. *Materials* **2019**, *12*, 2608. [CrossRef] [PubMed]
25. Liu, Z.; Tian, S.; Li, Q.; Wang, J.; Pu, J.; Wang, G.; Zhao, W.; Feng, F.; Qin, J.; Ren, L. Integrated dual-functional ormosil coatings with AgNPs@rGO nanocomposite for corrosion resistance and antifouling applications. *ACS Sustain. Chem. Eng.* **2020**, *8*, 6786–6797. [CrossRef]
26. Zhang, X.; Mikkelsen, Ø. Graphene oxide/silver nanocomposites as antifouling coating on sensor housing materials. *J. Clust. Sci.* **2021**, *33*, 627–635. [CrossRef]
27. Sousa-Cardoso, F.; Teixeira-Santos, R.; Mergulhão, F.J.M. Antifouling performance of carbon-based coatings for marine applications: A systematic review. *Antibiotics* **2022**, *11*, 1102. [CrossRef]
28. Yee, M.S.-L.; Khiew, P.-S.; Chiu, W.S.; Tan, Y.F.; Kok, Y.-Y.; Leong, C.-O. Green synthesis of graphene-silver nanocomposites and its application as a potent marine antifouling agent. *Colloids Surf. B Biointerfaces* **2016**, *148*, 392–401. [CrossRef]
29. Jiang, T.; Qi, L.; Qin, W. Improving the environmental compatibility of marine sensors by surface functionalization with graphene oxide. *Anal. Chem.* **2019**, *91*, 13268–13274. [CrossRef] [PubMed]
30. Parra, C.; Dorta, F.; Jimenez, E.; Henríquez, R.; Ramírez, C.; Rojas, R.; Villalobos, P. A nanomolecular approach to decrease adhesion of biofouling-producing bacteria to graphene-coated material. *J. Nanobiotechnol.* **2015**, *13*, 82. [CrossRef]
31. Krishnamoorthy, K.; Jeyasubramanian, K.; Premanathan, M.; Subbiah, G.; Shin, H.S.; Kim, S.-J. Graphene oxide nanopaint. *Carbon N. Y.* **2014**, *72*, 328–337. [CrossRef]
32. Balakrishnan, A.; Jena, G.; George, R.P.; Philip, J. Polydimethylsiloxane-graphene oxide nanocomposite coatings with improved anti-corrosion and anti-biofouling properties. *Environ. Sci. Pollut. Res.* **2020**, *28*, 7404–7422. [CrossRef]
33. Manderfeld, E.; Kleinberg, M.N.; Thamaraiselvan, C.; Koschitzki, F.; Gnutt, P.; Plumere, N.; Arnusch, C.J.; Rosenhahn, A. Electrochemically activated laser-induced graphene coatings against marine biofouling. *Appl. Surf. Sci.* **2021**, *569*, 150853. [CrossRef]

34. Fazli-Shokouhi, S.; Nasirpouri, F.; Khatamian, M. Epoxy-matrix polyaniline/*p*-phenylenediamine-functionalised graphene oxide coatings with dual anti-corrosion and anti-fouling performance. *RSC Adv.* **2021**, *11*, 11627–11641. [CrossRef] [PubMed]
35. Chambers, L.D.; Stokes, K.R.; Walsh, F.C.; Wood, R.J.K. Modern approaches to marine antifouling coatings. *Surf. Coat. Technol.* **2006**, *201*, 3642–3652. [CrossRef]
36. Palmer, J.; Flint, S.; Brooks, J. Bacterial cell attachment, the beginning of a biofilm. *J. Ind. Microbiol. Biotechnol.* **2007**, *34*, 577–588. [CrossRef]
37. Mostafaei, A.; Nasirpouri, F. Preparation and characterization of a novel conducting nanocomposite blended with epoxy coating for antifouling and antibacterial applications. *J. Coatings Technol. Res.* **2013**, *10*, 679–694. [CrossRef]
38. Hoge, J.; Leach, C. Epoxy resin infused boat hulls. *Reinf. Plast.* **2016**, *60*, 221–223. [CrossRef]
39. Faria, S.I.; Teixeira-Santos, R.; Gomes, L.C.; Silva, E.R.; Morais, J.; Vasconcelos, V.; Mergulhão, F.J.M. Experimental assessment of the performance of two marine coatings to curb biofilm formation of microfoulers. *Coatings* **2020**, *10*, 893. [CrossRef]
40. Oliveira, I.M.; Gomes, M.; Gomes, L.C.; Pereira, M.F.R.; Soares, O.S.G.P.; Mergulhão, F.J. Performance of graphene/polydimethylsiloxane surfaces against *S. aureus* and *P. aeruginosa* single- and dual-species biofilms. *Nanomaterials* **2022**, *12*, 355. [CrossRef]
41. Faria, S.I.; Teixeira-Santos, R.; Romeu, M.J.; Morais, J.; Vasconcelos, V.; Mergulhão, F.J. The relative importance of shear forces and surface hydrophobicity on biofilm formation by coccoid cyanobacteria. *Polymers* **2020**, *12*, 653. [CrossRef] [PubMed]
42. Romeu, M.J.; Lima, M.; Gomes, L.C.; de Jong, E.D.; Vasconcelos, V.; Pereira, M.F.R.; Soares, O.S.G.P.; Sjollema, J.; Mergulhão, F.J. The use of 3D optical coherence tomography to analyze the architecture of cyanobacterial biofilms formed on a carbon nanotube composite. *Polymers* **2022**, *14*, 4410. [CrossRef] [PubMed]
43. Whitehead, K.; Kelly, P.; Li, H.; Verran, J. Surface topography and physicochemistry of silver containing titanium nitride nanocomposite coatings. *J. Vac. Sci. Technol. B Nanotechnol. Microelectron. Mater. Process. Meas. Phenom.* **2010**, *28*, 180. [CrossRef]
44. Skovager, A.; Whitehead, K.; Siegumfeldt, H.; Ingmer, H.; Verran, J.; Arneborg, N. Influence of flow direction and flow rate on the initial adhesion of seven *Listeria monocytogenes* strains to fine polished stainless steel. *Int. J. Food Microbiol.* **2012**, *157*, 174–181. [CrossRef] [PubMed]
45. Ramos, V.; Morais, J.; Castelo-Branco, R.; Pinheiro, Â.; Martins, J.; Regueiras, A.; Pereira, A.L.; Lopes, V.R.; Frazão, B.; Gomes, D.; et al. Cyanobacterial diversity held in microbial biological resource centers as a biotechnological asset: The case study of the newly established LEGE culture collection. *J. Appl. Phycol.* **2018**, *30*, 1437–1451. [CrossRef]
46. Kotai, J. Instructions for the Preparation of Modified Nutrient Solution Z8 for Algae. *Nor. Inst. Water Res.* **1972**, *11*, 5.
47. Romeu, M.J.; Alves, P.; Morais, J.; Miranda, J.M.; de Jong, E.D.; Sjollema, J.; Ramos, V.; Vasconcelos, V.; Mergulhão, F.J.M. Biofilm formation behaviour of marine filamentous cyanobacterial strains in controlled hydrodynamic conditions. *Environ. Microbiol.* **2019**, *21*, 4411–4424. [CrossRef]
48. Boyer, J.N.; Kelble, C.R.; Ortner, P.B.; Rudnick, D.T. Phytoplankton bloom status: Chlorophyll a biomass as an indicator of water quality condition in the southern estuaries of Florida, USA. *Ecol. Indic.* **2009**, *9*, S56–S67. [CrossRef]
49. Porra, R.J.; Thompson, W.A.; Kriedemann, P.E. Determination of accurate extinction coefficients and simultaneous equations for assaying chlorophylls a and b extracted with four different solvents: Verification of the concentration of chlorophyll standards by atomic absorption spectroscopy. *Biochim. Biophys. Acta BBA Bioenerg.* **1989**, *975*, 384–394. [CrossRef]
50. Bakker, D.P.; Van der Plaats, A.; Verkerke, G.J.; Busscher, H.J.; Mei, H.C. Van der comparison of velocity profiles for different flow chamber designs used in studies of microbial adhesion to surfaces. *Appl. Environ. Microbiol.* **2003**, *69*, 6280–6287. [CrossRef]
51. Silva, E.R.; Tulcidas, A.V.; Ferreira, O.; Bayón, R.; Igartua, A.; Mendoza, G.; Mergulhão, F.J.; Faria, S.I.; Gomes, L.C.; Carvalho, S.; et al. Assessment of the environmental compatibility and antifouling performance of an innovative biocidal and foul-release multifunctional marine coating. *Environ. Res.* **2021**, *198*, 111219. [CrossRef] [PubMed]
52. Romeu, M.J.; Domínguez-Pérez, D.; Almeida, D.; Morais, J.; Araújo, M.J.; Osório, H.; Campos, A.; Vasconcelos, V.; Mergulhão, F.J. Hydrodynamic conditions affect the proteomic profile of marine biofilms formed by filamentous cyanobacterium. *NPJ Biofilms Microbiomes* **2022**, *8*, 80. [CrossRef] [PubMed]
53. Faria, S.; Gomes, L.C.; Teixeira-Santos, R.; Morais, J.; Vasconcelos, V.; Mergulhão, F. Developing new marine antifouling surfaces: Learning from single-strain laboratory tests. *Coatings* **2021**, *11*, 90. [CrossRef]
54. Romeu, M.J.L.; Domínguez-Pérez, D.; Almeida, D.; Morais, J.; Campos, A.; Vasconcelos, V.; Mergulhão, F.J.M. Characterization of planktonic and biofilm cells from two filamentous cyanobacteria using a shotgun proteomic approach. *Biofouling* **2020**, *36*, 631–645. [CrossRef]
55. Faria, S.; Teixeira-Santos, R.; Romeu, M.J.; Morais, J.; Jong, E.; Sjollema, J.; Vasconcelos, V.; Mergulhão, F.J. Unveiling the antifouling performance of different marine surfaces and their effect on the development and structure of cyanobacterial biofilms. *Microorganisms* **2021**, *9*, 1102. [CrossRef] [PubMed]
56. Romeu, M.J.; Domínguez-Pérez, D.; Almeida, D.; Morais, J.; Araújo, M.J.; Osório, H.; Campos, A.; Vasconcelos, V.; Mergulhão, F. Quantitative proteomic analysis of marine biofilms formed by filamentous cyanobacterium. *Environ. Res.* **2021**, *201*, 111566. [CrossRef] [PubMed]
57. Heydorn, A.; Nielsen, A.T.; Hentzer, M.; Sternberg, C.; Givskov, M.; Ersbøll, B.K.; Molin, S. Quantification of biofilm structures by the novel computer program COMSTAT. *Microbiology* **2000**, *146*, 2395–2407. [CrossRef]
58. Alves, P.; Gomes, L.C.; Rodríguez-Emmenegger, C.; Mergulhão, F.J. Efficacy of a poly (MeOEGMA) brush on the prevention of *Escherichia coli* biofilm formation and susceptibility. *Antibiotics* **2020**, *9*, 216. [CrossRef]

59. Zheng, S.; Bawazir, M.; Dhall, A.; Kim, H.-E.; He, L.; Heo, J.; Hwang, G. Implication of surface properties, bacterial motility, and hydrodynamic conditions on bacterial surface sensing and their initial adhesion. *Front. Bioeng. Biotechnol.* **2021**, *9*, 643722. [CrossRef]
60. Yuan, Y.; Lee, T.R. Contact angle and wetting properties. In *Surface Science Techniques*; Springer Series in surface sciences 51; Bracco, G., Holst, B., Eds.; Springer: Berlin/Heidelberg, Germany, 2013; Volume 51, pp. 3–34. ISBN 978-3-642-34242-4.
61. Doshi, B.; Sillanpää, M.; Kalliola, S. A review of bio-based materials for oil spill treatment. *Water Res.* **2018**, *135*, 262–277. [CrossRef]
62. Rafiee, J.; Mi, X.; Gullapalli, H.; Thomas, A.V.; Yavari, F.; Shi, Y.; Ajayan, P.M.; Koratkar, N.A. Wetting transparency of graphene. *Nat. Mater.* **2012**, *11*, 217–222. [CrossRef]
63. Chatterjee, S.; Nüesch, F.A.; Chu, B.T.T. Comparing carbon nanotubes and graphene nanoplatelets as reinforcements in polyamide 12 composites. *Nanotechnology* **2011**, *22*, 275714. [CrossRef]
64. Santos, C.M.; Mangadlao, J.; Ahmed, F.; Leon, A.; Advincula, R.C.; Rodrigues, D. Graphene nanocomposite for biomedical applications: Fabrication, antimicrobial and cytotoxic investigations. *Nanotechnology* **2012**, *23*, 395101. [CrossRef]
65. Zou, X.; Zhang, L.; Wang, Z.; Luo, Y. Mechanisms of the antimicrobial activities of graphene materials. *J. Am. Chem. Soc.* **2016**, *138*, 2064–2077. [CrossRef]
66. Mohammed, H.; Kumar, A.; Bekyarova, E.; Al-Hadeethi, Y.; Zhang, X.; Chen, M.; Ansari, M.S.; Cochis, A.; Rimondini, L. Antimicrobial mechanisms and effectiveness of graphene and graphene-functionalized biomaterials. A scope review. *Front. Bioeng. Biotechnol.* **2020**, *8*, 465. [CrossRef] [PubMed]
67. Staneva, A.D.; Dimitrov, D.K.; Gospodinova, D.N.; Vladkova, T.G. Antibiofouling activity of graphene materials and graphene-based antimicrobial coatings. *Microorganisms* **2021**, *9*, 1839. [CrossRef] [PubMed]
68. Sjollema, J.; Rustema-Abbing, M.; van der Mei, H.C.; Busscher, H.J. Generalized relationship between numbers of bacteria and their viability in biofilms. *Appl. Environ. Microbiol.* **2011**, *77*, 5027–5029. [CrossRef]
69. Song, C.; Yang, C.-M.; Sun, X.-F.; Xia, P.-F.; Qin, J.; Guo, B.-B.; Wang, S.-G. Influences of graphene oxide on biofilm formation of gram-negative and gram-positive bacteria. *Environ. Sci. Pollut. Res.* **2017**, *25*, 2853–2860. [CrossRef] [PubMed]
70. Ahmed, B.; Ameen, F.; Rizvi, A.; Ali, K.; Sonbol, H.; Zaidi, A.; Khan, M.S.; Musarrat, J. Destruction of cell topography, morphology, membrane, inhibition of respiration, biofilm formation, and bioactive molecule production by nanoparticles of Ag, ZnO, CuO, TiO_2, and Al_2O_3 toward beneficial soil bacteria. *ACS Omega* **2020**, *5*, 7861–7876. [CrossRef]
71. Fatima, N.; Qazi, U.Y.; Mansha, A.; Bhatti, I.A.; Javaid, R.; Abbas, Q.; Nadeem, N.; Rehan, Z.A.; Noreen, S.; Zahid, M. Recent developments for antimicrobial applications of graphene-based polymeric composites: A review. *J. Ind. Eng. Chem.* **2021**, *100*, 40–58. [CrossRef]
72. Radhi, A.; Mohamad, D.; Rahman, F.S.A.; Abdullah, A.M.; Hasan, H. Mechanism and factors influence of graphene-based nanomaterials antimicrobial activities and application in dentistry. *J. Mater. Res. Technol.* **2021**, *11*, 1290–1307. [CrossRef]
73. Azizi-Lalabadi, M.; Hashemi, H.; Feng, J.; Jafari, S.M. Carbon nanomaterials against pathogens; the antimicrobial activity of carbon nanotubes, graphene/graphene oxide, fullerenes, and their nanocomposites. *Adv. Colloid Interface Sci.* **2020**, *284*, 102250. [CrossRef]
74. Mountcastle, S.E.; Vyas, N.; Villapun, V.M.; Cox, S.C.; Jabbari, S.; Sammons, R.L.; Shelton, R.M.; Walmsley, A.D.; Kuehne, S.A. Biofilm viability checker: An open-source tool for automated biofilm viability analysis from confocal microscopy images. *NPJ Biofilms Microbiomes* **2021**, *7*, 44. [CrossRef] [PubMed]
75. Van den Driessche, F.; Rigole, P.; Brackman, G.; Coenye, T. Optimization of resazurin-based viability staining for quantification of microbial biofilms. *J. Microbiol. Methods* **2014**, *98*, 31–34. [CrossRef]
76. Ammarullah, M.I.; Santoso, G.; Sugiharto, S.; Supriyono, T.; Wibowo, D.B.; Kurdi, O.; Tauviqirrahman, M.; Jamari, J. Minimizing risk of failure from ceramic-on-ceramic total hip prosthesis by selecting ceramic materials based on tresca stress. *Sustainability* **2022**, *14*, 13413. [CrossRef]
77. Dingreville, R.; Karnesky, R.A.; Puel, G.; Schmitt, J.-H. Synergies between computational modeling and experimental characterization of materials across length scales. *J. Mater. Sci.* **2015**, *51*, 1176–1177. [CrossRef]

Article

Thickness, Adhesion and Microscopic Analysis of the Surface Structure of Single-Layer and Multi-Layer Metakaolin-Based Geopolymer Coatings

Martin Jaskevic [1,*], Jan Novotny [1], Filip Mamon [1], Jakub Mares [1] and Angelos Markopoulos [2]

[1] Faculty of Mechanical Engineering, J. E. Purkyne University in Usti nad Labem, Pasteurova 3334/7, 40001 Usti nad Labem, Czech Republic; jan.novotny@ujep.cz (J.N.); mamon@iic.cas.cz (F.M.); mares@iic.cas.cz (J.M.)

[2] School of Mechanical Engineering, National Technical University of Athens, Heroon Polytechniou 9, 15780 Athens, Greece; amark@mail.ntua.gr

* Correspondence: martin.jaskevic@ujep.cz

Abstract: This work is focused on creating coating layers made of a metakaolin-based geopolymer suspensions (GP)-formed Al matrix modified using H_3PO_4 acid with $Al(OH)_3$ in isopropyl alcohol, named GP suspension I, and H_3PO_4 acid with nano Al_2O_3 in isopropyl alcohol, named GP suspension J. The selected GP suspensions were applied on aluminum and steel underlying substrates as single-layer coatings and multi-layer coatings, where multi-layer coatings included three and five layers that were polymerized by a curing process. Curing was divided into two types with every layer curing process and final layer curing process. For both GP suspensions I and J, the effect of the number of layers and the type of substrate on adhesion was investigated. The prepared samples on underlying substrates were characterized on the microscopy analysis including SEM for high-resolution images and 3D laser confocal microscopy (CLSM) for the 3D visualization of the coatings structure. Microscopy analysis showed structural defects such as porosity, cracks and peeling, which increase with a greater number of applied layers. However, these defects were only evident on a micro scale and did not seem to be fatal for the performance of the surface stability. The EDS mapping of the prepared layer showed inhomogeneity in the distribution of elements caused by the brush application. A grid test and thickness measurement were performed to complete the microscopy analysis. The grid test confirmed a very high adhesion of GP coatings on the aluminum substrate with a rating of one (only in one case was there a rating of two) and a lower adhesion on the steel substrate with the most frequent rating of three (in one case, there were ratings of two and one). The thickness measurement proved a noticeably thicker thickness of the prepared layer on the Fe substrate compared to the Al substrate by 20%–30% in the case of suspension I and by 70%–10% in the case of suspension J. The thickness of the layer also showed a dependence on the method of application and curing, as a thicker layer was always achieved when curing after the final layer of the GP suspension, compared to curing after each applied layer. The resulting single-layer and multi-layer thicknesses ranged from approx. 7 to 30 μm for suspension I and from approx. 3 to 11 μm for suspension J. A non-linear increase in thickness was also evident from the thickness measurement data.

Keywords: geopolymers; single-layer coating; multi-layer coatings; aluminum; construction steel; microstructure characterization; layers thickness; grid test

Citation: Jaskevic, M.; Novotny, J.; Mamon, F.; Mares, J.; Markopoulos, A. Thickness, Adhesion and Microscopic Analysis of the Surface Structure of Single-Layer and Multi-Layer Metakaolin-Based Geopolymer Coatings. Coatings 2023, 13, 1731. https://doi.org/10.3390/coatings13101731

Academic Editor: Mohor Mihelčič

Received: 26 August 2023
Revised: 19 September 2023
Accepted: 29 September 2023
Published: 4 October 2023

Copyright: © 2023 by the authors. Licensee MDPI, Basel, Switzerland. This article is an open access article distributed under the terms and conditions of the Creative Commons Attribution (CC BY) license (https://creativecommons.org/licenses/by/4.0/).

1. Introduction

Synthetic inorganic polymers, also known as geopolymers (GP), consist of chains or networks of mineral molecules linked by covalent bonds [1].

GP are considered to be environmentally friendly materials that have the potential to be used as substitutes for ordinary Portland cement [2,3]. Geopolymers are produced from natural sources such as kaolinite or clays. Modern approaches also try to use industrial sources including fly ash, waste paper sludge or granulated blast furnace slag. Recycling these resources can have positive environmental impacts and reduce CO_2 production compared to Portland cement production [4].

GP, as a group of alkaline-activated materials, have a number of exceptional properties such as strength, resistance to acids and bases, fire resistance, good thermal stability and good adhesion to the underlying substrate [5].

Also, there are some possibilities to use GP as a coating material for metal and non-metal substrates. GP coating properties rely on the chemical composition of raw materials, followed by the roughness of the substrate or the Si/Al ratio [6,7]. Different types and concentrations of the acid could change the behavior and properties of the final GP coatings. For example, Shamala et. al. prepared GP coatings on wood substrates with various NaOH concentrations to find the best concentration for GP coatings [8]. The next factor for GP coatings depends on the water content, which affects the results of the GP coating thickness [9].

The preparation of geopolymers includes three basic phases. Dissolution is the first phase, in which Si and Al atoms transition from the basic raw material to the solution and complexes with hydroxide ions are formed. In the second phase, the condensation of monomers with mobile precursors follows, with a partial internal restructuring of the alkaline polysilicates. The third stage of the geopolymer formation process is the poly-condensation or the polymerization of monomers. Here, the polymer structure is formed, and the whole system solidifies. The product is an inorganic polymer structure. It is very complicated to analyze ongoing processes because they occur almost simultaneously [10].

The geopolymerization process is based on the reaction of reactive aluminosilicates supplemented with metakaolin or fly ash, which quickly dissolve in alkaline solutions in the presence of alkali hydroxides (NaOH/KOH). This creates tetrahedral units-connected polymeric precursors ($-SiO_4-AlO_4-$ or $-SiO_4-AlO_4-SiO_4-$ or $-SiO_4-AlO_4-SiO_4-SiO_4-$) forming amorphous geopolymer products with a 3D network structure [11,12]. Nergis reported that geopolymers also contain three types of pores formed by the arrangement of the OH− and Si groups (Si–OH), Si–O–Si groups, Si–O–Al groups and Si–O rings [13].

Some geopolymers are activated by acidic activators. The metakaolin-based geopolymer produced by using a phosphoric acid solution as an activator has a high compressive strength up to 93.8 MPa [14,15]. Another study also showed that acid-based geopolymers have a higher temperature resistance (up to 1450 °C) and better mechanical properties than alkali-based geopolymers [16]. However, GP have extremely good thermal stability and adhesion to the surfaces thought to be a kind of all-purpose material and potential tribological material [17].

Fire resistance is an interesting property of GP if the suspension contains a flame retardant. In our study, the stability of coatings with $Al(OH)_3$ is monitored for possible future fire protection applications on Fe and Al substrates. $Al(OH)_3$ as a flame retardant in a GP suspension was described on polystyrene and chipboard underlying substrates, and its positive influence has been proven [18].

The adhesion of the coating to the underlying substrate is an essential factor which is controlled by the surface treatments of the base metal as an underlying substrate [18]. Metal with a high surface roughness will have a higher adhesion strength with GP coating material compared to the polished metal substrate [9].

A general theory covering all relevant properties and parameters involved in the design and application of tribological coating composites is very far from being realized. Such a theory would have to treat the long chain of relations ranging from the coating deposition parameters to the tribological response of the coated component [19,20].

Obviously, a good adhesion to the substrate is a crucial property of most applications of coated components. Any adhesion test must superimpose an external stress field over the coating/substrate interface to cause a measurable adhesive failure. Since this stress field will depend on the geometry and type of loading (indentation, scratching, sliding, abrasion, impact, etc.) as well as on the elastic and plastic parameters of the coating and substrate, the resulting adhesion value will only be representative of the particular test from which it has been obtained [21].

There are many ways in which suspensions can be applied to a substrate surface. One of the simple methods of application, together with the satisfactory results of the final layer, is the application of the suspension with a brush. This method is very cheap, with minimal economic costs in creating a coating, and does not require deep knowledge and know-how in applying and creating a coating. The disadvantage is that it is not possible to accurately correct the achieved layer thickness, and also, the homogeneity of the resulting layer thickness fluctuates within a certain range. An airbrush appears to be another suitable method for applying the suspension. This method is also widely used, but its use is already economically and technologically more demanding. However, it can improve the homogeneity of the thickness of the resulting layer [22].

The aim of the work was to prepare and compare single- and multi-layer coatings on metal underlying substrates, aluminum and construction steel. It was important to explain what effect applying a thicker layer with a brush would have on the adhesion, overall surface quality and change in thickness compared to single-layer systems and various application/curing methods. The prepared GP coatings were characterized using confocal microscopy and SEM (scanning electron microscopy) to capture the microstructure and the visual quality of the coatings. These results were supplemented by a grid test and thickness measurement.

This work follows our previous research [23,24] and expands our knowledge about the selected single- and multi-layer GP coatings with different applications by brushes and some types of curing processes. We observed changes in the visual quality of the prepared GP coatings and changes in their mechanical properties.

2. Materials and Methods

The GP coating was selected based on our previous research [23,24]. The selected GP had good properties, which was a good starting point for the following complex research creating multi-layer GP coatings. In this work, we presented this advanced preparation of GP coatings compared with a multi-layered GP coating.

The chosen substrates were aluminum alloy EN-AW 6060 (AlMgSi$_{0.5}$) [25,26] and construction steel 1.0038 (according to EN 10025-2). These underlying substrates have been chosen as the most common alloys in all sectors of industry. The preparation of the substrate before the application of the geopolymer suspension consists only in removing gross impurities and degreasing the surface with an organic solvent (acetone). No other pre-treatment of the surface was applied; therefore, their natural oxide layers are found on the surface of the substrates.

2.1. Preparation of the Suspensions and Underlying Substrate

The preparation of the geopolymer suspension consists in mixing basic raw materials that have different phases (liquid and solid). In this research, there is a liquid component, phosphoric acid (H_3PO_4) and isopropylalkohol (iPrOH), for both GP suspensions and a solid component, metakaolin with $AlOH_3$ (for GP suspension I) and metakaolin with powder Al_2O_3 (for GP suspension J). The good homogenization of the resulting mixture after mixing the basic ingredients is very important. A laboratory homogenizer, AD300L-H, 10,000 RPM, was used for homogenization and mixing.

Geopolymer suspensions I and J were selected from previous research for their interesting properties on the Al substrate [23], where geopolymer I reached an average coating thickness of 2.7 μm and geopolymer J reached one of only 1.5 μm. Both suspensions had excellent adhesion to the Al substrate. Even the microhardness values of HV 0.1 achieved very good results (GP I 118.4 HV 0.1 and GP J 127.1 HV 0.1) compared to the underlying substrate Al 93.6 HV 0.1.

2.2. Roughness of Al Substrate EN AW-6060 and Fe Substrate 1.0038

For the application of conventional coatings with an organic or inorganic composition (most often for the anti-corrosion protection of metals) on metal surfaces, the roughness of the underlying substrate is very important [27,28]. For example, for aluminum alloys, the adhesion of such coatings is generally lower than that when applied to steel surfaces [29–33].

The roughness of the used metal substrates EN-AW 6060 and 1.0038 was measured using a Hommel Tester t1000 according to ISO 4287. The input values of the measurement were as follows: probe type T1E 2 µm/90°, compressive force 1.5 mN, traverse length 4.8 mm, traverse speed 0.5 mm/s and measurement range ±80 µm/0.01 µm. For the underlying metal substrates, a sheet with a thickness of 3 mm was used for both types. These sheets were processed by rolling and show a one-way orientation of the grooves that were created during the rolling process, as shown in the detail of the surface in Figure 1. This orientation is very well observed in the aluminum alloy. The roughness of the substrates was measured along the rolling direction A and also perpendicular to the rolling direction B, as shown in Figure 1. Table 1 shows the achieved surface roughness values in individual directions [23].

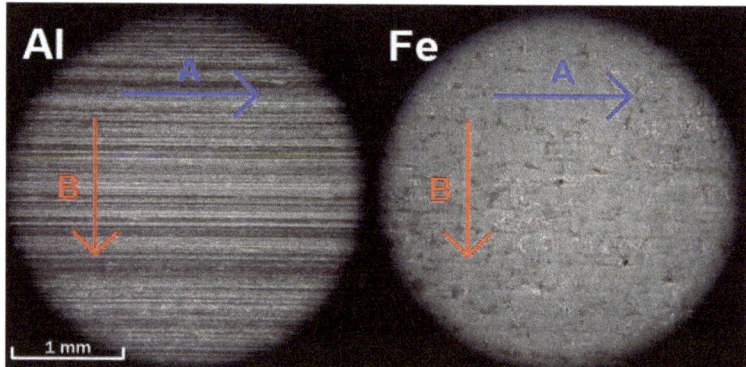

Figure 1. Surface detail of the EN-AW 6060 (Al) [23] and 1.0038 (Fe) underlying substrate with visible directional anisotropy and roughness measurement in the direction of rolling A and the perpendicular direction of rolling B.

Table 1. Surface roughness of the underlying substrate EN-AW 6060 and 1.0038.

Substrate	Al		Fe	
	Measurement Direction		Measurement Direction	
	A	B	A	B
R_a [µm]	0.206	0.841	1.091	0.874
R_z [µm]	1.117	5.410	6.225	5.098
R_{max} [µm]	1.488	6.710	6.850	6.613
R_t [µm]	1.603	6.958	8.448	7.538

R_a—arithmetic mean roughness; R_z—ten-point mean roughness; R_{max}—maximum roughness depth; R_t—maximum height of the profile.

2.3. Application of GP Suspensions

Application by brush was chosen, which is the simplest possible application of geopolymers with a sufficient resulting coating quality [23].

The geopolymer suspension was applied to the substrate by a brush, which is designed for water-based coatings. This method was chosen as in previous research [23], but with a different approach regarding the thickness of the coatings. In this case, we tried to prepare thicker layers with a brush compared to the previous research [23]. In previous research, it was found that by applying geopolymer suspensions with different compositions in one layer using a brush, very thin coatings with a thickness of up to approx. 20 μm (depending on the type of GP) with good adhesion can be achieved on the Al and Fe substrate. These thickness sizes were conditioned by the application of a very thin layer of the suspension with a brush; when applying the suspension, care must be taken to spread it very well over the surface of the substrate. The result was coatings that have a very good surface quality and very good adhesion [23,24]. Such application of GP in a thin layer is not complicated, but it requires concentration, and when applying it to larger or more fragmented surfaces, this procedure may no longer be followed exactly. A failure to follow the procedure can be caused by, for example, the human factor or even the brush application method itself, when this method is simple but not very accurate. The following research therefore simulates a process where the application procedure of a very thin layer is not followed, but the GP layer applied with a brush is thicker and a comparison is made of the effect on the properties (adhesion) and the appearance of the surface that the application of a thicker layer/layers will have.

Labeling explanations in Figure 2: GP—geopolymer suspension, X—type of geopolymer suspension (I or J), -/S—application and curing method (—curing after each applied layer; S—curing after three or five layers; see Figure 3), 1 L, 3 L, 5 L—number of layers applied (1 L—one layer, 3 L—three layers, 5 L—five layers). Our geopolymers were divided into four series, where two different GP coatings (I or J) were applied by brush on two metal substrates (Al or Fe).

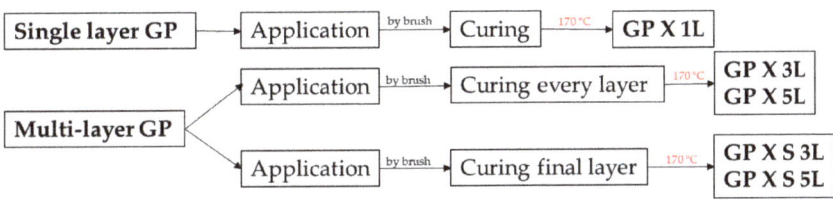

Figure 2. Scheme of the preparation of each GP coating.

In order for geopolymer suspensions to acquire their final properties after the application to the underlying substrate (Figure 2), chemical reactions, so-called geopolymerization, must occur in the mixture [24,34,35]. For the geopolymer mixture, geopolymerization occurs at elevated temperatures, in contrast to mixtures with a different composition, where geopolymerization can occur at lower temperatures [24,34,35]. For the selected geopolymer mixture, it is necessary to reach a certain minimum temperature, which was experimentally determined to be 170 °C, and to maintain the same conditions as in the previous research [23]. This temperature is an important parameter influencing the resulting quality of the coatings. This increase in the temperature is needed for the geopolymerization process [24].

Every series of GP has a different curing for the geopolymerization of these coatings shown in Figure 3. The geopolymers marked GP X 1L, GP X 3L and GP X 5L had a curing process with every layer immediately after the application. The GP marked GP X S 3L and GP X S 5L had a different approach to geopolymerization curing, with only single curing after the third or fifth layer of GP coatings. Between every layer, there were 24 h of drying, and after that, the next layer was applied. This approach changed the coatings' properties and their behavior (see the next chapters). This curing procedure was chosen in order to simplify the application of the suspensions to the substrate and at the same time reduce the resulting cost of creating multilayer coatings and analyze whether there is a difference between the layers when applying and curing after each layer compared to curing after applying the final layer.

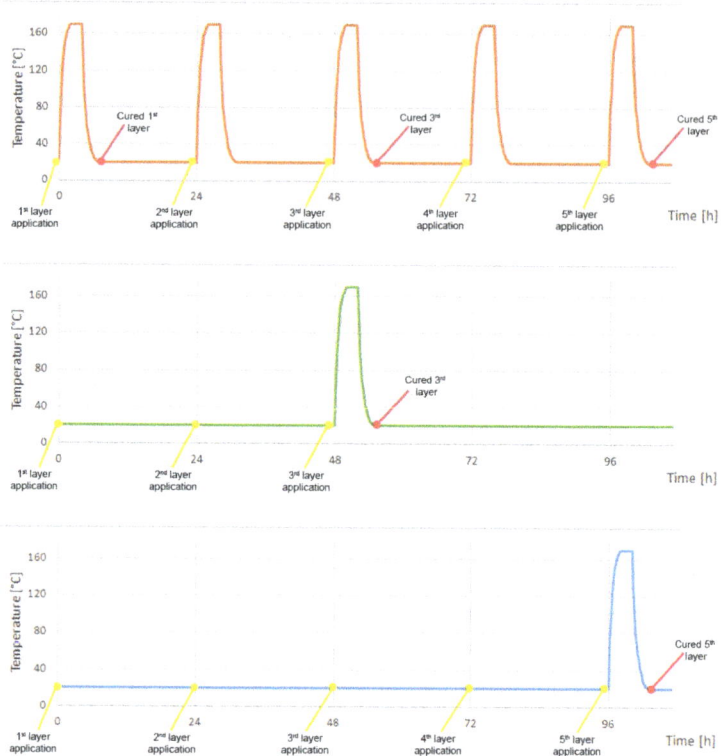

Figure 3. Curing for the geopolymerization of coatings.

2.4. Experimental Methods

All samples were prepared by manually painting the geopolymer suspensions onto the underlying substrates using a brush. The microstructure of the coatings was observed on an Olympus SZ61 optical microscope and then on a 3D laser confocal microscope LEXT OLS5000 SAF (CLSM) and on a Vega 3 scanning electron microscope (SEM) from Tescan company (Tescan, Vega 3, Brno, Czech Republic). The thickness of the geopolymer liners was measured with a DeFelsko PosiTector 6000 portable coating thickness meter for metal substrates with an FNS type probe, with a measurement range of 0–1500 µm accuracy ± (1 µm + 1%) for a coating thickness of 0–50 µm, according to ISO 2360. The adhesion of the geopolymer to the chosen substrates was analyzed by the grid test method, according to ISO 2409 [36]—specifically, with the Elcometer 1542 grid test set.

3. Results and Discussion

3.1. Microstructure Analysis of Geopolymer Coatings Using SEM and CLSM

3.1.1. Geopolymer I Al

Figure 4 shows:

- Sample I 1L Al: The surface was slightly rough and showed microporosity. A small number of cracks are visible.
- Sample I 3L Al: The surface was also slightly rough. On this surface, there is also visible microporosity, but in a smaller amount compared to the previous sample, and cracks emerge from the porosity, which are larger than those in the sample I 1L Al. However, there were visible cracks that were stable, and no flaking was present.

- Sample I 5L Al: The surface was very rough and contained large porosity that looked like bubbles.
- Sample I S 5L Al: The surface was slightly rough. It contained porosity that looked like bubbles, like the previous sample, but in a small amount. The cracks were smaller compared to those of sample I 3L Al and stable.
- Sample I S 5L Al: The surface was slightly rough, and it was different compared to that of the other sample coatings of this series. The surface was granular, without any porosity or cracks. We assume that this structure is not ideal, and from previous research, we can predict weak adhesion behavior.

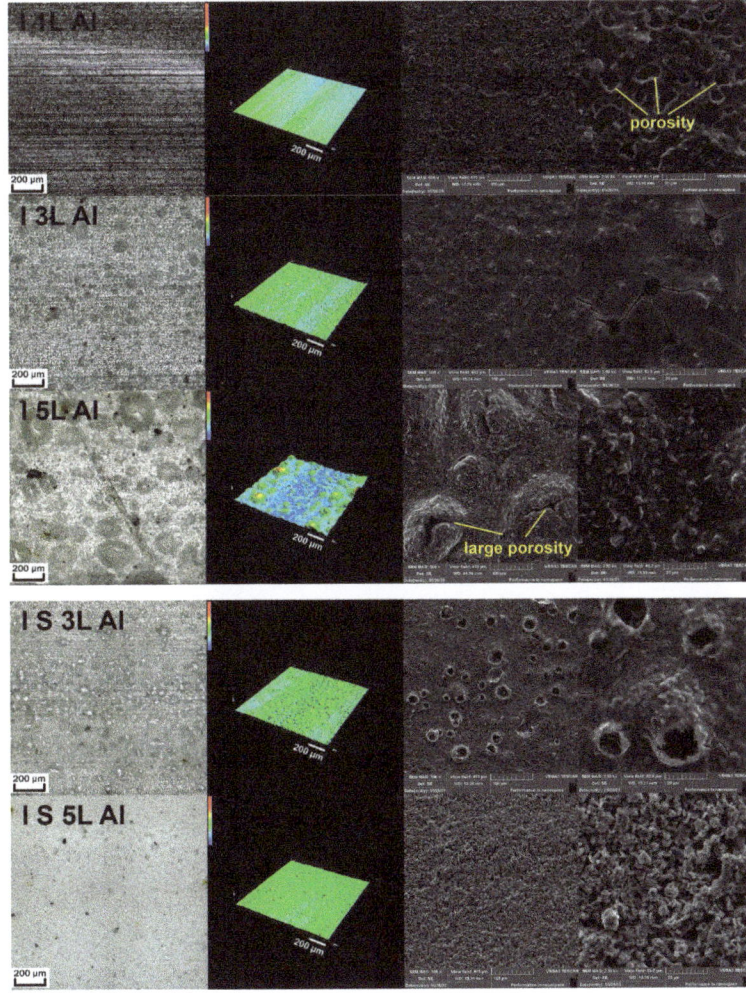

Figure 4. SEM and CLSM analysis of the surface of I geopolymer suspensions on the aluminum substrate.

Figure 4 shows the disparity in the surface structure between all of the analyzed samples. Even if they are coatings created from the same type of geopolymer suspension, which differ only in the number of created layers or in the curing method, the resulting surfaces are completely different. All coatings except for I S 5L Al exhibit some form of porosity. The I S 5L Al coating was granular and did not show any form of porosity. All coatings are stable, the surface shows no flaking and the cracks are minimal and distributed homogeneously over the entire surface.

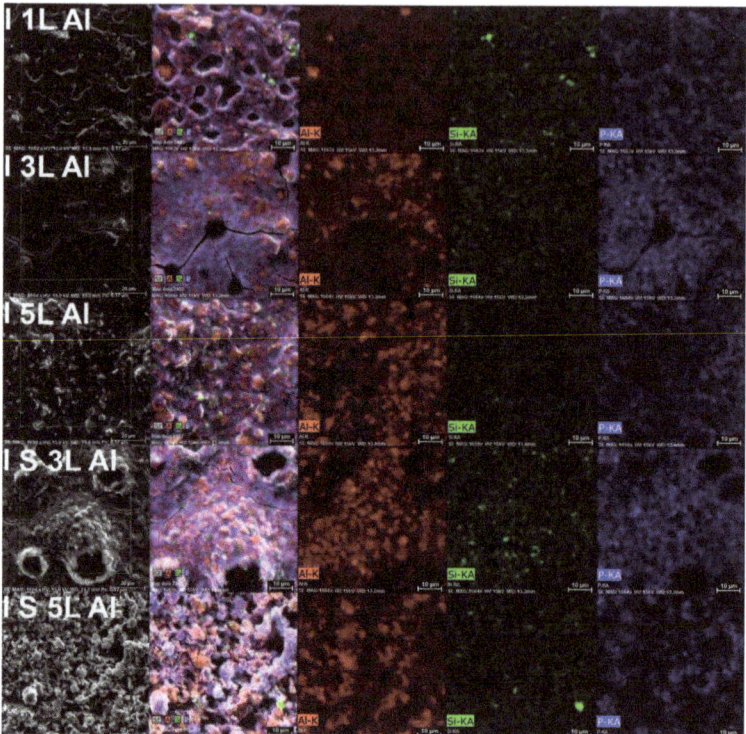

Figure 5. EDS mapping analysis of the surface of I geopolymer suspensions on the aluminum substrate.

The Figure 5 showed that except for the porosity positions, where, naturally, the presence of the elements analyzed using the EDS method was lower, it is evident that the elements Al, Si and P were distributed homogeneously, except for the sample I S 5L Al with a granular structure. It was observed that Al is more represented in the structure compared to Si, as the main elements of the geopolymer system.

3.1.2. Geopolymer I Fe

Figure 6 shows:
- Sample I 1L Fe: The coating surface of this sample is uniform, with a couple of small cracks. There is no visible porosity.
- Sample I 3L Fe: This surface is similar to sample I 1L Fe. It is evident that there is not much difference between a single-layer and multi-layer system, but we can observe that the cracks tend to coalesce to form long cracks.
- Sample I 5L Fe: This coating surface is the same as previous coatings on a steel substrate. However, it is completely without cracks.
- Sample I S 5L Fe: This sample is very different compared to the sample I 3L Fe. The surface is granular, with fine cracks.
- Sample I S 5L Fe: Even with this sample, the surface is different compared to that of the sample I 5L Fe, and a continuing trend can be seen with the layers applied in a different way (Figure 3). The layer is very fragmented and completely granular.

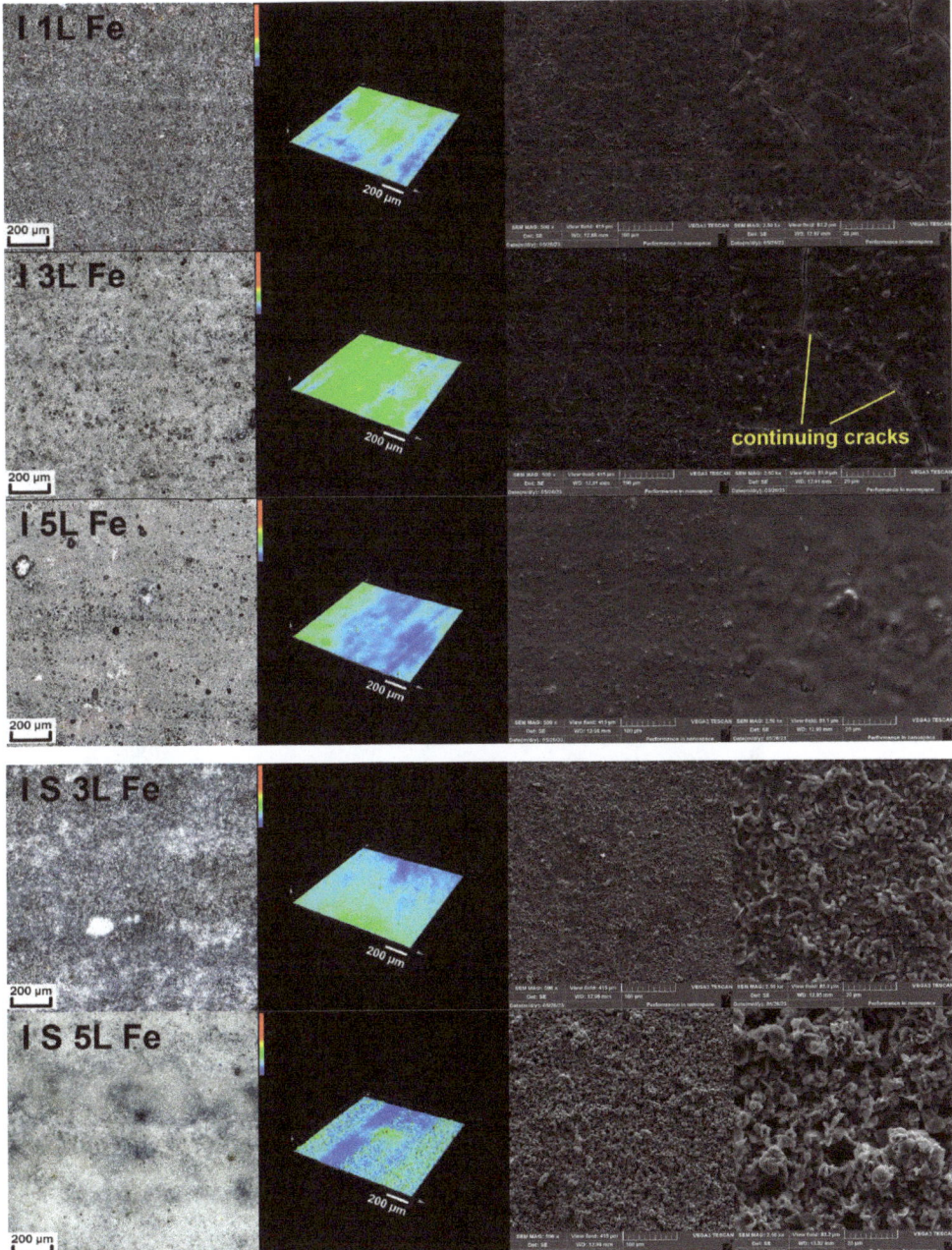

Figure 6. SEM and CLSM analysis of the surface of I geopolymer suspensions on the steel substrate.

In Figure 6, it was observed that no form of porosity occurs anymore in the coatings on the steel substrate. For the I S 5L Fe sample, a granular surface structure was again observed, just like the sample on the aluminum substrate in Figure 4. A lesser extent of the granular structure of the coating was also observed in the I S 5L Fe sample.

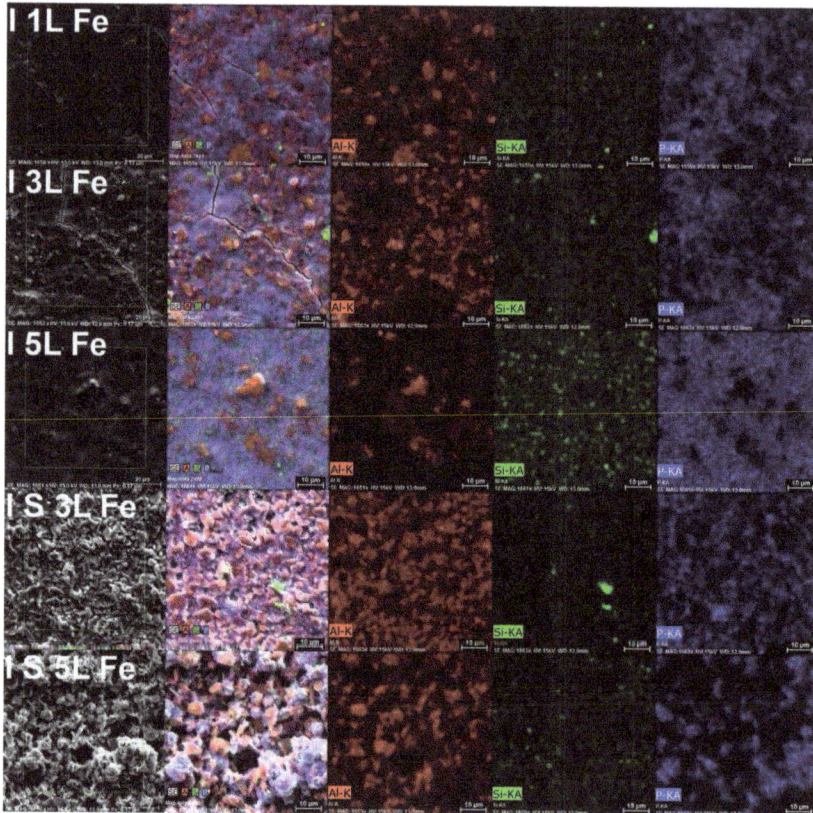

Figure 7. EDS mapping analysis of the surface of I geopolymer suspensions on the steel substrate.

Inhomogeneity was evident in all samples, even in samples without a granular structure. Figure 7 showed that the samples without a granular structure (I 1L Fe, I 3L Fe and I 5L Fe) have visible lumps on the surface, with an absence of P and a higher concentration of Al. The reason for this inhomogeneity was not studied, but it is probably a behavior of the geopolymerization and a property of the geopolymer. Application suspension by brush can also have an effect, since this phenomenon is also observed in the same suspension on the Al substrate, as described above. Si is distributed homogeneously.

3.1.3. Geopolymer J Al

Figure 8 shows the following:

- Sample J 1L Al: The surface had a very fine and smooth surface structure. It was stable, without visible cracks or flaking. There were visible lines after painting with a brush.
- Sample J 3L Al: There were also visible lines after painting with a brush; however, there were cracks on places with the thickest coating, which represents brighter places on the SEM picture.
- Sample J 5L Al: This surface was not as smooth as previous surfaces. There were visibly larger cracks than in the previous sample, which were visible mainly in places with a thicker layer of coating, but no flaking.
- Sample J S 3L Al: This surface was also not smooth. Cracks were visible.
- Sample J S 5L Al: The surface was slightly rough. On this surface, there were visible, large cracks. Here, there was also visible flaking for the first time.

Figure 8. SEM and CLSM analysis of the surface of J geopolymer suspensions on the aluminum substrate.

The GP J coatings on the aluminum substrate show different surface structures compared to the GP I coatings in Figure 4. In the GP J coatings, there were not any signs of porosity or granular structures. The J S 5L Al sample showed a considerable number of cracks. The same J 5L Al multi-layer sample, where the layers were cured sequentially, showed fewer cracks, and spalling was present only in the sample J S 5L Al.

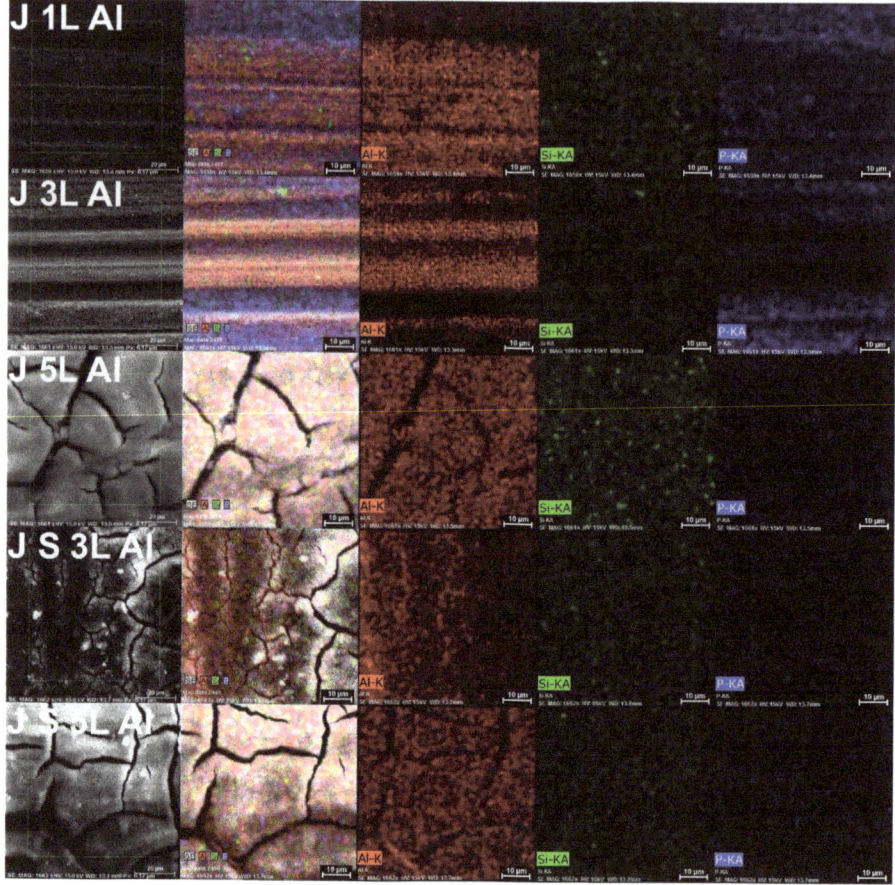

Figure 9. EDS mapping analysis of the surface of J geopolymer suspensions on the aluminum substrate.

The inhomogeneity of the distribution of Al and P elements which could be seen on Figure 9 was clearly visible on the J 1L Al and J 3L Al samples in the lines, which were created after the application of the geopolymer by brush. These lines were characteristic of a thinner layer of the geopolymer coating. The lines were enriched more with phosphorous, and a smaller presence of aluminum was evident.

3.1.4. Geopolymer J Fe

Figure 10 shows:

- Sample J 1L Fe: The surface had a very fine and smooth surface structure. Cracks were regular and visible only in the detail.
- Sample J 3L Fe: The surface was also very fine, with a smooth structure. Cracks were larger but without flaking.
- Sample J 5L Fe: The cracks of this multi-layer system were larger, without flaking.
- Sample J S 3L Fe: The cracks were extensive. No flaking was observed, but the space of the cracks was large and seemed unstable. More cracks were situated in the places with a thicker coating (after the brush application).
- Sample J S 5L Fe: The surface was rough. The cracks were extensive and homogeneously distributed over the surface, not only in places with a thicker coating.

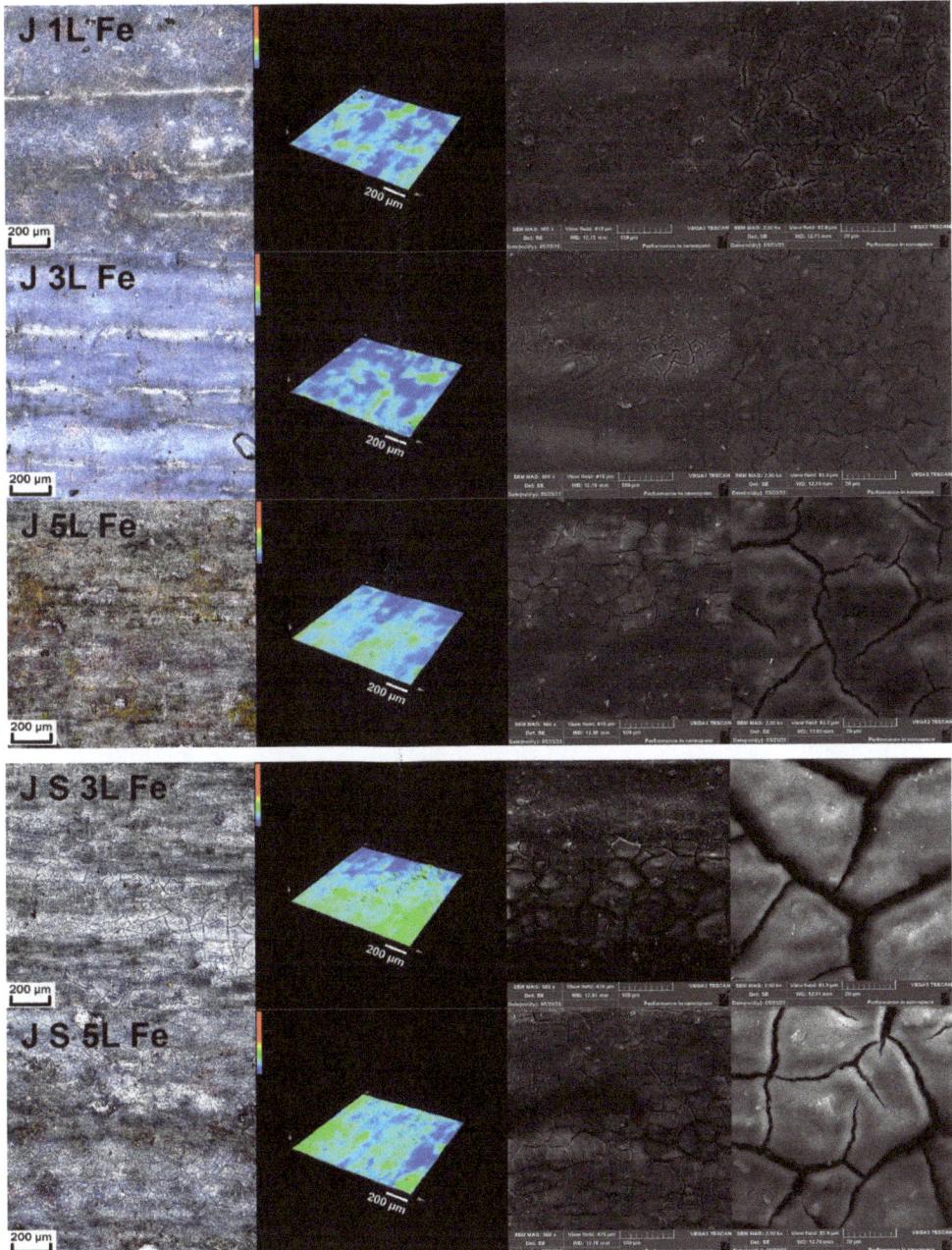

Figure 10. SEM and CLSM analysis of the surface of J geopolymer suspensions on the steel substrate.

Here, in Figure 10, cracks were observed in all of the measured samples. Their intensity increased gradually with the increasing number of layers. No porosity or granular structure of the coating was visible. The surfaces for all samples are very similar to the surfaces on the aluminum substrate.

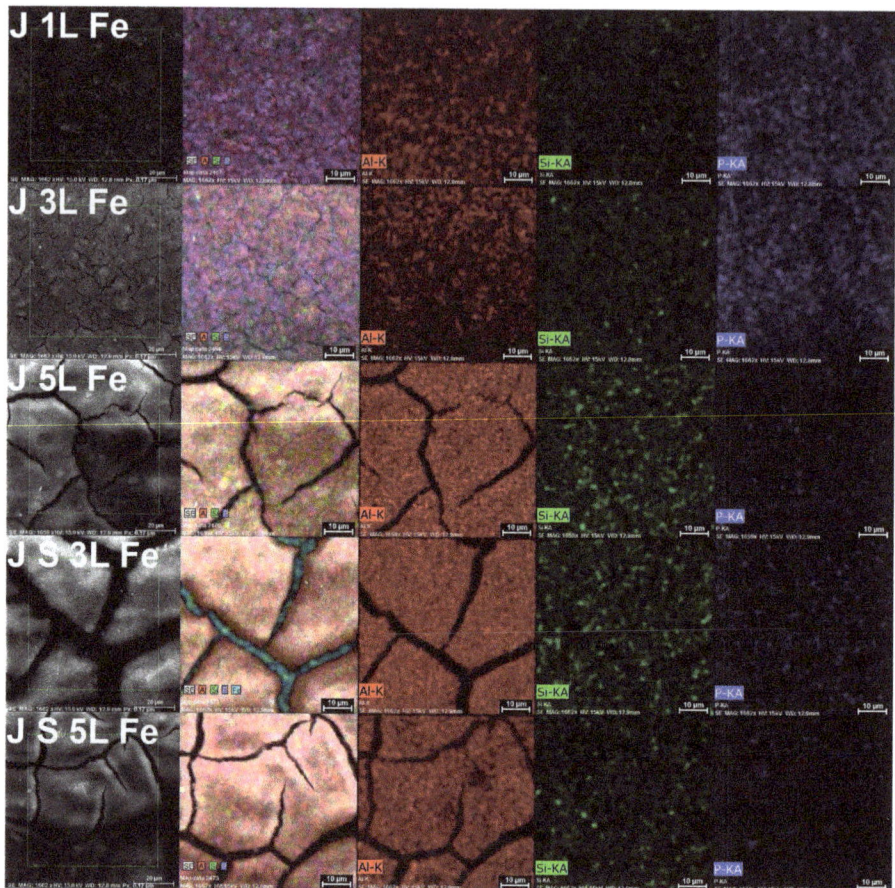

Figure 11. EDS mapping analysis of the surface of J geopolymer suspensions on the steel substrate.

EDS analysis seen on Figure 11 showed a homogeneous distribution of all elements. Homogeneity was the most pronounced of all the samples examined. No representation of elements was visible in the area of the cracks. The sample J S 3L Fe showed that, through the cracks, the steel substrate was visible and was in contact with atmospheric conditions. These coatings cannot provide full chemical protection of the substrate. This sample was chosen to present this phenomenon. The beam intensity of the SEM microscope could not penetrate the entire coating [37].

3.2. Analysis of the Adhesion of the Geopolymer Layer by the Grid Test

The achieved results of the grid test are shown in Figure 12, and the evaluation of the results is shown in Table 2. Table 2 shows the evaluated results of the grid test of various GP on Al and Fe substrates. GP I and J on the Al substrate achieved a rating of one, but only GP I S 5L had a rating of two, most likely due to too large of a layer of the GP suspension in combination with the method of application and curing. GP I on the Fe substrate achieved a worse rating of three, due to the presence of a natural oxide layer on the surface of the substrate and its peeling, together with the coating. GP J on the Al substrate achieved a rating of one for all samples and layers. GP J on the Fe substrate showed a better value of the grid test, with a rating of one for J 5L and a rating of two for J S 3L. The other GP J had a rating of three.

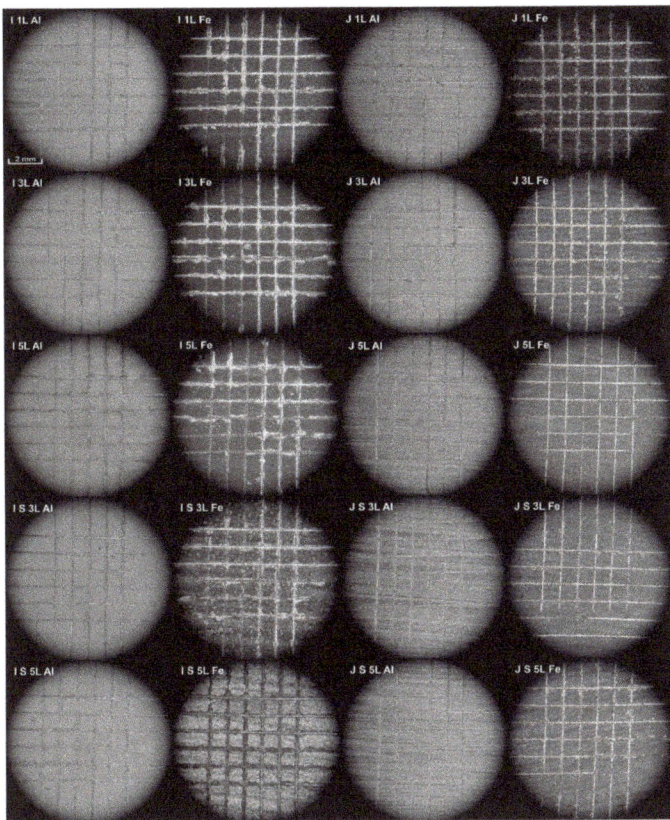

Figure 12. Grid test of GP coatings on the Al and Fe substrates.

Table 2. Grid test rates of GP coatings on the Al and Fe substrates.

Substrate	I 1L	I 3L	I 5L	I S 5L	I S 5L	J 1L	J 3L	J 5L	J S 3L	J S 5L
Al	1	1	1	1	2	1	1	1	1	1
Fe	3	3	3	3	3	3	3	1	2	3

The results of the GP I 1L Al and GP J 1L Al + Fe coatings corresponded with our previous research [23,24], but GP I 1L Fe had worse ratings than those in previous research. Due to the fact that no pre-treatment of the surface before the application of GP suspensions took place on the substrates (except for surface degreasing), there are natural oxides on both types of substrates. As can be seen from the grid test, the natural Al_2O_3 oxide layer on the Al substrate does not have a negative effect on the adhesion of the GP coatings to the surface. The Fe substrate is also covered with a natural oxide layer that is thicker than that on the Al substrate. According to the XRD analysis, it was found that Fe_3O_4, Fe_2O_3 and FeO oxides are found on the surface of the Fe substrate. The lower adhesion of GP coatings on the steel substrate is apparently caused by this oxide layer, which does not have high adhesion to the substrate itself (the steel layer located below this layer) and thus peels off from the surface together with the GP coating.

3.3. Analysis of the Thickness of the Geopolymer Layer

A comparison of the measured values of the thicknesses of all layers for GP I on the aluminum and steel substrates can be seen in Figure 13. Single-layer coatings reach almost the same thickness for both types of substrates (11% difference). The resulting thickness of I 1L Al is approximately 2.7 times higher than the thickness achieved in previous research [23]. All samples on both types of substrates show an almost linear increase in the layer thickness with an increasing number of layers. It can be seen from the graph that overall thicker layers were achieved on the aluminum substrate than on the steel substrate (about 20%–30%). What is interesting is the comparison of the multi-layer sample I 3L with I S 5L and that of I 5L with I S 5L, where a clear trend can be observed, where the samples cured after each layer reach an overall lower thickness than the samples cured after the last-applied layer.

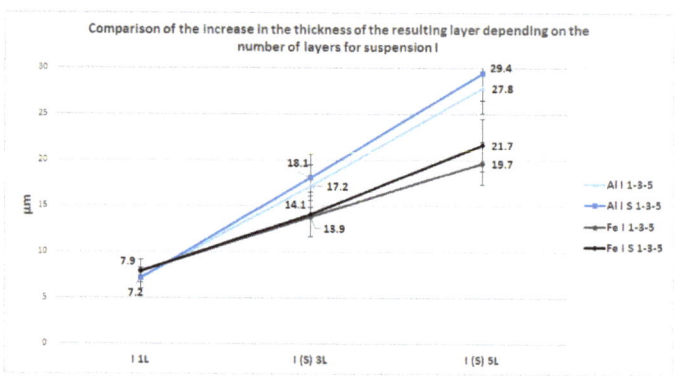

Figure 13. Comparison of the increase in the thickness of the resulting layer depending on the number of layers for suspension I.

The thicknesses of the GP J layers on both types of substrates are shown in Figure 14. As we can see, the J 1L Al single-layer sample reaches a thickness of 3.4 µm, which, as in the rare case of GP I, is an increase in thickness compared to previous research [23] by 2.3 times. Sample J 1L Fe has a greater thickness for this geopolymer, almost twice that of the aluminum substrate. All the layers on the Fe substrate reach greater thicknesses by about 70%–110% compared to the Al substrate, which is the opposite trend to that of GP I. The comparison of the multi-layer sample J 3L with J S 3L and that of J 5L with J S 5L again confirm the previous trend, where the samples cured after each layer reach a lower thickness than the samples cured after the last layer (except for J 3L and J S 3L on the Al substrate, where the thickness is almost identical).

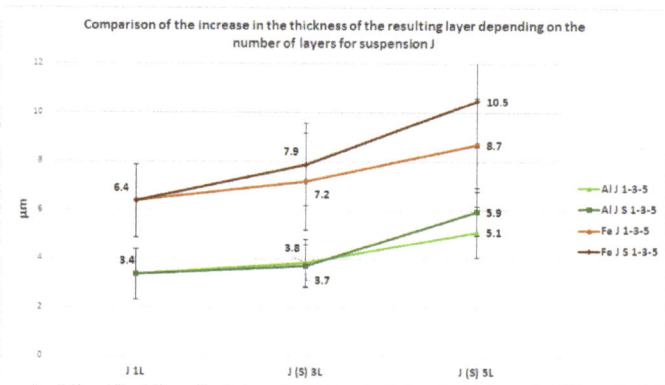

Figure 14. Comparison of the increase in the thickness of the resulting layer depending on the number of layers for suspension J.

4. Conclusions

The basis of this research was the creation of geopolymer coatings on aluminum and construction steel substrates, which were applied to the surface using a brush. A thicker layer of the suspension was already created on the substrates during the application itself, which was supposed to verify the properties and surface of the thicker layers created in this way, which follows previous research that, on the contrary, was focused on creating very thin layers [23,24]. Two types of geopolymer suspensions, I and J, were selected from the previous research. Furthermore, GP suspensions were applied in multiple layers, three and five, and two different methods of curing were used (see Section 2.2).

The formed coatings showed a certain porosity, and, above all, they are prone to the formation of cracks. These cracks are created naturally during curing by the emission of water from the volume and the different thermal expansion of the geopolymeric suspension and the underlying substrate. Cracks generally increased with the increasing number of layers. However, in most cases, the cracks did not appear to affect the cohesion of the coating or the adhesion of the coatings to the substrate surface. Thus, the porosity, rough surface and cracks do not have to reduce the resulting properties or the use of the coating, and in some cases, they can be beneficial. A rough or cracked surface can show better adhesion, e.g., when using glue in glued joints, when a larger surface is needed for the good adhesion of the joint [37]. A rough surface can help improve part handling and increase safety [38]. A jagged, rough or cracked surface change the optical properties of the surface so that, for example, there are no reflections of light from the surface [39]. Self-lubricating systems appear to be a very suitable application for this type of surface, where the surface of the part is provided with this coating, which contains many cracks and capillaries into which the lubricant is applied, and then gets between the functional surfaces and thus affects the tribological properties, e.g., by reducing friction, which leads to an increase in the life of the component and a reduction in the need for maintenance [40–42].

The application of suspensions with a brush is economical, but according to microscopic analysis, it is evident that it introduces a certain inhomogeneity into the coating, whether it is the fluctuating thickness of the layer, which then causes cracking, or the inhomogeneity in the distribution of Al, Si and P elements in the coating.

The grid test confirmed the very high adhesion of GP coatings on the aluminum substrate, independent of the thickness, the number of layers and the method of curing. On the steel substrate, the adhesion of the coatings was lower and partly dependent on the above-mentioned variables.

Observing the thickness of the layer on the underlying substrate is an important aspect. Thicker layers protect the underlying substrate better against corrosion and mechanical wear. As can be seen from the electron microscopy images, cracks on the surface increased in size with the increasing number of layers. Cracks of the underlying substrate are undesirable in the case of the application of GP anti-corrosion suspensions, as they reduce corrosion resistance (the corrosive environment with cracks can reach the underlying substrate). Large cracks are undesirable because they lead to the peeling of the GP layer from the underlying substrate, which was observed in some cases.

Finding a balance between the layer thickness and cracks is essential for future applications.

By applying a thicker layer of the GP suspension when applied with a brush, a final, thicker layer could be created, which, according to analyses, had no negative effect on the quality of the resulting surface. By applying additional layers, thicker layers could be created, which could affect other properties (mechanical, chemical) of the resulting surface. The joint thickness is further influenced by both the composition of the geopolymer suspension and the method of the application and curing of single layers in multi-layer samples.

To increase the homogeneity of the distribution of individual elements in the resulting layers and further improve the quality of the surface, in terms of, e.g., reducing the roughness or reducing the heterogeneity of the layer thickness in different places of the surface, which is caused by an application with a brush, and, thus, to reduce the formation, number and size of cracks and fissures, which are apparently caused by too thick of an applied suspension (again, in certain places), it would be necessary to change the method of application of the suspension. As a suitable solution to these inhomogeneities during the application, the suspension could be applied by spraying (air brush) or possibly even by means of a roller. These application methods (mainly the air brush) are more complicated and complex. Other forms of applications, e.g., dipping, are not suitable due to the need to maintain a very small thickness of the applied suspension. Moreover, the search for another application method contradicts the basic assumption of cheap and simple painting.

This research shows that it is possible to create multi-layer geopolymer coatings that achieve a good surface quality and adhesion on various underlying substrates while maintaining a relatively small thickness. This can be advantageous, for example, for functional components, where after the application of several layers of GP suspensions, there will be no large dimensional change, and the possible functionality is thus not affected. To further increase the quality of the surface, it would be advisable, for example, to change the method of application of the suspension in order to achieve a more even coverage of the surface of the substrate, which will ultimately affect the homogeneity of the resulting layers. The lower adhesion of the layers on the steel substrate could probably be solved by a suitable mechanical (or even chemical) pre-treatment of the surface, when the surface oxide layer is removed. Adhesion is excellent with the aluminum substrate, and the oxide layer does not seem to negatively affect adhesion. Furthermore, it is possible to focus on the final temperature when curing the layers. In this research, it was applied at a curing temperature of 170 °C, but it is possible to change this temperature further, which can affect the resulting mechanical properties of the applied layers. Lowering the resulting temperature and shortening the holding time will have a positive effect on production costs, but at the same time, the temperature must not be too low in order for the applied layers to properly geopolymerize. On the contrary, a higher temperature can further positively affect the mechanical properties of the layers, e.g., by increasing the surface hardness (mainly in the case of GP suspensions with Al_2O_3 content). Continuing that research and the previous research [23,24], the focus may be on the analysis of the mechanical properties of multilayer coatings, such as the microhardness of the surface of the layers or the tribological properties [40,42] (possibly the corrosion resistance of the coatings [43]).

Author Contributions: Conceptualization, M.J. and F.M.; Data curation, M.J. and J.N.; Formal analysis, F.M., J.M. and A.M.; Funding acquisition, M.J. and J.N.; Investigation, M.J., F.M. and J.M.; Methodology, M.J., F.M. and J.M.; Project administration, M.J. and F.M.; Resources, M.J. and F.M. and J.M.; Supervision, J.N.; Validation, M.J. and J.N.; Visualization, M.J., F.M. and J.M.; Writing—original draft, M.J., F.M. and J.M.; Writing—review and editing, M.J., F.M., J.M. and A.M. All authors have read and agreed to the published version of the manuscript.

Funding: This research was funded by NANOTECH ITI II. No. CZ.02.1.01/0.0/0.0/18_069/0010045 and the internal UJEP Grant Agency (UJEP-SGS-2022-48-002-2).

Data Availability Statement: Data sharing is not applicable to this article.

Conflicts of Interest: The authors declare no conflict of interest.

References

1. Matsimbe, J.; Dinka, M.; Olukanni, D.; Musonda, I. Geopolymer: A systematic review of methodologies. *Materials* **2022**, *15*, 6852. [CrossRef] [PubMed]
2. Rong, X.; Wang, Z.; Xing, X.; Zhao, L. Review on the Adhesion of Geopolymer Coatings. *ACS Omega* **2021**, *8*, 5108–5112. [CrossRef] [PubMed]
3. Singh, N.B.; Middendorf, B. Geopolymers as an alternative to Portland cement: An overview. *Constr. Build. Mater.* **2020**, *237*, 117455. [CrossRef]
4. Amritphale, S.S.; Bhardwaj, P.; Gupta, R. Advanced Geopolymerization Technology. In *Geopolymers and Other Geosynthetics*; IntechOpen: London, UK, 2020.
5. He, R.; Dai, N.; Wang, Z. Thermal and Mechanical Properties of Geopplymers Exposed to High Temperature: A Literature Review. *Adv. Civ. Eng.* **2020**, *2020*, 1–17.
6. Duxson, P.; Mallicoat, S.W.; Lukey, G.C.; Kriven, W.M.; Deventer, J.S.J.V. The effect of alkali and Si/Al ratio on the development of mechanical properties of me-takaolin-based geopolymers. *Colloids Surf. A Physicochem. Eng. Asp.* **2007**, *292*, 8–20. [CrossRef]

7. Duxson, P.; Lukey, G.C.; van Deventer, J.S.J. Thermal evolution of metakaolin geopolymers: Part 1—Physical evolution. *J. Non-Cryst. Solids* **2006**, *352*, 2186–2200. [CrossRef]
8. Ramasamy, S.; Abdullah, M.M.B.; Kamarudin, H.; Yue, H.; Jin, W. Improvement of Kaolin Based Geopolymer Coated Wood Substrates for Use in NaOH Molarity. *Mater. Sci. Forum* **2019**, *967*, 241–249.
9. Temuujin, J.; Minjigmaa, A.; Rickard, W.; Lee, M.; Williams, I.; Riessen, A. Fly ash based geopolymer thin coatings on metal substrates and its thermal evaluation. *J. Hazard. Mater.* **2010**, *180*, 748–752. [CrossRef]
10. Vickers, L.; Riessen, A.; Rickard, W.D.A. *Fire-Resistant Geopolymers: Role of Fibres and Fillers to Enhance Thermal Properties*; Briefs in Materials; Springer: New York, NY, USA, 2015; ISBN 978-981-287-310-1.
11. Zhang, Y.; Li, Z.; Sun, W.; Li, W. Setting and Hardening of Geopolymeric Cement Pastes In-corporated with Fly Ash. *ACI Mater. J.* **2009**, *106*, 405–412.
12. Duxson, P.; Fernández-Jiménez, A.; Provis, J.; Lukey, G.C.; Palomo, A.; Deventer, J.S.J.V. Geopolymer Technology: The Current State of the Art. *J. Mater. Sci.* **2007**, *42*, 2917–2933. [CrossRef]
13. Nergis, D.B.; Vizureanu, P.; Ardelean, I.; Sandu, A.; Corbu, O.; Matei, E. Revealing the Influence of Microparticles on Geopolymers' Synthesis and Porosity. *Materials* **2020**, *13*, 3211. [CrossRef]
14. Shuai, Q.; Xu, Z.; Yao, Z.; Chen, X.; Jiang, Z.; Peng, X.; An, R.; Li, Y.; Jiang, X.; Li, H. Fire resistance of phosphoric acid-based geopolymer foams fabricated from metakaolin and hydrogen peroxide. *Mater. Lett.* **2019**, *263*, 127228. [CrossRef]
15. Kouamo, H.T.; Rüscher, C.H. Mechanical and microstructural properties of metakaolin-based geopolymer cements from sodium waterglass and phosphoric acid solution as hardeners: A comparative study. *Appl. Clay Sci.* **2017**, *140*, 81–87.
16. Celerier, H.; Jouin, J.; Gharzouni, A.; Mathivet, V.; Sobrados, I.; Tessier-Doyen, N.; Rossignol, S. Relation between working properties and structural properties from 27Al, 29Si and 31P NMR and XRD of acid-based geopolymers from 25 to 1000 °C. *Mater. Chem. Phys.* **2019**, *228*, 293–302. [CrossRef]
17. Wang, H.; Li, H.; Yan, F. Synthesis and tribological behavior of etakaolinite-based geopolymer composites. *Mater. Lett.* **2005**, *59*, 3976–3981. [CrossRef]
18. Troconis, B.C.R.; Frankel, G.S. Effect of Roughness and Surface Topography on Adhesion of PVB to AA2024-T3 using the Blister Test. *Surf. Coat. Technol.* **2013**, *236*, 531–539. [CrossRef]
19. Hogmark, S.; Hedenqvist, P.; Jacobson, S. Tribological properties of thin hard coatings—Demands and evaluation. *Surf. Coat. Technol.* **1997**, *90*, 247–257. [CrossRef]
20. Jadon, V.K.; Kumar, S. Effect of substrate surface conditions on tribological behaviour of machine element coating. *Aust. J. Mech. Eng.* **2020**, *20*, 1000–1007.
21. Hogmark, S.; Jacobson, S.; Larsson, M. Design and evaluation of tribological coatings. *Wear* **2000**, *246*, 20–33. [CrossRef]
22. Mao, Y.; Biasetto, L.; Colombo, P. Metakaolin-based geopolymer coatings on metals by airbrush spray deposition. *J. Coat. Technol. Res.* **2020**, *17*, 991–1002. [CrossRef]
23. Novotny, J.; Jaskevic, M.; Mamon, F.; Mares, J.; Houska, P. Manufacture and Characterization of Geopolymer Coatings Deposited from Suspensions on Aluminium Substrates. *Coatings* **2022**, *12*, 1695. [CrossRef]
24. Mares, J.; Mamon, F.; Jaskevic, M.; Novotny, J. Adhesion of Various Geopolymers Coatings on Metal Substrates. *Manuf. Technol.* **2023**, *23*, 81–87. [CrossRef]
25. *AW, E.N. 6060*; AlMgSi. AlMgSi. Vydavatelství Úřadu pro Normalizaci a Měření: Prague, Czech Republic, 1978.
26. ALLOY DATA SHEET EN-AW 6060. Available online: N.A.2019/07/NEDAL_Datasheet-6060.pdf (accessed on 25 October 2022).
27. Kanaboyana, N.; Hanchate, S.R.; Ghorpade, V.G. Durability properties of geopolymer concrete produced with recycled coarse aggregates and quarry stone dust. *Nat. Volatiles Essent* **2021**, *8*, 10450–10459.
28. Ruzaidi, C.; Al Bakri, A.; Binhusain, M.; Salwa, M.; Alida, A.; Faheem, M.; Azlin, S. Study on Properties and Morphology of Kaolin Based Geopolymer Coating on Clay Substrates. *Key Eng. Mater.* **2013**, *594*, 540–545. [CrossRef]
29. Zainal, F.; Fazill, M.; Hussin, K.; Rahmat, A.M.; Al Bakri, A.; Wazien, W. Effect of Geopolymer Coating on Mild. *Solid State Phenom.* **2018**, *273*, 175–180. [CrossRef]
30. Zhang, X.; Yao, A.; Chen, L. A Review on the Immobilization of Heavy Metals with Geopolymers. *Adv. Mater. Res.* **2013**, *634*, 173–177. [CrossRef]
31. Lingyu, T.; Dongpo, H.; Jianing, Z.; Hongguang, W. Durability of geopolymers and geopolymers concretes: A review. *Rev. Adv. Mater. Sci.* **2021**, *60*, 1–14. [CrossRef]
32. Bhardwaj, P.; Gupta, R.; Deshmukh, K.; Mishra, D. Optimization studies and characterization of advanced geopolymer coatings for the fabrication of mild steel substrate by spin coating technique. *Indian J. Chem. Technol.* **2021**, *28*, 59–67.
33. Jiang, C.; Wang, A.; Bao, X.; Chen, Z.; Ni, T.; Wang, Z. Protective Geopolymer Coatings Containing Multi-Componential Precursors: Preparation and Basic Properties Characterization. *Materials* **2020**, *13*, 3448. [CrossRef]
34. Zhu, C.; Guo, Y.; Wen, Z.; Zhou, Y.; Zhang, L.; Wang, Z.; Fang, Y.; Long, W. Hydrophobic Modification of a Slag-based Geopolymer Coating. *IOP Conf. Ser. Earth Environ. Sci.* **2021**, *783*, 012015. [CrossRef]
35. Gupta, R.; Tomar, A.S.; Mishra, D.; Sanghi, S.K. Multifaceted geopolymer coating: Material development, characterization and study of long term anti-corrosive properties. *Microporous Mesoporous Mater.* **2021**, *317*, 110995. [CrossRef]
36. *ISO 2409:2020(en)*; Paints and Varnishes-Pull of Test for Adhesion. ISO: Geneva, Switzerland, 2020.
37. Rieutord, F.; Moriceau, H.; Beneyton, R.; Capello, L.; Morales, C.; Charvet, A.-M. Rough Surface Adhesion Mechanisms for Wafer Bonding. *Electrochem. Soc.* **2006**, *3*, 205.

38. Beketov, A.; Khalimova, S. Impact of Roughness and Friction Properties of Road Surface of Urban Streets on the Traffic Safety. *Commun. Sci. Lett. Univ. Zilina* **2023**, *25*, 51–63. [CrossRef]
39. Kohli, J.T.; Nguyen, K.; Zhang, L. Anti-Glare Surface and Method of Making. US8992786B2,09.04.2015.
40. Kumar, A.; Kumar, M.; Tailor, S. Self-lubricating composite coatings: A review of deposition techniques and material advancement. *Mater. Today Proc.* **2023**, *11*, 302. [CrossRef]
41. Zhang, G.; Cai, W.; Wei, X.; Yin, Y. Percolation and Supply Behavior of Lubricant on Porous Self-Lubricating Material. *Adv. Eng. Mater.* **2023**, *25*, 12. [CrossRef]
42. Wang, X.L.; Yang, L.Y.; Wang, S. Research and Development of Self-Lubricating Bearing Materials. *Adv. Mater. Res.* **2013**, *651*, 198–203. [CrossRef]
43. Omer, L.; Gomaa, M.; Sufe, W.H.; Elsayed, A.A.; Elghazaly, H.A. Enhancing corrosion resistance of RC pipes using geopolymer mixes when subjected to aggressive environment. *J. Eng. Appl. Sci.* **2022**, *69*, 3. [CrossRef]

Disclaimer/Publisher's Note: The statements, opinions and data contained in all publications are solely those of the individual author(s) and contributor(s) and not of MDPI and/or the editor(s). MDPI and/or the editor(s) disclaim responsibility for any injury to people or property resulting from any ideas, methods, instructions or products referred to in the content.

Article

Nanostructures and Thin Films of Poly(Ethylene Glycol)-Based Surfactants and Polystyrene Nanocolloid Particles on Mica: An Atomic Force Microscopy Study

John Walker [1], Andrew B. Schofield [2] and Vasileios Koutsos [1,*]

[1] School of Engineering, Institute for Materials and Processes, The University of Edinburgh, Robert Stevenson Road, Edinburgh EH9 3FB, UK
[2] SUPA, School of Physics and Astronomy, The University of Edinburgh, Edinburgh EH9 3FD, UK; a.b.schofield@ed.ac.uk
* Correspondence: vasileios.koutsos@ed.ac.uk; Tel.: +44-(0)131-6508704

Abstract: We studied the nanostructures and ultrathin films resulting from the deposition and adsorption of polystyrene nanocolloidal particles and methoxy poly(ethylene glycol) methacrylate surfactants on mica surfaces from mixed suspensions in water. The samples were prepared by droplet evaporation and dip coating and imaged with atomic force microscopy. Topography and phase imaging revealed a significant richness in morphological features of the deposited/adsorbed films. We observed uniform ultrathin films and extended islands of the surfactant oligomers indicating their self-assembly in monolayers and multilayers, while the polystyrene nanocolloids were embedded within the surfactant structures. Droplet evaporation resulted in the migration of particles towards the edges of the droplet leaving an intricate network of imprints within the surfactant film. Dip coating induced the formation of extended nanocolloid clusters with colloidal crystalline structuring.

Keywords: nanocolloids; nanoparticles; nanostructures; monolayers; polymers; surfaces; thin films; mica; poly(ethylene-glycol); atomic force microscopy

1. Introduction

The self-assembly of polymers on surfaces has emerged as an elegant and inexpensive method to form nanostructures and thin films on surfaces for various applications, from functional devices [1] and nanophotonics [2] to smart nanostructured materials [3] and biomedicine [4]. Various polymer architectures have been employed including homopolymers [5], block-copolymers [6–8], random copolymers [9,10], hyperbranched polymers [11], and star polymers [12,13]. One area that is still relatively unexplored is the self-assembly of polymeric colloidal [14,15] and in particular nanocolloidal particles [16] on surfaces and the formation of associated nanostructures and thin films. This has the potential to achieve nanostructures and nanopatterns for advanced products such as high-density, defect-free data storage devices, sensors, and biochips made for a substantially lower cost than at the present and using more environmentally friendly materials and processes (suspensions of particles in water) [17–20]. In order for these colloidal particles to maintain their dispersion in a liquid suspension and not sediment out by flocculation, they must have repulsive interactions to counteract the ubiquitous attractive van der Waals forces. Colloidal particles are stabilized against this attraction either by electrostatic repulsion from surface charges [21] or with steric repulsions originating from anchored polymers or surfactants [22]. While an electrostatic interaction provides the basis of colloidal stability [23,24] by repulsion between like surface charges, it is limited because it can be used only with polar liquid mediums and it is susceptible to flocculations with extremes in pH, salt concentration, and temperature. Surfactant-based colloidal particles can be produced to provide the necessary interparticle repulsive forces in either polar or apolar mediums depending on the particular application such as targeted drug delivery [25,26] and smart biosensors based on selective surfactant films [27]. One such surfactant is methoxy poly(ethylene glycol) methacrylate (MeOPEGMA). It consists mainly of poly(ethylene glycol) (PEG) or poly(ethylene oxide) (PEO); both terms signify the same chemical, but historically PEG has tended to refer to oligomers and polymers with a molecular mass below 20,000 g/mol and PEO to polymers with a molecular mass above 20,000 g/mol. PEG is widely used as a stabilizing agent in colloid water suspensions due to its water-solubility and its low toxicity, which make it an ideal candidate for biomedical applications [28]. Functionalisation of a colloidal surface by PEG can be achieved by combining the PEG with a suitable co-monomer which attaches to the colloidal surface, anchoring the PEG chain and thus creating a polymeric layer of steric repulsion.

The advent of scanning probe microscopy (SPM) techniques such as atomic force microscopy (AFM) has given us the ability to image and manipulate nanometre-sized particles. Apart from the high-resolution topographical imaging of a material surface, AFM [29] can also provide information on the physical properties of the material under investigation in the form of phase information [30,31] due to its direct interaction with the surface at the atomic level. The phase image contrast originates from the material's mechanical and adhesive properties, facilitating an indirect distinction between different materials/chemical species on the surface of a sample that would otherwise be undetected using topography imaging alone.

In this work, suspensions of polystyrene nanocolloidal particles copolymerised with methoxy poly(ethylene glycol) methacrylate (MeOPEGMA) surfactants were used. We deposited/adsorbed the suspension material onto mica by droplet evaporation and dip coating. In both cases, nanostructures, (sub-)monolayers, and ultrathin films were formed and studied with AFM topography and phase imaging in air. We observed the formation of highly ordered crystalline structuring of the polystyrene nanocolloids and evidence of self-assembled ultrathin films of the surfactant which originated from the nanocolloidal suspension as suspended excess material. To the best of our knowledge this is the first time that such a system has been investigated and the results we present are of an exploratory nature.

2. Experimental

The colloidal particles are made from the polymerisation of styrene molecules with a non-ionic co-monomer/stabilizer, methoxy poly(ethylene glycol) methacrylate (MeOPEGMA), in a process outlined by Ottewill et al. [32]. The MeOPEGMA polymer consists of a 2000 g/mol molecular weight PEG terminated on one end with a methoxy group. On the other end is a methyl methacrylate group which co-polymerises with the styrene and chemically links the PEG stabilizer to the particles. This results in the formation of a colloid particle with a polystyrene core and PEG polymer on its surface, which provides steric repulsion against aggregation/flocculation (PS-PEG). A 1 litre, 3-necked round-bottom flask was taken and into the 3 inlets were placed a water-cooled condenser, a nitrogen gas source, and a stirrer with a PTFE blade which was rotated at 350 rpm. The flask was placed in an oil bath which was kept at 60 °C to maintain temperature equilibrium. Into the flask were poured 475 mL of distilled water, 40.04 g (0.77 mol/dm^3) of distilled styrene, and 6.82 g of a 60% MeOPEGMA solution in water (4.1×10^{-3} mol/dm^3), which had been obtained from the Ottewill laboratory. This mixture was stirred for 20 min to allow it to come to temperature equilibrium with the oil bath. Whilst this was happening, in a separate vial the initiator solution of 0.098 g (1.12×10^{-3} mol/dm^3) of ascorbic acid, 0.22 g (3.56×10^{-3} mol/dm^3) of a 27.5% solution of hydrogen peroxide, and 22.1 g of distilled water was mixed. Once the 20 min equilibrium time had elapsed the initiator solution was added to the flask and the chemical reaction started and it was allowed to run for 6 h. The final latex particles were cleaned by centrifugation and resuspension in clean, distilled water. This process was repeated several times. The average particle diameter was measured to be 41 ± 10 nm by dynamic light scattering (DLS).

The samples for imaging were prepared in two different manners: droplet evaporation and dip coating. Mica sheets (Fisher), pre-cut to 11 mm squares, were freshly cleaved using a scalpel along the lateral plane of the sheet inside a fume cupboard to minimise airborne pollutants from contaminating the surface. The colloidal suspensions were agitated for 30 s to ensure the minimisation of any sedimentation that may have occurred. Droplet samples were created by applying 0.16 mL of the suspension using a pipette onto the freshly cleaved mica surface. The liquid droplet completely covered the mica surface and its contact line was 'pinned' onto the mica edge. Subsequently, the sample was placed into a drying box at room temperature and pressure, in an effort to avoid heterogeneous particles from contaminating the sample. These were left until the droplet had visibly evaporated, after which the sample was placed in a glass covered Petri dish and moved into an oven at 60 °C for 1 h to ensure that any residual suspension liquid had evaporated. The dip-coated samples were prepared by submersion of the freshly-cleaved mica into a vial of the colloidal suspension in the upright position and leaving it for a predetermined incubation time. The sample was then removed and rinsed gently with deionised water followed by drying from a nitrogen jet. The sample was then placed in a glass covered Petri dish and moved into an oven at 60 °C for 1 h to ensure complete drying. If the samples were not imaged immediately after preparation, they were sealed with parafilm until imaging.

Imaging was performed using a molecular imaging PicoSPM AFM (Agilent Technologies, Santa Clara, CA, USA) operating in tapping mode (intermediate contact with the sample) using MikroMasch tips with a nominal spring constant and resonant frequency of 40 N/m and 170 kHz, respectively. The nominal tip radius was quoted as 10 nm. We operated in a light-tapping mode, where the contact amplitude approached the free oscillation amplitude of the cantilever, to minimise tip–sample interaction forces. The cantilevers were oscillated at 5% below their natural resonance frequency. For post-processing of the AFM images, the scanning probe image processor (SPIP, version 5.1.0, Image Metrology) was used. All images presented are representative of the corresponding preparation protocol and were post-processed by simple flattening alone, unless otherwise stated.

3. Results

Droplet evaporation: On imaging a droplet-prepared sample near its centre, we were presented with a topography of islands of various sizes with fractal-like edges, surrounded by an interconnected series of channels (Figure 1). The phase image shows a very high contrast between the islands and the surrounding network, indicating that there is a difference in the mechanical/adhesive properties between the two, i.e., indicating a different material. A height profile of the surface (Figure 1d) reveals a uniform island thickness of approximately 13 nm. Figure 1 shows typical AFM data from a scan area towards the centre of the sample. Moving towards the periphery of the sample, we start to observe topographical changes.

In Figure 2, we can see that the topography shows the existence of particles towards the top of the scan image. We also observe the same network of interconnected channels previously observed in Figure 1. The middle of the sample exhibits some unusual trail lines, which are clearly present on the phase image. The depth of these lines is approximately 13 nm. Towards the bottom of the topography image we see some isolated holes. The width of the isolated holes and trail lines are in the range of 40 nm.

Dip coating—incubation time 1 h: Figure 3 shows a typical example of an AFM image for a dip-coated sample for an incubation time of 1 h. We observe the presence of numerous clusters of the particles, with varying heights across the surface. We also present a selection of profiles taken across the topography image giving an indication that their heights have some preferred values: the taller clusters are approximately twice the height of the lower clusters. This is clearly demonstrated by frequency histograms of the z-heights (Figure 4) obtained by grain analysis of the cluster formations. It shows that the maximal z-height ranges have peaks at 30 nm and 60–80 nm. The area coverage has a peak value of approximately 10^5 nm^2.

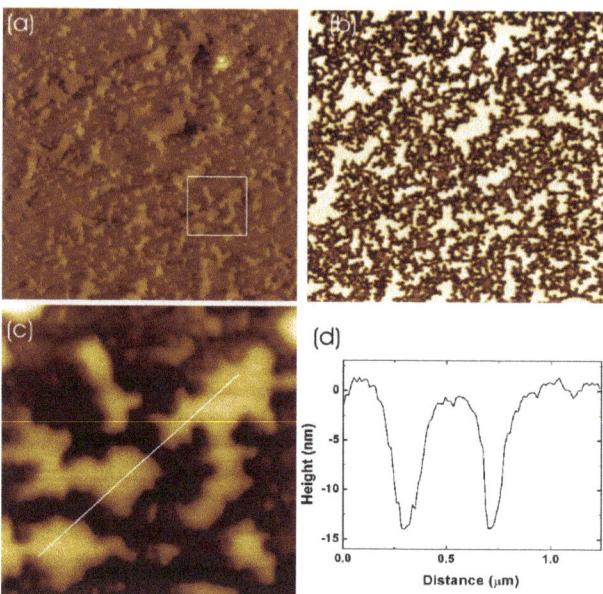

Figure 1. (a) 6.6 × 6.6 μm² AFM topography scan of the droplet-prepared sample (z-height scale ≈ 50 nm) with (b) associated phase image (voltage range 4.87 V), and (c) magnified section with (d) height profile.

Figure 2. 6.6 × 6.6 μm² AFM topography scan of the droplet-prepared sample (**left**) with its corresponding phase image (**right**) (z-height scale ≈ 70 nm and phase voltage scale ≈ 5.2 V).

Dip coating—incubation time 72 h: Increasing the submersion time to 72 h resulted in clearly visible regular nanocolloid arrangements on the mica substrate of various lateral sizes (Figure 5). We can also observe that the surrounding areas of the particle clusters have an extended border of uniform height. This layer has been measured to be approximately 12 nm thick (Figure 6) and can be found surrounding all particle adsorption sites. The phase image suggests that there is a distinct difference between the physical properties of this plateau layer and the underlying substrate. The structuring of the particles has been observed to be highly crystalline (Figure 6c). The particles are in a close-packed hexagonal arrangement leading to very well ordered structure, as indicated by fast Fourier transform (FFT). A characteristic length of 57 nm between the particles was evaluated from the FFT graph.

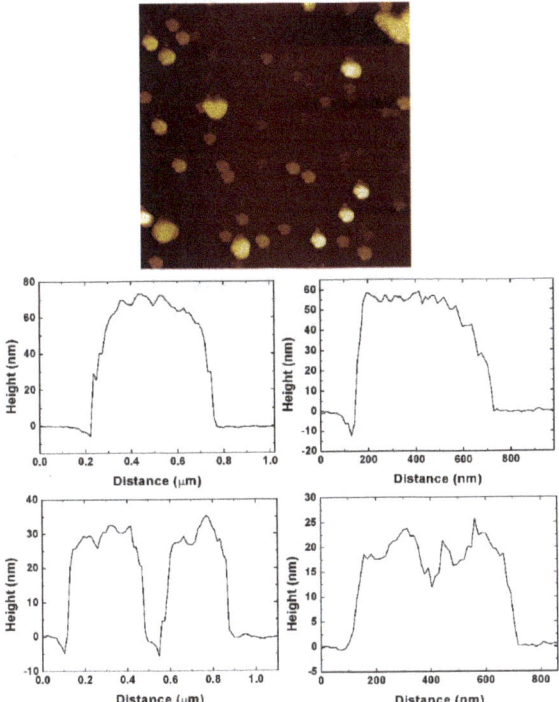

Figure 3. 6.6 × 6.6 µm² AFM topography scan of the 1 h dip-coated sample (**top**) (z-height scale ≈ 80 nm) with representative height profiles of several colloidal clusters showing their relative height to the surrounding surface (**bottom**).

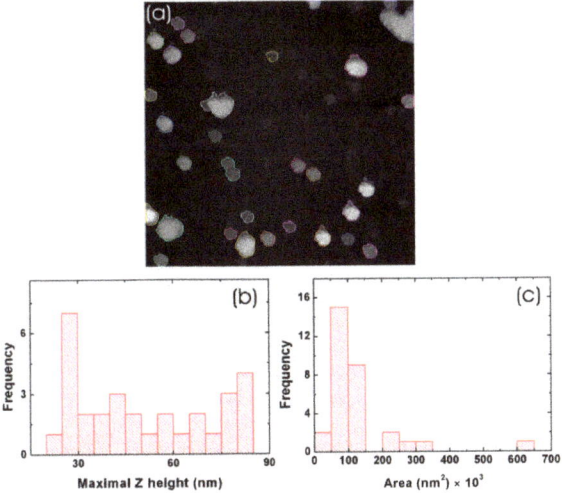

Figure 4. (**a**) Grain analysis of the 6.6 × 6.6 µm² topography scan of the sample submerged for 1 h in PS–PEG. (**b**) Maximal z-height distribution relative to the surrounding surface. (**c**) Cluster area coverage distribution.

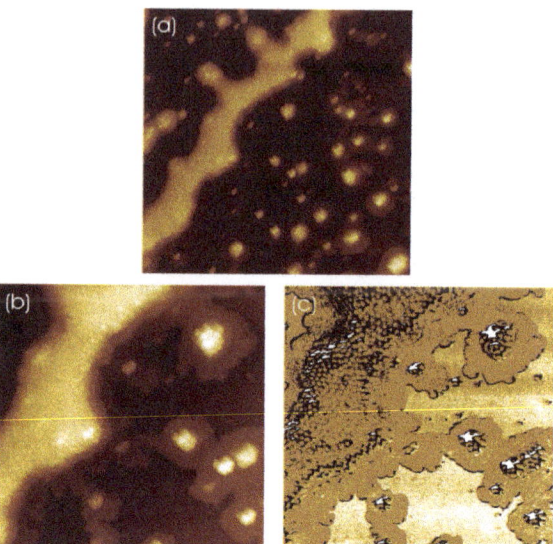

Figure 5. (a) 6.6 × 6.6 μm² AFM topography scan; (b) 3 × 3 μm² zoom scan and (c) its corresponding phase image showing high contrast for a sample submerged for 72 h in the suspension (z-height scale ≈ 100 nm for both topography scans, phase voltage scale ≈ 4.7 V).

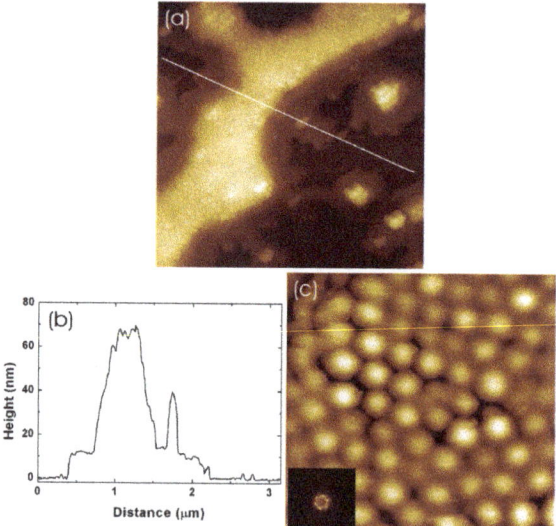

Figure 6. (a) 3 × 3 μm² AFM topography scan (z-height scale ≈ 100 nm); (b) height profile (indicated by white line on image) showing a relative height of approx. 70 nm for the large structure, approx. 40 nm for a small isolated cluster, and a surrounding plateau layer of approx. 12 nm. (c) Zoom of larger structure (image sharpened) showing hexagonal crystalline structuring of the particles along with (insert) its 2D FFT.

Dip coating—incubation time 240 h: After a submersion time of 10 days, scattered protrusions across the topography image were observed (Figure 7). Height profiles of the protrusions relative to the surrounding surface give a range of elevations. Grain analysis

(Figure 8) of the topography images shows that there are well-defined peaks in the relative z-heights of the protrusions at approximately 15, 25, 40, and 70 nm.

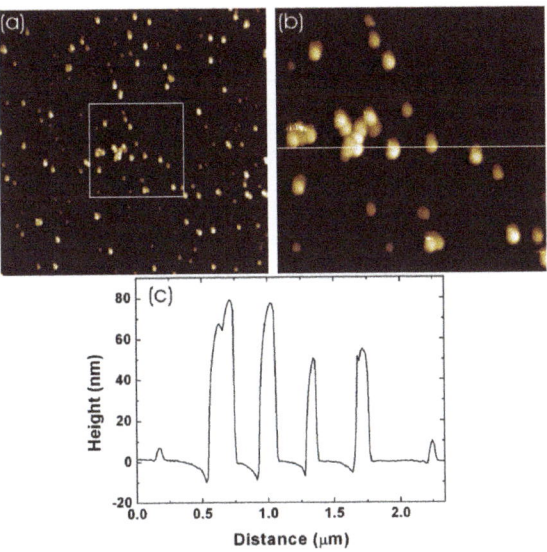

Figure 7. (a) 6.6 × 6.6 μm² AFM topography scan of the sample submerged for 240 h in the suspension (z-height scale ≈ 80 nm). (b) Zoom section and (c) corresponding height profile (indicated by white line on image).

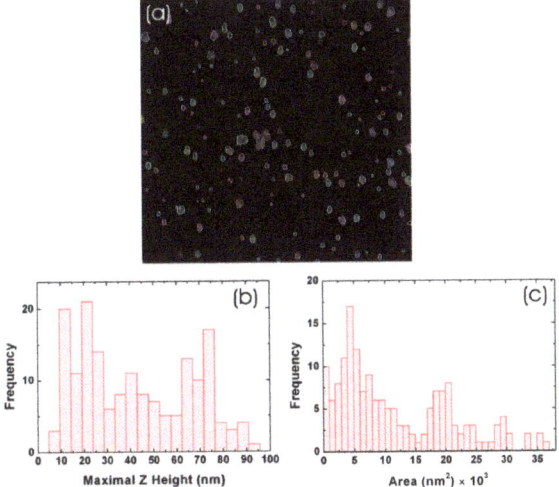

Figure 8. (a) Grain analysis of the 6.6 × 6.6 μm² AFM topography scan of the sample submerged for 240 h in the suspension; (b) z-height frequency histogram relative to the surrounding surface and (c) area coverage frequency histogram of protrusions.

The area coverage of the protrusions shows peaks at 5×10^3, 19×10^3, and 29×10^3 nm². As we would have expected, increased submersion times create higher coverage of material on this sample than seen in the 72 h dip-coating experiments. We surmise that some sort of film layer has formed on the surface of the substrate of the 240 h samples. To investigate further, the tip was raised and the AFM reconfigured for contact mode. The set-point voltage to determine the tip contact force was raised to provide a sufficiently high enough cantilever tip-point force on the surface of the sample that would start to penetrate the film layer. The tip then scanned a square raster of approx. 1×1 μm^2 for 30 min that would effectively "dig" a hole into the surface through the film layer. The tip was then withdrawn and a tapping mode scan carried out over 4×4 μm^2 to image the dig site. As is shown in Figure 9, the dig area is clearly marked and appears to have material built up on either side. The hole is quite clean and well defined. Measurements (Figure 9d) determined the depth of the hole to be approximately 32 nm over the majority of its area. The material to the sides varies considerably in height, spanning 100–300 nm. In contrast, the material further away from the hole that was present before the digging is approximately 70 nm in height (relative to the bottom of the dig site), which is in agreement with the larger protrusions observed in Figure 7. It is interesting to note that the phase information in Figure 9 suggests that the film may be even thicker than what we dug out, as there is no contrast and thus little discernable difference in the physical properties between the surface and the bottom of the dig site, indicating the same material.

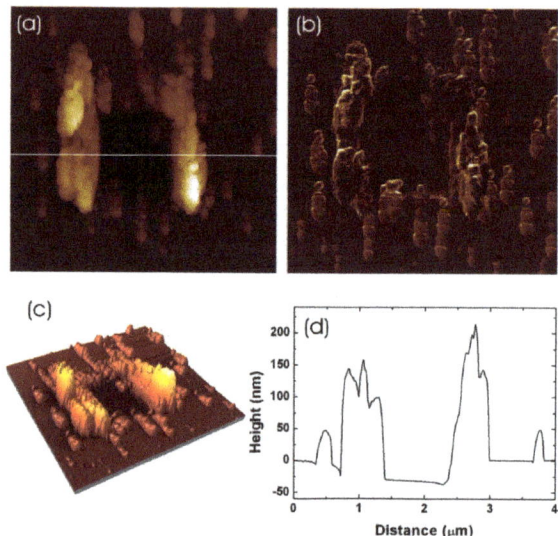

Figure 9. (a) 4×4 μm^2 AFM topography scan of dig site on 240 h submersion sample; (b) phase image; (c) 3D rendering of the topography and (d) height profile of the dig site as indicated by white line on the topography image (z-height scale ≈ 300 nm, phase voltage scale ≈ 4.6 V).

4. Discussion

Our observations (Figure 1) suggest that a uniform ultrathin film of the excess PEG-based surfactant in the colloidal suspension was formed on the mica surface during the droplet evaporation. The surfactant islands' borders form an intricate pattern of circular domains of the size of the nanocolloidal particles. It seems that both PS-PEG nanocolloids and PEG-based surfactant molecules were initially deposited on the surface but upon the slow evaporation of the droplet the subsequent convection current flows within the droplet induced desorption and redeposition of the particles towards the periphery of the sample (Figure 2). This is aided by the steric repulsion that exists between the surfactant-coated mica substrate and the PS-PEG nanocolloids. Additional evidence for this hypothesis comes from the trail lines appearing towards the perimeter of the sample (Figure 2). These lines could be caused by the PS-PEG particle being dragged as opposed to lifted due to the thinning of the droplet's contact line as it nears complete evaporation of the liquid. The existence of particles at the perimeter also suggests mass transport within the droplet.

The initiation of convection flows within the droplet as it began to evaporate caused the nanocolloids that were initially deposited on the sample surface to be lifted or dragged along to the extremities. The mechanism of movement of suspended material towards the periphery is well documented in other experimental work as the "coffee stain effect" [33], where a pinned droplet will stream liquid towards the perimeter in order to replace the evaporating liquid carrying within it any particulates in suspension [34].

The apparent thickness of the thin film is approximately 13 nm, which is compatible with a dense self-assembled monolayer (SAM) of the MeOPEGMA oligomer that has a contour length of about 16 nm [35]. We note that we have imaged in ambient conditions; due to the humidity in the environment and to the fact that both mica and PEG have a high affinity for water, there is the opportunity for the PEG SAM to be swollen [12]. The MeOPEGMA oligomer terminates at one end with a methyl methacrylate group which is hydrophobic while PEG is hydrophilic. If we consider that the PEG component of the surfactant has a very high affinity for water then there would be no reason for it to adsorb onto the surface for any of the submersion samples [36,37]; the methacrylate termination group, however, will be in a bad solvent when in water, and so will prefer to align to the mica surface [38]. If this adsorption configuration occurred at a high enough adsorption density it would have the effect of creating a self-assembled monolayer of surfactant chains in a brush regime. We note that when the samples are removed from the water suspension the hydrophilic mica retains an ultrathin layer of water and could attract the PEG [39,40], changing the orientation of MeOPEGMA.

In the dip-coated samples, we have also observed a layer of approx. 12 nm surrounding the nanocolloid structures (Figure 6). In these samples we also see some evidence of slight steps in the surrounding layer edge which could come from capillary forces as the liquid around the nanocolloid structure evaporates. We note that there is an evaporation process of the remaining ultrathin water domains at the micro/nanoscale for the dip-coated samples even if most of the liquid has been dried by nitrogen injection [12]. The surfactants exposed at the edge of the plateau will subside and tilt in some form creating small nanometre-size steps. Parallel layering to the sample substrate cannot be excluded.

The clustering observed in the 1 h submersion samples (Figure 3) showed some interesting configurations. Grain analysis showed that the z-heights relative to the surrounding surface had quite well-defined peaks at 30 and 70 nm, suggesting that there is evidence of bilayering occurring in the cluster structure. This is reinforced by the area histogram which has a very strong peak in the region of the $\approx 90 \times 10^3$ nm^2, indicating that there seems to be a preferred size for the area occupied by the cluster. The (≈ 10 nm) lower than expected height values could be the result of a thin film of surfactant having already formed on the mica substrate (approx. 12–13 nm). Thus, the height peaks compare well for a particle monolayer and bilayer since the particle size of the PS-PEG particles is ≈ 41 nm.

The 240 h incubation samples were submerged for the longest time in this study and they had the highest material coverage of all the samples under investigation, as shown by the digging experiment (Figure 9). The depth of the dig site (32 nm) indicates that there is a bilayer, if not multilayer, of the surfactant present on the mica substrate. The hydrophilic nature of the PEG combined with the hydrophobic methyl methacrylate termination would facilitate such multilayer lamellar structuring. This configuration could explain the range of protrusion sizes for these samples.

Further evidence for the multilayer surfactant formation comes from the grain analysis (Figure 8). The z-height histograms indicate reoccurring peaks relative to the surrounding surface at 15 nm, 25 nm, 40 nm, and 70 nm. Peaks also appear on the area measurements at 5×10^3, 19×10^3, and 29×10^3 nm^2. This multimodal distribution suggests that there are regular separations in the z-heights, the differences are in the region of 15–30 nm which would fit well with our proposed lamellar structuring of the surfactant. If this is the case, then particles have become trapped on different layers of the surfactant; the multimodal distribution in the area measurements promotes this idea; particles at different, but well defined, depths will protrude through to the top layer creating a series of defined protrusions on the surface.

The FFT of the particle structures for the 72 h incubation time sample showed highly ordered crystalline structuring of the PS-PEG particles (Figure 6). It revealed that the characteristic length between the particles was 57 nm. The larger than expected length could be due to the presence of the dry surfactant layer between them. The self-assembled crystalline structures were formed due to capillary/meniscus forces towards the final steps of the drying/evaporation; at this point the necessary interparticle forces required to draw the particles closely together were generated [41,42]. A meniscus must exist between the particles of the suspension liquid in order to generate the capillary forces when the particles are still partially submerged in liquid at this point.

5. Conclusions

We found that the PS-PEG colloidal suspension containing an excess of PEG-based surfactant molecules produced mixed monolayers, (sub-)monolayers, and ultrathin films on the mica surface. Our AFM study indicates that the PEG-based surfactant organised into self-assembled monolayers and multilayer ultrathin films. Flow currents during the droplet evaporation led to the desorption and redeposition of PS-PEG colloidal nanoparticles into the periphery of the sample, leaving the formation of an imprint of the particle deposition towards the centre of the samples and thus creating a 'fingerprint' of the nanocolloid initial structuring. The dip-coated samples showed a rich range of nanostructures including highly ordered domains of the PS-PEG particles at low incubation times and particles embedded in a multilayer surfactant ultrathin film at higher incubation times. To the best of our knowledge this system has not been studied before and our AFM study indicates the presence of very interesting nanostructures and nanopatterns which could be relevant for many applications.

Author Contributions: Conceptualization, V.K. and J.W.; methodology, V.K. and J.W.; software, J.W.; validation, J.W., V.K. and A.B.S.; formal analysis, J.W.; investigation, J.W.; resources, V.K. and A.B.S.; data curation, J.W.; writing—original draft preparation, J.W.; writing—review and editing, J.W., V.K. and A.B.S.; visualization, J.W.; supervision, V.K.; project administration, V.K.; funding acquisition, J.W. and V.K. All authors have read and agreed to the published version of the manuscript.

Funding: This research was funded by EPSRC, EP/P500206/1, DTA—University of Edinburgh.

Institutional Review Board Statement: Not applicable.

Informed Consent Statement: Not applicable.

Data Availability Statement: Data available upon request.

Acknowledgments: J.W. acknowledges financial support from EPSRC (DTA—University of Edinburgh). For the purpose of open access, the author has applied a Creative Commons Attribution (CC BY) licence to any Author Accepted Manuscript version arising from this submission.

Conflicts of Interest: The authors declare that they have no conflict of interest.

References

1. Pinto-Gómez, C.; Pérez-Murano, F.; Bausells, J.; Villanueva, L.G.; Fernández-Regúlez, M. Directed Self-Assembly of Block Copolymers for the Fabrication of Functional Devices. *Polymers* **2020**, *12*, 2432. [CrossRef] [PubMed]
2. Kulkarni, A.A.; Doerk, G.S. Thin film block copolymer self-assembly for nanophotonics. *Nanotechnology* **2022**, *33*, 292001. [CrossRef]
3. Kim, J.H.; Jin, H.M.; Yang, G.G.; Han, K.H.; Yun, T.; Shin, J.Y.; Jeong, S.-J.; Kim, S.O. Smart Nanostructured Materials based on Self-Assembly of Block Copolymers. *Adv. Funct. Mater.* **2020**, *30*, 1902049. [CrossRef]
4. Li, J.; Wang, J.; Chen, L.; Dong, Y.; Chen, H.; Nie, G.; Li, F. Self-assembly of DNA molecules at bio-interfaces and their emerging applications for biomedicines. *Nano Res.* **2023**. [CrossRef]
5. Glynos, E.; Chremos, A.; Camp, P.J.; Koutsos, V. Surface Nanopatterning Using the Self-Assembly of Linear Polymers on Surfaces after Solvent Evaporation. *Nanomanuf. Metrol.* **2022**, *5*, 297–309. [CrossRef]

6. Cummins, C.; Lundy, R.; Walsh, J.J.; Ponsinet, V.; Fleury, G.; Morris, M.A. Enabling future nanomanufacturing through block copolymer self-assembly: A review. *Nano Today* **2020**, *35*, 100936. [CrossRef]
7. Karayianni, M.; Pispas, S. Block copolymer solution self-assembly: Recent advances, emerging trends, and applications. *J. Polym. Sci.* **2021**, *59*, 1874–1898. [CrossRef]
8. Brassat, K.; Lindner, J.K.N. Nanoscale Block Copolymer Self-Assembly and Microscale Polymer Film Dewetting: Progress in Understanding the Role of Interfacial Energies in the Formation of Hierarchical Nanostructures. *Adv. Mater. Interfaces* **2020**, *7*, 1901565. [CrossRef]
9. McClements, J.; Shaver, M.P.; Sefiane, K.; Koutsos, V. Morphology of Poly(styrene-co-butadiene) Random Copolymer Thin Films and Nanostructures on a Graphite Surface. *Langmuir* **2018**, *34*, 7784–7796. [CrossRef]
10. McClements, J.; Buffone, C.; Shaver, M.P.; Sefiane, K.; Koutsos, V. Poly(styrene-co-butadiene) random copolymer thin films and nanostructures on a mica surface: Morphology and contact angles of nanodroplets. *Soft Matter* **2017**, *13*, 6152–6166. [CrossRef]
11. Zhou, Y.; Huang, W.; Liu, J.; Zhu, X.; Yan, D. Self-Assembly of Hyperbranched Polymers and Its Biomedical Applications. *Adv. Mater.* **2010**, *22*, 4567–4590. [CrossRef] [PubMed]
12. Glynos, E.; Chremos, A.; Petekidis, G.; Camp, P.J.; Koutsos, V. Polymer-like to Soft Colloid-like Behavior of Regular Star Polymers Adsorbed on Surfaces. *Macromolecules* **2007**, *40*, 6947–6958. [CrossRef]
13. Mendrek, B.; Oleszko-Torbus, N.; Teper, P.; Kowalczuk, A. Towards next generation polymer surfaces: Nano- and microlayers of star macromolecules and their design for applications in biology and medicine. *Prog. Polym. Sci.* **2023**, *139*, 101657. [CrossRef]
14. Zhang, J.; Sun, Z.; Yang, B. Self-assembly of photonic crystals from polymer colloids. *Curr. Opin. Colloid Interface Sci.* **2009**, *14*, 103–114. [CrossRef]
15. van Dommelen, R.; Fanzio, P.; Sasso, L. Surface self-assembly of colloidal crystals for micro- and nano-patterning. *Adv. Colloid Interface Sci.* **2018**, *251*, 97–114. [CrossRef] [PubMed]
16. MacFarlane, L.R.; Shaikh, H.; Garcia-Hernandez, J.D.; Vespa, M.; Fukui, T.; Manners, I. Functional nanoparticles through π-conjugated polymer self-assembly. *Nat. Rev. Mater.* **2021**, *6*, 7–26. [CrossRef]
17. Martin, C.P.; Blunt, M.O.; Vaujour, E.; Fahmi, A.; D'Aléo, A.; De Cola, L.; Vögtle, F.; Moriarty, P. Chapter 1 Self-Organised Nanoparticle Assemblies: A Panoply of Patterns. In *Studies in Multidisciplinarity*; Krasnogor, N., Gustafson, S., Pelta, D.A., Verdegay, J.L., Eds.; Elsevier: Amsterdam, The Netherlands, 2008; Volume 5, pp. 1–20.
18. Mutch, K.J.; Koutsos, V.; Camp, P.J. Deposition of Magnetic Colloidal Particles on Graphite and Mica Surfaces Driven by Solvent Evaporation. *Langmuir* **2006**, *22*, 5611–5616. [CrossRef] [PubMed]
19. Xia, Y.; Gates, B.; Yin, Y.; Lu, Y. Monodispersed Colloidal Spheres: Old Materials with New Applications. *Adv. Mater.* **2000**, *12*, 693–713. [CrossRef]
20. Zbonikowski, R.; Mente, P.; Bończak, B.; Paczesny, J. Adaptive 2D and Pseudo-2D Systems: Molecular, Polymeric, and Colloidal Building Blocks for Tailored Complexity. *Nanomaterials* **2023**, *13*, 855. [CrossRef]
21. Scherer, C.; Figueiredo Neto, A. Ferrofluids: Properties and Applications. *Braz. J. Phys.* **2005**, *35*, 718–727. [CrossRef]
22. Ghodbane, J.; Denoyel, R. Competitive adsorption between non-ionic polymers and surfactants on silica. *Colloids Surf. A Physicochem. Eng. Asp.* **1997**, *127*, 97–104. [CrossRef]
23. Verwey, E.J.W.; Overbeek, J.T.G. *Theory of the Stability of Lyophobic Colloids*; Elsevier Publishing Co.: New York, NY, USA; Amsterdam, The Netherlands; London, UK; Brussels, Belgium, 1948; p. 205.
24. Derjaguin, B.; Landau, L. Theory of the stability of strongly charged lyophobic sols and of the adhesion of strongly charged particles in solutions of electrolytes. *Prog. Surf. Sci.* **1993**, *43*, 30–59. [CrossRef]
25. Drummond, C.J.; Fong, C. Surfactant self-assembly objects as novel drug delivery vehicles. *Curr. Opin. Colloid Interface Sci.* **1999**, *4*, 449–456. [CrossRef]
26. Torchilin, V.P. Structure and design of polymeric surfactant-based drug delivery systems. *J. Control. Release* **2001**, *73*, 137–172. [CrossRef] [PubMed]
27. Velev, O.D.; Kaler, E.W. In Situ Assembly of Colloidal Particles into Miniaturized Biosensors. *Langmuir* **1999**, *15*, 3693–3698. [CrossRef]
28. Otsuka, H.; Nagasaki, Y.; Kataoka, K. Self-assembly of poly(ethylene glycol)-based block copolymers for biomedical applications. *Curr. Opin. Colloid Interface Sci.* **2001**, *6*, 3–10. [CrossRef]
29. Butt, H.J.; Berger, R.; Bonaccurso, E.; Chen, Y.; Wang, J. Impact of atomic force microscopy on interface and colloid science. *Adv. Colloid Interface Sci.* **2007**, *133*, 91–104. [CrossRef]
30. Schmitz, I.; Schreiner, M.; Friedbacher, G.; Grasserbauer, M. Phase imaging as an extension to tapping mode AFM for the identification of material properties on humidity-sensitive surfaces. *Appl. Surf. Sci.* **1997**, *115*, 190–198. [CrossRef]
31. Bar, G.; Thomann, Y.; Brandsch, R.; Cantow, H.J.; Whangbo, M.H. Factors Affecting the Height and Phase Images in Tapping Mode Atomic Force Microscopy. Study of Phase-Separated Polymer Blends of Poly(ethene-co-styrene) and Poly(2,6-dimethyl-1,4-phenylene oxide). *Langmuir* **1997**, *13*, 3807–3812. [CrossRef]
32. Ottewill, R.H.; Satgurunathan, R. Nonionic latices in aqueous media part 1. Preparation and characterization of polystyrene latices. *Colloid Polym. Sci.* **1987**, *265*, 845–853. [CrossRef]
33. Deegan, R.D.; Bakajin, O.; Dupont, T.F.; Huber, G.; Nagel, S.R.; Witten, T.A. Capillary flow as the cause of ring stains from dried liquid drops. *Nature* **1997**, *389*, 827–829. [CrossRef]

34. Askounis, A.; Sefiane, K.; Koutsos, V.; Shanahan, M.E.R. Structural transitions in a ring stain created at the contact line of evaporating nanosuspension sessile drops. *Phys. Rev. E* **2013**, *87*, 012301. [CrossRef] [PubMed]
35. Wang, T.; Xu, J.; Qiu, F.; Zhang, H.; Yang, Y. Force spectrum of a few chains grafted on an AFM tip: Comparison of the experiment to a self-consistent mean field theory simulation. *Polymer* **2007**, *48*, 6170–6179. [CrossRef]
36. Chai, L.; Klein, J. Role of Ion Ligands in the Attachment of Poly(ethylene oxide) to a Charged Surface. *J. Am. Chem. Soc.* **2005**, *127*, 1104–1105. [CrossRef] [PubMed]
37. Geke, M.O.; Shelden, R.A.; Caseri, W.R.; Suter, U.W. Ion Exchange of Cation-Terminated Poly(ethylene oxide) Chains on Mica Surfaces. *J. Colloid Interface Sci.* **1997**, *189*, 283–287. [CrossRef]
38. Kumaki, J.; Nishikawa, Y.; Hashimoto, T. Visualization of Single-Chain Conformations of a Synthetic Polymer with Atomic Force Microscopy. *J. Am. Chem. Soc.* **1996**, *118*, 3321–3322. [CrossRef]
39. Glynos, E.; Pispas, S.; Koutsos, V. Amphiphilic Diblock Copolymers on Mica: Formation of Flat Polymer Nanoislands and Evolution to Protruding Surface Micelles. *Macromolecules* **2008**, *41*, 4313–4320. [CrossRef]
40. Kalloudis, M.; Glynos, E.; Pispas, S.; Walker, J.; Koutsos, V. Thin Films of Poly(isoprene-b-ethylene Oxide) Diblock Copolymers on Mica: An Atomic Force Microscopy Study. *Langmuir* **2013**, *29*, 2339–2349. [CrossRef] [PubMed]
41. Kralchevsky, P.A.; Denkov, N.D. Capillary forces and structuring in layers of colloid particles. *Curr. Opin. Colloid Interface Sci.* **2001**, *6*, 383–401. [CrossRef]
42. Kralchevsky, P.A.; Nagayama, K. Capillary forces between colloidal particles. *Langmuir* **1994**, *10*, 23–36. [CrossRef]

Disclaimer/Publisher's Note: The statements, opinions and data contained in all publications are solely those of the individual author(s) and contributor(s) and not of MDPI and/or the editor(s). MDPI and/or the editor(s) disclaim responsibility for any injury to people or property resulting from any ideas, methods, instructions or products referred to in the content.

Article

Crystallization of Poly(ethylene oxide)-Based Triblock Copolymers in Films Swollen-Rich in Solvent Vapors

Iulia Babutan [1,2], Otto Todor-Boer [3], Leonard Ionut Atanase [4,5], Adriana Vulpoi [1] and Ioan Botiz [1,2,*]

1. Interdisciplinary Research Institute on Bio-Nano-Sciences, Babeș-Bolyai University, 400271 Cluj-Napoca, Romania; iulia.babutan@ubbcluj.ro (I.B.); adriana.vulpoi@ubbcluj.ro (A.V.)
2. Faculty of Physics, Babeș-Bolyai University, 400084 Cluj-Napoca, Romania
3. INCDO-INOE 2000, Research Institute for Analytical Instrumentation, 400293 Cluj-Napoca, Romania; otto.todor@icia.ro
4. Department of Biomaterials, Faculty of Medical Dentistry, "Apollonia" University of Iasi, 700511 Iasi, Romania; leonard.atanase@univapollonia.ro
5. Academy of Romanian Scientists, 050045 Bucharest, Romania
* Correspondence: ioan.botiz@ubbcluj.ro

Abstract: In this study, we employed a polymer processing method based on solvent vapor annealing in a confined environment to swell-rich thin films of polybutadiene-*b*-poly(2-vinylpyridine)-*b*-poly(ethylene oxide) triblock copolymers and to promote their crystallization. As revealed by optical and atomic force microscopy, thin films of triblock copolymers containing a rather short crystalline poly(ethylene oxide) block that was massively obstructed by the other two blocks were unable to crystallize following the spin-casting process, and their further swelling in solvent vapors was necessary in order to produce polymeric crystals displaying a dendritic morphology. In comparison, thin films of triblock copolymers containing a much longer poly(ethylene oxide) block that was less obstructed by the other two blocks were shown to crystallize into dendritic structures right after the spin-casting procedure, as well as upon rich swelling in solvent vapors.

Keywords: block copolymers; thin films; solvent vapor annealing; polymer crystallization; atomic force microscopy

1. Introduction

The wide diversity of polymeric properties has its first source in the nature of polymers' soft material component units named monomers [1–11]. The capability of long polymer chains to adopt, after specific processing conditions, a multitude of conformational arrangements at multiple length scales, represents a second consistent source that nourishes the development of new and/or enhanced polymer properties in thin films and on various surfaces of interest, in solutions and in the solid state [2,9,12–16]. Obviously, the optimized polymer properties can be advantageously employed to design and produce various functional devices, develop new technologies, and engineer high-impact applications [7,17–27].

There is a broad range of prominent processing methods that can efficiently alter the microstructure of various polymer chains and tune their molecular arrangements at micrometer and nanometer scales [10,14,28–34]. Such methods can be used to align polymer molecules in thin films during [35–37] or after [15,33,38–43] their fabrication, as well as in solutions [44–46] and in the solid state [14,47,48], and may include the selection of solvent quality [49] and polymer concentration [50], addition of non-solvents [51], employement of convective forces [31,34], thermal annealing [40], pressure [52] or mechanical stretching [53], and use of space confinements [54,55] or annealing in solvent vapors [39,41]. The aforementioned methods mainly rely on physical processes such as self-assembly [56–58] or crystallization [40,41,59–62]. In particular, the crystallization process becomes important when employed to control the mechanical [63,64], optoelectronic [64,65] or other [64,66] properties of polymeric materials. Interestingly, while the confinement of polymer chains has been reported to exhibit significant impact on the kinetics of crystallization [67] of block copolymers (BCPs), the solvent vapor annealing procedure has proved to be critical in the ordering of a large variety of BCPs [68–70] inclusively when investigating highly complex molecules [39].

In this study, we relied on a processing approach that takes advantage of both confinement and solvent vapor annealing (C-SVA) and further induces crystallization of BCP systems based on rather short crystallizable poly(ethylene oxide) (PEO) blocks. The PEO blocks are composed of only about 104 and 154 monomer units, with their corresponding number average molecular weights being determined as $M_{nPEO104}$ = 4600 g/mol and $M_{nPEO154}$ = 6800 g/mol, respectively [71]. We show that although the aforementioned blocks are capable of crystallizing at temperatures below 40 °C [72], triblock copolymers based on such blocks will not crystallize into large dendritic crystals under spin-casting conditions, unless further processed by utilizing the improved C-SVA approach [73,74] (note that polymer chains containing much larger numbers of ethylene oxide monomers are known to crystallize following spin casting or thermal annealing processes at room temperature or various higher crystallization temperatures; such macromolecules are often used as model systems for studying general concepts of the crystallization process [38,75]). Compared to other SVA-based studies [39,41,42], this improved approach was recently shown to be capable of self-assembling various BCP systems into highly ordered nanostructures [73,74]. It consists of a sample chamber of reduced depth and well-regulated temperature and a "bubbling" system able to inject precise amounts of solvent vapors inside the sample chamber (Figure 1). With this design, the rich-swelling of BCP films is possible without encountering weak variations in the sample temperature during the swelling-deswelling processes, and thus without experiencing unwanted fluctuations in the swollen-film thickness that can generate film defects upon drying [73].

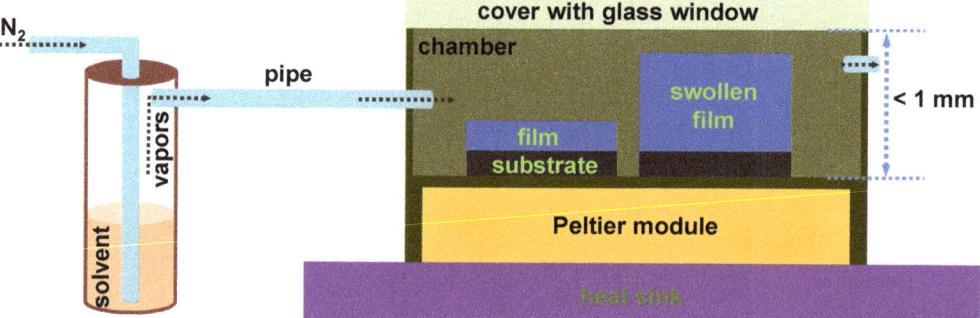

Figure 1. Schematic representation of the experimental setup used to process thin BCP films. In this setup, the Peltier module was connected both to a temperature controller and a PT100 sensor. Meanwhile, the heat sink was coupled to a fan. At the same time, the pipe transporting solvent vapors was connected to a nitrogen-based "bubbling" system depicted on the left. The scheme shows both the unswollen (on the **left**) and swollen (on the **right**) states of a BCP film. Note that the dimensions are not drawn at scale.

2. Materials and Methods

The polymer system employed in this study was a polybutadiene-*b*-poly(2-vinylpyridine)-*b*-poly(ethylene oxide) (PB-*b*-P2VP-*b*-PEO) triblock copolymer (see its chemical structure depicted in Figure 2). We employed four different BCP systems displaying number average molecular weights of M_n = 20,500 g/mol (PB$_{100}$-*b*-P2VP$_{100}$-*b*-PEO$_{104}$), M_n = 28,200 g/mol (PB$_{185}$-*b*-P2VP$_{108}$-*b*-PEO$_{154}$), M_n = 76,000 g/mol (PB$_{348}$-*b*-P2VP$_{252}$-

b-PEO$_{697}$) and M$_n$ = 26,400 g/mol (PB$_{66}$-b-P2VP$_{69}$-b-PEO$_{356}$), respectively [71]. These polymer systems were synthesized by living anionic polymerization in tetrahydrofuran (THF) in the presence of cumyl potassium (PIK) as initiator, and through the employment of well-established procedures [71,76,77].

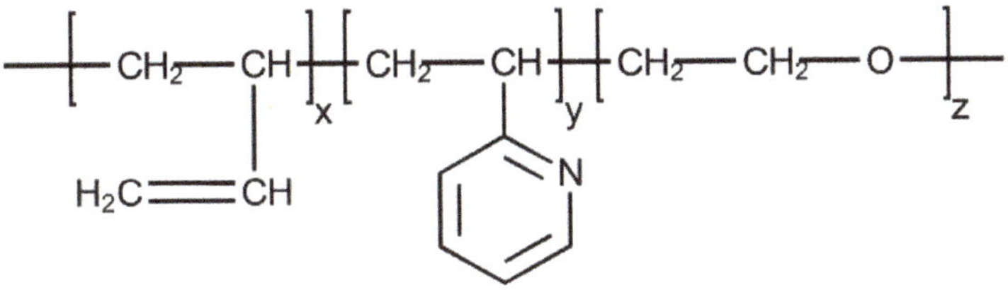

Figure 2. Chemical structure of the PB$_x$-b-P2VP$_y$-b-PEO$_z$ triblock copolymer employed in this study. Here, x, y and z are the degrees of polymerization corresponding to each of the three blocks (x is representing the values of 66, 100, 185 and 348, y is representing the values of 69, 100, 108 and 252, and z is representing the values of 356, 104, 154 and 697, respectively).

The reagent used for the preparation of copolymer solutions was toluene (98%), purchased from the Chemical Company (Iasi, Romania). Toluene was employed because it is a good solvent for both PEO and PB. Copolymer solutions were prepared by dissolving 10 mg of copolymer powder in 1 mL of toluene. The process was followed by gentle stirring to further stimulate dissolution and homogenization. Afterwards, polymeric solutions were annealed at 70 °C in a silicon oil bath (ONE 7-45, Schwabach, Germany) for 30 min to complete the dissolution of copolymers in toluene.

Thin triblock copolymer films of a thickness of 79 ± 5 nm (this value was determined after scratching a film and measuring the corresponding height profile with the atomic force microscopy/AFM) were obtained by spin casting copolymer solutions onto solid silicon substrates using a WS-650mz23nppb spin coater from Laurell Technologies Corporation (North Wales, PA, USA). Films were deposited at a speed of 2000 rpm for 30 s. Type 4PO/5-10/380 ± 15/SSP/TTV < 5 silicon substrates were acquired from Siegert Wafer GmbH (Aachen, Germany) and were subjected to UV-ozone treatment for 20 min (in a PSD Pro Series-Digital UV Ozone System from Novascan; Boone, IA, USA) before their further use.

For the swelling and deswelling of BCP films via their exposure to toluene vapors in a quasi-confined environment (C-SVA), we have used a home-made equipment consisting of an aluminum chamber with a high-performance Peltier element (15.4 V/8.5 A from Stonecold) placed beneath (see Figure 1). The 100 W powerful Peltier module permitted the setup to exhibit a maximum temperature difference ΔT between the two sides of about 58 °C. The temperature of the Peltier module (i.e., the temperature of the bottom of the sample chamber and thus, the temperature of the film; note that the bottom of the chamber was thermally separated from the rest of the chamber by design) could be regulated by a temperature controller (model TCM U 10 A from Electron Dynamics Ltd.; Southampton, UK) that received feedback from a PT100 temperature sensor located in the chamber, in the vicinity of the BCP film. The PT100 sensor was continuously communicating the film temperature to the controller. The latter could change the strength and direction of the electric current depending on whether it needed to cool or heat the system (electricity was provided by a 12 V/10 A power supply). Moreover, an aluminum heat sink and a fan were placed on the other side of the Peltier module, which helped to equalize the ΔT temperature. The controller was connected to computer software that used proportional integral derivative technology to accurately set the temperature within the desired time. With the above-described device, it was possible to control the sample temperature with a precision of 0.01 °C and to keep it constant over time. Furthermore, the time required to reach a specific temperature could be varied within the seconds–hours range. This allowed us to finely tune the rate at which the temperature changed and to avoid weak variations in the sample temperature, which may appear when reaching a specific temperature setpoint during a swelling or deswelling procedure. Finally, note that the sample chamber was saturated with toluene vapors using a nitrogen-based "bubbling" system that was connected to a flow meter that allowed the amount of vapors to be regulated.

The following experimental procedure was used to swell/de-swell thin BCP films in a quasi-confined environment saturated with toluene vapors. A PB-b-P2VP-b-PEO triblock copolymer film was placed in the sample chamber. While the chamber was heated up to 40 °C, the desired quantity of toluene vapors was bubbled inside. Next, the film temperature was set to 15 °C. While the temperature was decreasing at a rate of 0.3 °C/s, at around 22 °C the toluene vapors started to condense gradually on the surface of the film and the latter started to swell and change its color. This change in the interference colors was associated with the changes in the film thickness and could be used to monitor the thickness of the latter in its swollen state (an interference color–film thickness calibration can be generated before the start of the swelling experiments by measuring the thickness of many BCP films using the AFM technique and then associating each thickness to the corresponding film color observed under the optical microscope; see additional details on the procedure elsewhere [39,41,78]). At 15 °C there were enough toluene vapors condensed on the film to transform it into a quasi-two-dimensional (2D) "solution" with a polymer concentration (c_p) of about 7% ± 3% (this concentration value was calculated as the ratio between the initial film thickness and the thickness of the swollen film; see details in [39,41,78]). After about 30 s at this low c_p, we reversed the process and initiated the deswelling process when the film temperature was increased slowly back to 40 °C, at a rate of only 0.01 °C/s. During this time, toluene vapors began to gradually evaporate and the film slowly returned to its original thickness, but with its surfaces covered with newly induced crystalline structures.

For the acquisition of AFM images, a system from Molecular Devices and Tools for Nano Technology (NT-MDT) mounted on an Olympus IX71 optical microscope was used in noncontact (tapping) mode. The AFM measurements were conducted utilizing high resolution Noncontact Golden Silicon probes from NT-MDT. These probes had a tip radius of curvature smaller than 10 nm and a tip height in the range of 14–16 µm. Moreover, they were coated with Au on the detector side cantilever. The latter had a length of 125 ± 5 µm, and displayed a resonance frequency in the range of 187–230 kHz and a nominal force constant ranging between 1.45–15.1 N/m. The AFM images (256 × 256 lines) were obtained using a scanning speed of about 1–2 µm/s and a setpoint ranging between 9 and 12 V. The setpoint was adjusted so that a very soft tapping regime was obtained. The optical microscopy micrographs were acquired using a KERN OKN-177 optical microscope operating in reflection mode.

3. Results and Discussion

The main objective of this study was to reveal the impact of C-SVA processing on the crystallization of thin BCP films. To achieve this objective, we firstly used a triblock PB_{100}-b-$P2VP_{100}$-b-PEO_{104} copolymer system containing one crystalline [40] PEO block. Because the PEO block was rather short (it comprised only 104 repeating monomers) and comparable to the lengths of the other two constituent PB and P2VP blocks (each containing 100 corresponding monomers), we expected that this triblock copolymer would face difficulties in crystallizing (i.e., to nucleate and grow into single crystals) due to unavoidable obstructions exerted by the non-crystalline blocks. Indeed, many attempts to nucleate and crystallize thin PB_{100}-b-$P2VP_{100}$-b-PEO_{104} films from the melt at various temperatures below 40 °C were unsuccessful (although partial melting of such films was observable under the optical microscope at 67 °C and above, the subsequent crystallization in the temperature range of 26–38 °C would not occur, possibly due to insufficiently increased chain mobility combined with the inhibitions generated by the non-crystalline blocks). Moreover, crystallization would also be difficult under normal, high-speed spin-casting conditions that would kinetically trap the films in a rather disordered state, as there is only limited time for the BCP to initiate nucleation and subsequent crystallization prior to the loss of solvent. This latter statement was confirmed by the results presented in Figure 3. As we can observe in the optical micrograph in Figure 3b, there were no crystals forming during and/or after the spin-casting process, besides some irregular and randomly distributed elongated objects (of a length of 10–20 µm, as indicated by the yellow dotted arrows; a part of such a quasi-flat object was further indicated by the dotted shape in Figure 3d). The AFM height images presented in Figure 3d,f further demonstrated that the rest of the film surface was covered with bright-colored domains a few tens of nanometers tall, surrounding some "empty" and dark-colored irregular regions. Nonetheless, these latter regions appeared brighter in the AFM phase image in Figure 3h, indicating that at the bottom they were made of rather stiffer structures, possibly some poorly developed aggregates or crystalline objects. Instead, the taller bright-colored domains visible in Figure 3f were composed of a rather soft, amorphous material, as indicated by their darker appearance in the AFM phase image (Figure 3h). Here, some rather circular sub-50 nm nanostructures, emphasized in Figure 3j by the dotted circular shapes, could nonetheless be observed.

In comparison, the as-cast PB_{100}-b-$P2VP_{100}$-b-PEO_{104} BCP film that was further processed with the C-SVA method exhibited a morphology composed of essentially single crystals randomly distributed over the whole surface and displaying a dendritic structure (Figures 3a,c and 4a). This observation proved that processing with the C-SVA method favored and induced crystallization, even though the crystalline PEO block was short and well obstructed by the other two blocks and otherwise did not form crystals. Moreover, the dendritic crystals shown in Figure 4a nucleated and grew during the C-SVA processing, within the range of 17–20 °C while the c_p was determined to increase from about 20% to around 40% (note that the four crystals numbered in their centers are "deformed" due to their partial coalescence).

All the above experimental observations suggested that dendritic crystals only formed when polymer chains experienced high mobilities conferred by the rather diluted quasi-2D "film-solution" regime. This was not the case during the thermal annealing, when chain mobilities are expected to be lower than those in solutions or during the spin-casting process, when such mobilities are rapidly decreasing with the dramatic evaporation of the solvent. Clearly, the dendritic crystals are a result of (partial) alignments of, at least, PEO molecular chains (i.e., their crystallization). The growth of these crystals was driven not only by the transport of the molecules to the crystal (which is ultimately dictated by the concentration of polymer molecules in rich-swollen "film-solutions" and most probably by their diffusion rate [75]), but also by the probability of a polymer chain attaching to the surface of a crystal. Considering that the dendritic crystals formed in about 5 min (while increasing the sample temperature from about 17 °C to 20 °C, with a rate of 0.01 °C/s) in a regime where the molecules most likely experienced high mobilities that favored the diffusion of molecules, we tentatively conclude that the corresponding attachment probability was rather high and most probably dominated by the diffusion-limited aggregation that finally led to a dendritic morphology [75].

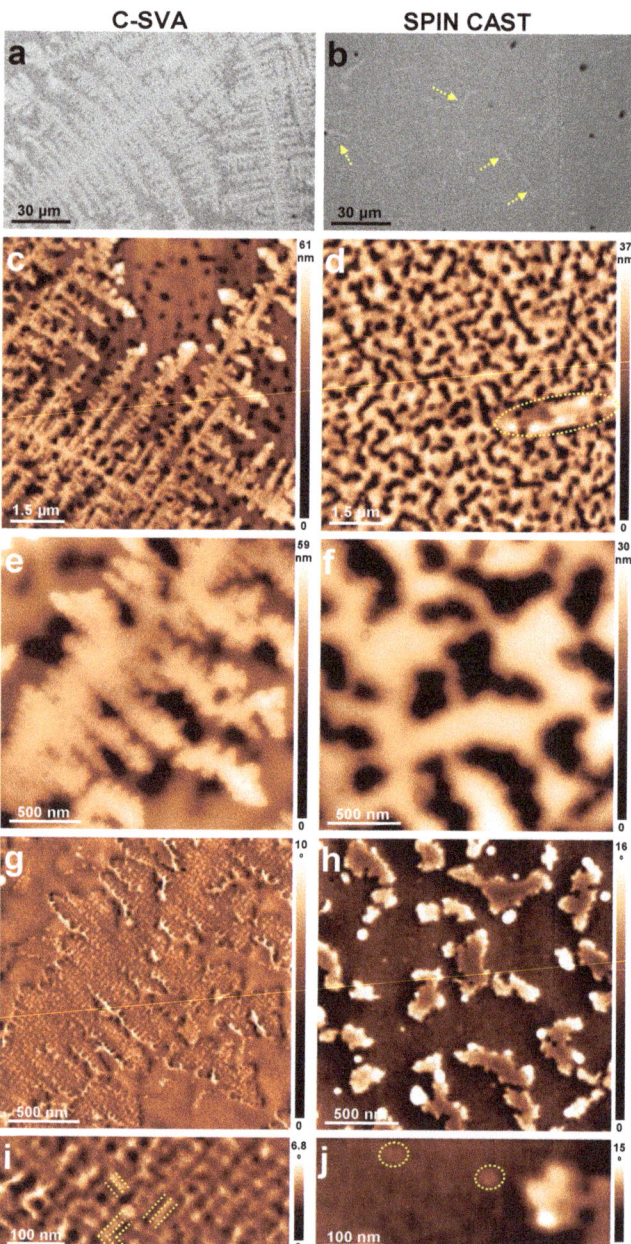

Figure 3. Optical micrographs (**a,b**) and AFM height (**c–f**) and phase (**g–j**) images depicting the surface microstructure observed in a thin film of PB$_{100}$-*b*-P2VP$_{100}$-*b*-PEO$_{104}$ after (**a,c,e,g,i**) and before (**b,d,f,h,j**) its exposure to toluene vapors in a rather confined sample chamber. While height (**e**) and phase (**g**) images each represent a zoom-in of a region depicted in (**c**), images in (**f,h**) correspond to a zoom-in of a region shown in (**d**). Moreover, images presented in (**i,j**) are each a zoom-in of the images portrayed in (**g,h**), respectively. The only purpose of the dotted arrows, shapes and lines is to guide the eye.

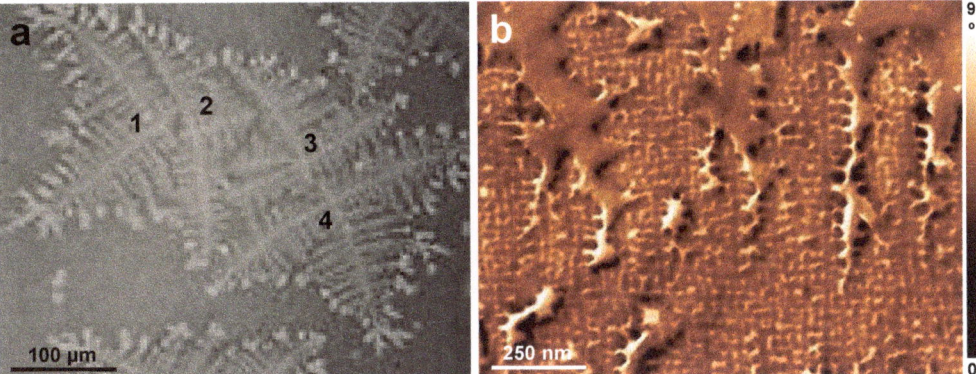

Figure 4. (**a**) Four (partially) coalesced single crystals of dendritic structures observed under the optical microscope in a PB_{100}-b-$P2VP_{100}$-b-PEO_{104} thin film after its rich-exposure to toluene vapors. (**b**) Zoom-in and a 230° rotation of the AFM phase micrograph shown in Figure 3g realized in order to better emphasize the existence of 16 ± 2 nm large substructures.

Furthermore, the average thickness of the dendritic crystal presented in Figure 3c,e was extracted from multiple AFM cross-section profiles and was estimated to be 26 ± 5 nm. This value seemed to match the thickness of a PEO crystalline lamella, if considering that the maximum length of a crystalline PEO chain in its fully extended conformation is almost 29 nm (according to the literature, the dimension of one ethylene oxide monomer is 0.2783 nm [79]). Nonetheless, we do not know whether the crystal extended deeper within the 79 nm thick film, below the surface probed by the AFM. Additionally, the semicrystalline P2VP block could possibly extend the lamellar thickness by another 25 nm (100 × 0.25 nm [80,81]). Therefore, it is not possible to conclude whether the observed crystals were made of folded, tilted or fully extended chains without a further structural analysis employing X-ray experiments. Nonetheless, by comparing the crystal with its surrounding areas in the AFM phase image shown in Figure 3g, we observed that the color of the crystalline regions appeared to be slightly lighter than that of the surrounding areas, pointing towards a stiffer material. This was in accordance with the expectation that a crystalline material should be stiffer than its uncrystallized counterpart. A further analysis on the crystal morphology revealed not only that the dendritic crystals were composed, as expected, of a multitude of orthogonal branches, but also that these branches exhibited fine, rather orthogonal substructures of a lateral dimension of 16 ± 2 nm (see the dotted lines in the AFM image of Figure 3i and the high magnification of substructures in Figure 4b). To corelate this value with the dimension of the polymer chains and their precise arrangements within the crystal, a full structural analysis based on X-ray measurements will be needed in the future.

In order to further demonstrate that the C-SVA approach can efficiently induce and promote the crystallization process even under unfavorable conditions (i.e., when the PEO block is well obstructed by the other two blocks), we have further increased the length of all blocks, while maintaining a similar ratio between the number of ethylene oxide monomers and the number of total butadiene and 2-vinylpyridine monomers (~0.52). In this case, the resulting PB_{185}-b-$P2VP_{108}$-b-PEO_{154} BCP system possessed a longer PEO crystalline chain, but was still well obstructed by the other two blocks. Figure 5 shows structures obtained in a PB_{185}-b-$P2VP_{108}$-b-PEO_{154} film after and before its processing in toluene vapors. While the optical micrograph presented in Figure 5a and corresponding to the processed film clearly emphasized the presence of crystals of a seaweed dendritic morphology [82,83] with growing tips splitting intermittently during crystallization, the optical micrograph shown in Figure 5b and recorded for the unprocessed film suggested that there were no signs of crystals. In this latter case, the film surface was covered with irregular and randomly distributed structures of a few micrometers in size. These are better visualized in Figure 5d as bright, irregular aggregates indicated by the dotted arrows. In between these aggregates, the surface was covered with sub-500 nm structures that exhibited various irregular shapes and were randomly distributed on the surface. One such structure is pointed out in Figure 5f,h by the dotted square shapes, and further magnified in Figure 5j. These structures appeared to be composed of roundish substructures of molecular dimension (32 ± 3 nm in diameter; see the dotted circular green shapes in Figure 5j). The regions in between the sub-500 nm structures were also covered with spherical objects of a molecular diameter (see the dotted circular yellow shapes in Figure 5j). Knowing that PB_{185}-b-$P2VP_{108}$-b-PEO_{154} BCP displays, under specific conditions, a micellar nature [84,85], with micelles that could be exhibiting a total hydrodynamic radius of up to 28.5 nm [84], we cannot exclude the possibility that the as-cast morphology of this BCP system is based on randomly distributed 32 ± 3 nm large micellar objects.

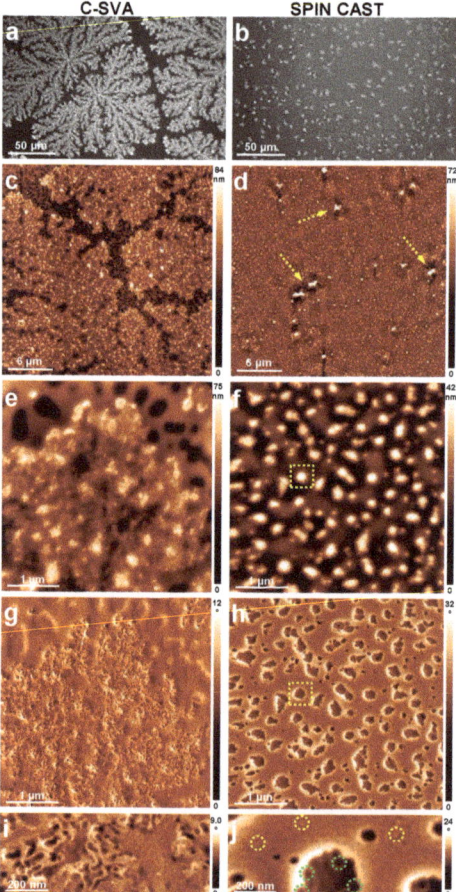

Figure 5. Optical micrographs (**a**,**b**) and AFM height (**c**–**f**) and phase (**g**–**j**) images depicting the surface microstructure observed in a thin film of PB_{185}-b-$P2VP_{108}$-b-PEO_{154} after (**a**,**c**,**e**,**g**,**i**) and before (**b**,**d**,**f**,**h**,**j**) its exposure to toluene vapors. While height (**e**) and phase (**g**) images each represent a zoom-in of a region depicted in (**c**), images in (**f**,**h**) correspond to a zoom-in of a region portrayed in (**d**). Moreover, micrographs presented in (**i**,**j**) are each a zoom-in of the images shown in (**g**,**h**), respectively. The purpose of the dotted shapes is for the eye guiding only.

Instead, when the same BCP system was processed by utilizing the C-SVA method, the morphology was composed of crystalline seaweed dendrites of an average height of 29 ± 7 nm (determined by evaluating several cross-sections of the dendrites shown in Figure 5c). As it can be observed in Figure 5e, the dendrites had a rather irregular shape and were surrounded by a rather porous, yet amorphous morphology. Obviously, the latter displayed a darker color in the AFM phase micrograph when compared to the dendritic crystal (Figure 5g). Interestingly, a further magnification of the dendritic area revealed the existence of both folded "stripe"-like and spherical substructures of a few tens of nanometers in lateral dimension (Figure 5i). In conclusion, the comparison between the two morphologies of the PB_{185}-b-$P2VP_{108}$-b-PEO_{154} film obtained before and after its exposure to toluene vapors, revealed again that the crystallization process was induced only in the diluted quasi-2D "film-solution" regime when the C-SVA method was utilized.

Finally, in order to compare the resulting crystalline structures when they are also generated during the spin-casting process, we have massively increased the crystalline PEO block within the BCP system (the ratio between the number of ethylene oxide monomers and the number of total butadiene and 2-vinylpyridine monomers was increased to more than 1.16; in this case the PEO block was expected to be less obstructed by the other two constituent blocks when undergoing crystallization). In Figure 6 we compare a film of PB_{345}-b-$P2VP_{252}$-b-PEO_{697}, after and before being processed using the C-SVA method. When the BCP system contained 697 ethylene oxide monomer units, the crystallization process was spotted right after the spin-casting procedure. In this case, the BCP film exhibited a uniform morphology fully covered with densely packed dendritic crystalline structures (23 ± 5 nm in height; Figure 6b,d). Similarly, the film that was processed with the C-SVA method also displayed a surface covered with dendritic crystalline structures, but of a height about 33 ± 7 nm (Figure 6a,c; note here that the dendritic structures formed during the initial spin casting process were dissolved during the exposure of the film to toluene vapors in the "film-solution" configuration and then re-crystallized). In this latter case, some "empty" regions, most probably depleted of chain molecules by the crystallization process, were observed in between the dendrites. Clearly, the dendrites grown in the C-SVA processed film were larger than those grown in the as-spin-cast film (compare Figure 6c with Figure 6d). Moreover, the former dendrites were visibly covered with various polymer decorations of an average height of 20 ± 5 nm (Figure 6e,g). This was not the case for the as-spin-cast dendrites, which only displayed a rather uniform surface (Figure 6f). Interestingly, the decorations were composed of spherical substructures that had an average diameter ranging from ~30 nm to ~55 nm (Figure 6g) and displayed a rather soft texture (as inferred from the AFM phase micrograph shown in Figure 6i). These spherical structures failed to develop during the spin-casting process (Figure 6h,j). In conclusion, when dealing with BCPs composed of rather long crystalline PEO blocks, the crystallization process occurred both after the spin casting and after the C-SVA processing. Nonetheless, only in the latter case were the crystalline dendrites larger, better defined, and displaying decorations composed of spherical soft structures of molecular dimensions.

Similar results to those reported in Figure 6 were also obtained for the PB_{66}-b-$P2VP_{69}$-b-PEO_{356} triblock copolymer. Although composed of a shorter PEO block that contained only 356 monomer units, this BCP also exhibited crystalline structures in both C-SVA processed and unprocessed films (Figure 7). Nonetheless, in this case the ratio between the number of ethylene oxide monomers and the number of total butadiene and 2-vinylpyridine monomers was more than 2.63, as the PB and P2VP blocks were reduced to only 66 and 69 monomers, respectively. Therefore, there was not much obstruction of the crystallization process that could have been exerted by the latter two PB and P2VP blocks. Moreover, the fine substructures that can be seen in Figure 7i,j and that form the dendritic crystals shown in Figure 7a,b, are displaying a rather lamellar appearance with lateral dimensions of 18 ± 3 nm in width. Thus, these substructures are not spherical as those observed in Figure 6i,j for the BCP containing a much longer PEO block.

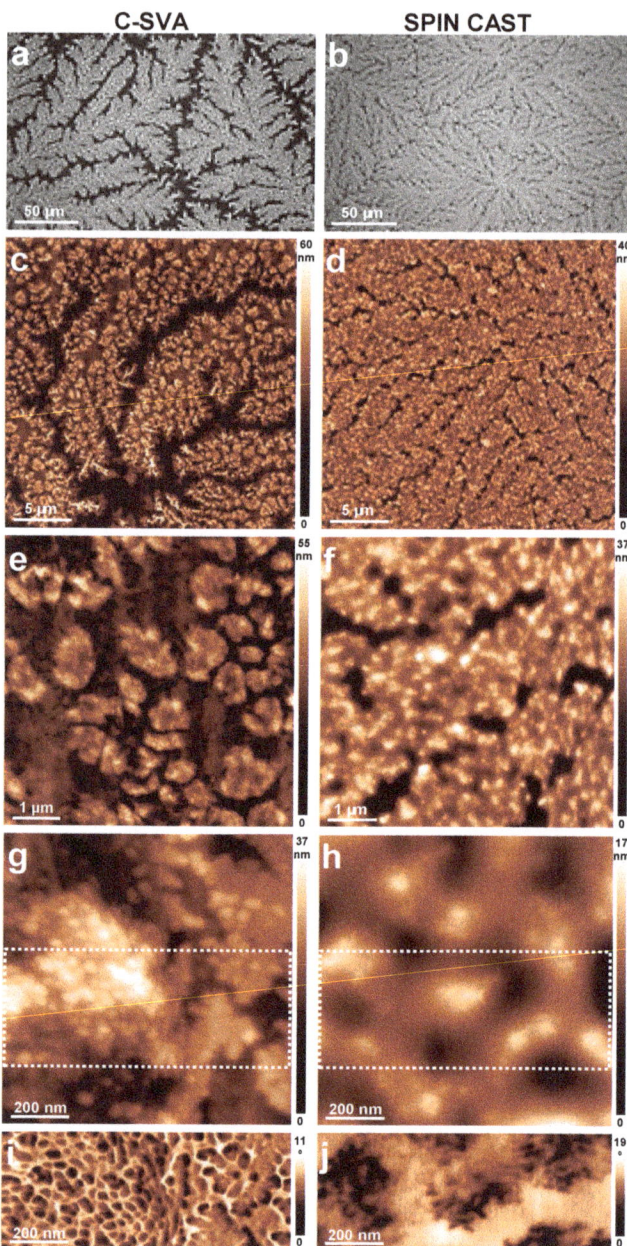

Figure 6. Optical micrographs (**a,b**) and AFM height (**c–h**) and phase (**i,j**) images depicting the surface microstructure observed in a thin film of PB$_{348}$-*b*-P2VP$_{252}$-*b*-PEO$_{697}$ after (**a,c,e,g,i**) and before (**b,d,f,h,j**) its exposure to toluene vapors in a confined sample chamber. Height images in (**e,f**) each represent a zoom-in of regions depicted in (**c,d**), respectively. Similarly, the height images in (**g,h**) each correspond to a zoom-in of regions portrayed in (**e,f**), respectively. Moreover, phase images shown in (**I,j**) are each a zoom-in corresponding to the regions delimited by dotted shapes in (**g,h**), respectively.

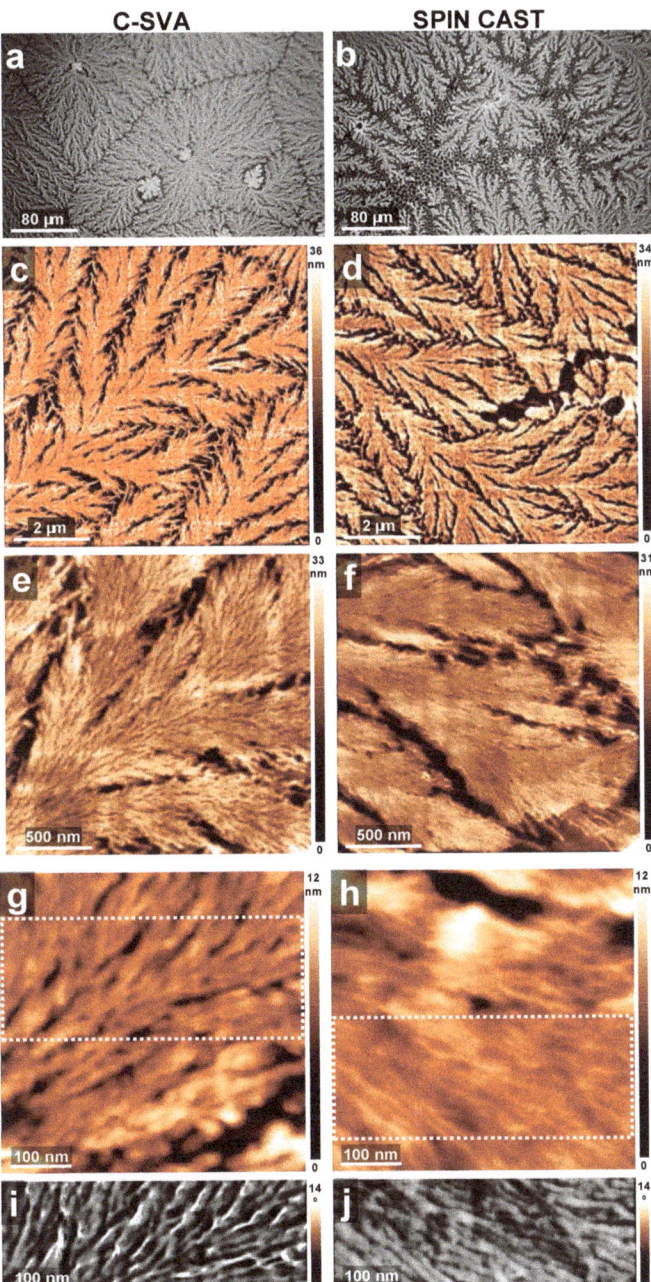

Figure 7. Optical micrographs (**a**,**b**) and AFM height (**c**–**h**) and phase (**i**,**j**) images depicting the surface microstructure observed in a thin film of PB$_{66}$-b-P2VP$_{69}$-b-PEO$_{356}$ after (**a**,**c**,**e**,**g**,**i**) and before (**b**,**d**,**f**,**h**,**j**) its rich-swelling in toluene vapors. Height images in (**e**,**f**) each represent a zoom-in of regions depicted in (**c**,**d**), respectively. Similarly, the height images in (**g**,**h**) each correspond to a zoom-in of regions portrayed in (**e**,**f**), respectively. Moreover, phase micrographs shown in (**i**,**j**) are each a zoom-in corresponding to the regions delimited by dotted shapes in (**g**,**h**), respectively.

4. Conclusions

We have used a polymer processing approach based on solvent vapor annealing in a space-confined environment in order to induce crystallization in thin films of PB-*b*-P2VP-*b*-PEO triblock copolymers that contained a rather short crystalline PEO and which would not crystallize otherwise. Indeed, the obtained optical microscopy and AFM results have shown that the PB-*b*-P2VP-*b*-PEO films based on short PEO blocks (i.e., 104–154 monomer units) that were well hindered by the other two constituent blocks led to crystals of (seaweed) dendritic morphology only following their generous swelling in solvent vapors. As expected, the BCP films based on a much longer PEO crystalline block (i.e., 356–697 monomers) that were less hindered by the other two constituent blocks underwent crystallization inclusively under normal spin-casting conditions.

Author Contributions: Conceptualization, I.B. (Iulia Babutan), O.T.-B., L.I.A. and I.B. (Ioan Botiz); Methodology, O.T.-B., L.I.A., A.V. and I.B. (Ioan Botiz); Formal analysis, I.B. (Iulia Babutan), O.T.-B. and I.B. (Ioan Botiz); Investigation, I.B. (Iulia Babutan) and I.B. (Ioan Botiz); Writing—original draft, I.B. (Iulia Babutan); Writing—review & editing, L.I.A., A.V. and I.B. (Ioan Botiz); Supervision, I.B. (Ioan Botiz); Project administration, A.V.; Funding acquisition, A.V. All authors have read and agreed to the published version of the manuscript.

Funding: A.V. and I.B. acknowledge the financial support of the Romanian National Authority for Scientific Research and Innovation, CNCS-UEFISCDI, project no. PN-III-P1-1.1-TE-2021-0388.

Institutional Review Board Statement: Not applicable.

Informed Consent Statement: Not applicable.

Data Availability Statement: The data presented in this study are available on request from the corresponding author.

Conflicts of Interest: The authors declare no conflict of interest.

References

1. Ronca, S. Chapter 10—Polyethylene. In *Brydson's Plastics Materials*, 8th ed.; Gilbert, M., Ed.; Butterworth-Heinemann: Oxford, UK; pp. 247–278, ISBN 978-0-323-35824-8.
2. Babutan, I.; Lucaci, A.-D.; Botiz, I. Antimicrobial Polymeric Structures Assembled on Surfaces. *Polymers* **2021**, *13*, 1552. [CrossRef] [PubMed]
3. Moohan, J.; Stewart, S.A.; Espinosa, E.; Rosal, A.; Rodríguez, A.; Larrañeta, E.; Donnelly, R.F.; Domínguez-Robles, J. Cellulose Nanofibers and Other Biopolymers for Biomedical Applications. A Review. *Appl. Sci.* **2020**, *10*, 65. [CrossRef]
4. Pattanashetti, N.A.; Heggannavar, G.B.; Kariduraganavar, M.Y. Smart Biopolymers and Their Biomedical Applications. *Procedia Manuf.* **2017**, *12*, 263–279. [CrossRef]
5. Smith, M.; Kar-Narayan, S. Piezoelectric Polymers: Theory, Challenges and Opportunities. *Int. Mater. Rev.* **2022**, *67*, 65–88. [CrossRef]
6. Yao, H.; Fan, Z.; Cheng, H.; Guan, X.; Wang, C.; Sun, K.; Ouyang, J. Recent Development of Thermoelectric Polymers and Composites. *Macromol. Rapid Commun.* **2018**, *39*, 1700727. [CrossRef]
7. Tarcan, R.; Handrea-Dragan, M.; Leordean, C.-I.; Cioban, R.C.; Kiss, G.-Z.; Zaharie-Butucel, D.; Farcau, C.; Vulpoi, A.; Simon, S.; Botiz, I. Development of Polymethylmethacrylate/Reduced Graphene Oxide Composite Films as Thermal Interface Materials. *J. Appl. Polym. Sci.* **2022**, *139*, e53238. [CrossRef]
8. Foster, D.P.; Majumdar, D. Critical Behavior of Magnetic Polymers in Two and Three Dimensions. *Phys. Rev. E* **2021**, *104*, 024122. [CrossRef]
9. Botiz, I.; Durbin, M.M.; Stingelin, N. Providing a Window into the Phase Behavior of Semiconducting Polymers. *Macromolecules* **2021**, *54*, 5304–5320. [CrossRef]
10. Botiz, I.; Astilean, S.; Stingelin, N. Altering the Emission Properties of Conjugated Polymers. *Polym. Int.* **2016**, *65*, 157–163. [CrossRef]
11. Nguyen, H.V.-T.; Jiang, Y.; Mohapatra, S.; Wang, W.; Barnes, J.C.; Oldenhuis, N.J.; Chen, K.K.; Axelrod, S.; Huang, Z.; Chen, Q.; et al. Bottlebrush Polymers with Flexible Enantiomeric Side Chains Display Differential Biological Properties. *Nat. Chem.* **2022**, *14*, 85–93. [CrossRef]
12. Song, X.; Gan, B.; Qi, S.; Guo, H.; Tang, C.Y.; Zhou, Y.; Gao, C. Intrinsic Nanoscale Structure of Thin Film Composite Polyamide Membranes: Connectivity, Defects, and Structure–Property Correlation. *Environ. Sci. Technol.* **2020**, *54*, 3559–3569. [CrossRef] [PubMed]

13. Shi, Y.; Guo, H.; Qin, M.; Wang, Y.; Zhao, J.; Sun, H.; Wang, H.; Wang, Y.; Zhou, X.; Facchetti, A.; et al. Imide-Functionalized Thiazole-Based Polymer Semiconductors: Synthesis, Structure–Property Correlations, Charge Carrier Polarity, and Thin-Film Transistor Performance. *Chem. Mater.* **2018**, *30*, 7988–8001. [CrossRef]
14. Botiz, I. Prominent Processing Techniques to Manipulate Semiconducting Polymer Microstructure. *J. Mater. Chem. C* **2023**, *11*, 364–405. [CrossRef]
15. Botiz, I.; Freyberg, P.; Leordean, C.; Gabudean, A.-M.; Astilean, S.; Yang, A.C.-M.; Stingelin, N. Emission Properties of MEH-PPV in Thin Films Simultaneously Illuminated and Annealed at Different Temperatures. *Synth. Met.* **2015**, *199*, 33–36. [CrossRef]
16. Peng, Z.; Stingelin, N.; Ade, H.; Michels, J.J. A Materials Physics Perspective on Structure-Processing-Function Relations in Blends of Organic Semiconductors. *Nat. Rev. Mater.* **2023**, 1–17. [CrossRef]
17. Dimov, I.B.; Moser, M.; Malliaras, G.G.; McCulloch, I. Semiconducting Polymers for Neural Applications. *Chem. Rev.* **2022**, *122*, 4356–4396. [CrossRef] [PubMed]
18. de Leon, A.C.C.; da Silva, Í.G.M.; Pangilinan, K.D.; Chen, Q.; Caldona, E.B.; Advincula, R.C. High Performance Polymers for Oil and Gas Applications. *React. Funct. Polym.* **2021**, *162*, 104878. [CrossRef]
19. Pham, Q.-T.; Chern, C.-S. Applications of Polymers in Lithium-Ion Batteries with Enhanced Safety and Cycle Life. *J. Polym. Res.* **2022**, *29*, 124. [CrossRef]
20. Yarali, E.; Baniasadi, M.; Zolfagharian, A.; Chavoshi, M.; Arefi, F.; Hossain, M.; Bastola, A.; Ansari, M.; Foyouzat, A.; Dabbagh, A.; et al. Magneto-/ Electro-responsive Polymers toward Manufacturing, Characterization, and Biomedical/Soft Robotic Applications. *Appl. Mater. Today* **2022**, *26*, 101306. [CrossRef]
21. Angel, N.; Li, S.; Yan, F.; Kong, L. Recent Advances in Electrospinning of Nanofibers from Bio-Based Carbohydrate Polymers and Their Applications. *Trends Food Sci. Technol.* **2022**, *120*, 308–324. [CrossRef]
22. He, Y.; Kukhta, N.A.; Marks, A.; Luscombe, C.K. The Effect of Side Chain Engineering on Conjugated Polymers in Organic Electrochemical Transistors for Bioelectronic Applications. *J. Mater. Chem. C* **2022**, *10*, 2314–2332. [CrossRef] [PubMed]
23. Handrea-Dragan, M.; Botiz, I. Multifunctional Structured Platforms: From Patterning of Polymer-Based Films to Their Subsequent Filling with Various Nanomaterials. *Polymers* **2021**, *13*, 445. [CrossRef] [PubMed]
24. Mohanty, A.K.; Wu, F.; Mincheva, R.; Hakkarainen, M.; Raquez, J.-M.; Mielewski, D.F.; Narayan, R.; Netravali, A.N.; Misra, M. Sustainable Polymers. *Nat. Rev. Methods Primers* **2022**, *2*, 1–27. [CrossRef]
25. Chohan, J.S.; Boparai, K.S.; Singh, R.; Hashmi, M.S.J. Manufacturing Techniques and Applications of Polymer Matrix Composites: A Brief Review. *Adv. Mater. Process. Technol.* **2022**, *8*, 884–894. [CrossRef]
26. Sun, C.; Zhu, C.; Meng, L.; Li, Y. Quinoxaline-Based D–A Copolymers for the Applications as Polymer Donor and Hole Transport Material in Polymer/Perovskite Solar Cells. *Adv. Mater.* **2022**, *34*, 2104161. [CrossRef]
27. Ritsema van Eck, G.C.; Chiappisi, L.; de Beer, S. Fundamentals and Applications of Polymer Brushes in Air. *ACS Appl. Polym. Mater.* **2022**, *4*, 3062–3087. [CrossRef]
28. Kirillova, A.; Yeazel, T.R.; Asheghali, D.; Petersen, S.R.; Dort, S.; Gall, K.; Becker, M.L. Fabrication of Biomedical Scaffolds Using Biodegradable Polymers. *Chem. Rev.* **2021**, *121*, 11238–11304. [CrossRef] [PubMed]
29. Fischer, F.S.U.; Tremel, K.; Saur, A.-K.; Link, S.; Kayunkid, N.; Brinkmann, M.; Herrero-Carvajal, D.; Navarrete, J.T.L.; Delgado, M.C.R.; Ludwigs, S. Influence of Processing Solvents on Optical Properties and Morphology of a Semicrystalline Low Bandgap Polymer in the Neutral and Charged States. *Macromolecules* **2013**, *46*, 4924–4931. [CrossRef]
30. Baklar, M.A.; Koch, F.; Kumar, A.; Buchaca Domingo, E.; Campoy-Quiles, M.; Feldman, K.; Yu, L.; Wöbkenberg, P.; Ball, J.; Wilson, R.M.; et al. Solid-State Processing of Organic Semiconductors. *Adv. Mater.* **2010**, *22*, 3942–3947. [CrossRef]
31. Botiz, I.; Codescu, M.-A.; Farcau, C.; Leordean, C.; Astilean, S.; Silva, C.; Stingelin, N. Convective Self-Assembly of π-Conjugated Oligomers and Polymers. *J. Mater. Chem. C* **2017**, *5*, 2513–2518. [CrossRef]
32. Peterson, G.W.; Lee, D.T.; Barton, H.F.; Epps, T.H.; Parsons, G.N. Fibre-Based Composites from the Integration of Metal–Organic Frameworks and Polymers. *Nat. Rev. Mater.* **2021**, *6*, 605–621. [CrossRef]
33. Chowdhury, M.; Sajjad, M.T.; Savikhin, V.; Hergué, N.; Sutija, K.B.; Oosterhout, S.D.; Toney, M.F.; Dubois, P.; Ruseckas, A.; Samuel, I.D.W. Tuning Crystalline Ordering by Annealing and Additives to Study Its Effect on Exciton Diffusion in a Polyalkylthiophene Copolymer. *Phys. Chem. Chem. Phys.* **2017**, *19*, 12441–12451. [CrossRef] [PubMed]
34. Todor-Boer, O.; Petrovai, I.; Tarcan, R.; Vulpoi, A.; David, L.; Astilean, S.; Botiz, I. Enhancing Photoluminescence Quenching in Donor–Acceptor PCE11:PPCBMB Films through the Optimization of Film Microstructure. *Nanomaterials* **2019**, *9*, 1757. [CrossRef] [PubMed]
35. Li, Q.-Y.; Yao, Z.-F.; Lu, Y.; Zhang, S.; Ahmad, Z.; Wang, J.-Y.; Gu, X.; Pei, J. Achieving High Alignment of Conjugated Polymers by Controlled Dip-Coating. *Adv. Electron. Mater.* **2020**, *6*, 2000080. [CrossRef]
36. Xiao, M.; Kang, B.; Lee, S.B.; Perdigão, L.M.A.; Luci, A.; Warr, D.A.; Senanayak, S.P.; Nikolka, M.; Statz, M.; Wu, Y.; et al. Anisotropy of Charge Transport in a Uniaxially Aligned Fused Electron-Deficient Polymer Processed by Solution Shear Coating. *Adv. Mater.* **2020**, *32*, 2000063. [CrossRef]
37. Basu, A.; Niazi, M.R.; Scaccabarozzi, A.D.; Faber, H.; Fei, Z.; Anjum, D.H.; Paterson, A.F.; Boltalina, O.; Heeney, M.; Anthopoulos, T.D. Impact of P-Type Doping on Charge Transport in Blade-Coated Small-Molecule:Polymer Blend Transistors. *J. Mater. Chem. C* **2020**, *8*, 15368–15376. [CrossRef]
38. Grozev, N.; Botiz, I.; Reiter, G. Morphological Instabilities of Polymer Crystals. *Eur. Phys. J. E* **2008**, *27*, 63–71. [CrossRef]

39. Botiz, I.; Grozev, N.; Schlaad, H.; Reiter, G. The Influence of Protic Non-Solvents Present in the Environment on Structure Formation of Poly(γ-Benzyl-L-Glutamate in Organic Solvents. *Soft Matter* **2008**, *4*, 993–1002. [CrossRef]
40. Darko, C.; Botiz, I.; Reiter, G.; Breiby, D.W.; Andreasen, J.W.; Roth, S.V.; Smilgies, D.M.; Metwalli, E.; Papadakis, C.M. Crystallization in Diblock Copolymer Thin Films at Different Degrees of Supercooling. *Phys. Rev. E* **2009**, *79*, 041802. [CrossRef]
41. Jahanshahi, K.; Botiz, I.; Reiter, R.; Thomann, R.; Heck, B.; Shokri, R.; Stille, W.; Reiter, G. Crystallization of Poly(γ-Benzyl L-Glutamate) in Thin Film Solutions: Structure and Pattern Formation. *Macromolecules* **2013**, *46*, 1470–1476. [CrossRef]
42. Jahanshahi, K.; Botiz, I.; Reiter, R.; Scherer, H.; Reiter, G. Reversible Nucleation, Growth, and Dissolution of Poly(γ-Benzyl l-Glutamate) Hexagonal Columnar Liquid Crystals by Addition and Removal of a Nonsolvent. *Cryst. Growth Des.* **2013**, *13*, 4490–4494. [CrossRef]
43. Singh, M.; Agrawal, A.; Wu, W.; Masud, A.; Armijo, E.; Gonzalez, D.; Zhou, S.; Terlier, T.; Zhu, C.; Strzalka, J.; et al. Soft-Shear-Aligned Vertically Oriented Lamellar Block Copolymers for Template-Free Sub-10 Nm Patterning and Hybrid Nanostructures. *ACS Appl. Mater. Interfaces* **2022**, *14*, 12824–12835. [CrossRef] [PubMed]
44. Zhang, Q.; Weber, C.; Schubert, U.S.; Hoogenboom, R. Thermoresponsive Polymers with Lower Critical Solution Temperature: From Fundamental Aspects and Measuring Techniques to Recommended Turbidimetry Conditions. *Mater. Horiz.* **2017**, *4*, 109–116. [CrossRef]
45. Macedo, A.S.; Carvalho, E.O.; Cardoso, V.F.; Correia, D.M.; Tubio, C.R.; Fidalgo-Marijuan, A.; Botelho, G.; Lanceros-Méndez, S. Tailoring Electroactive Poly(Vinylidene Fluoride-co-Trifluoroethylene) Microspheres by a Nanoprecipitation Method. *Mater. Lett.* **2020**, *261*, 127018. [CrossRef]
46. Lin, X.; Liu, R.; Ding, C.; Deng, J.; Guo, Y.; Long, S.; Li, L.; Li, M. Modulation of Microstructure and Charge Transport in Polymer Monolayer Transistors by Solution Aging. *Chin. J. Chem.* **2021**, *39*, 3079–3084. [CrossRef]
47. Khalil, Y.; Hopkinson, N.; Kowalski, A.; Fairclough, J.P.A. Characterisation of UHMWPE Polymer Powder for Laser Sintering. *Materials* **2019**, *12*, 3496. [CrossRef]
48. Jana, A.; Selvaraj, S.; Subramani, K. A Novel Technique for the Development of Acetabular Cup by Cold Isostatic Compaction and Sintering of UHMWPE Powder with Optimized Processing Parameters. *Polym. Eng. Sci.* **2021**, *61*, 2536–2556. [CrossRef]
49. Khan, A.L.T.; Sreearunothai, P.; Herz, L.M.; Banach, M.J.; Köhler, A. Morphology-Dependent Energy Transfer within Polyfluorene Thin Films. *Phys. Rev. B* **2004**, *69*, 085201. [CrossRef]
50. Danesh, C.D.; Starkweather, N.S.; Zhang, S. In Situ Study of Dynamic Conformational Transitions of a Water-Soluble Poly (3-Hexylthiophene) Derivative by Surfactant Complexation. *J. Phys. Chem. B* **2012**, *116*, 12887–12894. [CrossRef]
51. Adachi, T.; Tong, L.; Kuwabara, J.; Kanbara, T.; Saeki, A.; Seki, S.; Yamamoto, Y. Spherical Assemblies from π-Conjugated Alternating Copolymers: Toward Optoelectronic Colloidal Crystals. *J. Am. Chem. Soc.* **2013**, *135*, 870–876. [CrossRef]
52. Guha, S.; Chandrasekhar, M.; Scherf, U.; Knaapila, M. Tuning Structural and Optical Properties of Blue-Emitting Polymeric Semiconductors. *Phys. Status Solidi B* **2011**, *248*, 1083–1090. [CrossRef]
53. Tung, K.-P.; Chen, C.-C.; Lee, P.; Liu, Y.-W.; Hong, T.-M.; Hwang, K.C.; Hsu, J.H.; White, J.D.; Yang, A.C.-M. Large Enhancements in Optoelectronic Efficiencies of Nano-Plastically Stressed Conjugated Polymer Strands. *ACS Nano* **2011**, *5*, 7296–7302. [CrossRef] [PubMed]
54. Loo, Y.L.; Register, R.A.; Ryan, A.J. Modes of Crystallization in Block Copolymer Microdomains: Breakout, Templated, and Confined. *Macromolecules* **2002**, *35*, 2365–2374. [CrossRef]
55. Bao, J.; Dong, X.; Chen, S.; Lu, W.; Zhang, X.; Chen, W. Confined Crystallization, Melting Behavior and Morphology in PEG-b-PLA Diblock Copolymers: Amorphous versus Crystalline PLA. *J. Polym. Sci.* **2020**, *58*, 455–465. [CrossRef]
56. MacFarlane, L.R.; Shaikh, H.; Garcia-Hernandez, J.D.; Vespa, M.; Fukui, T.; Manners, I. Functional Nanoparticles through π-Conjugated Polymer Self-Assembly. *Nat. Rev. Mater.* **2021**, *6*, 7–26. [CrossRef]
57. Stevens, C.A.; Kaur, K.; Klok, H.-A. Self-Assembly of Protein-Polymer Conjugates for Drug Delivery. *Adv. Drug Deliv. Rev.* **2021**, *174*, 447–460. [CrossRef]
58. Scanga, R.A.; Reuther, J.F. Helical Polymer Self-Assembly and Chiral Nanostructure Formation. *Polym. Chem.* **2021**, *12*, 1857–1897. [CrossRef]
59. Kos, P.I.; Ivanov, V.A.; Chertovich, A.V. Crystallization of Semiflexible Polymers in Melts and Solutions. *Soft Matter* **2021**, *17*, 2392–2403. [CrossRef]
60. Jin, F.; Yuan, S.; Wang, S.; Zhang, Y.; Zheng, Y.; Hong, Y.; Miyoshi, T. Polymer Chains Fold Prior to Crystallization. *ACS Macro Lett.* **2022**, *11*, 284–288. [CrossRef]
61. Sheng, J.; Chen, W.; Cui, K.; Li, L. Polymer Crystallization under External Flow. *Rep. Prog. Phys.* **2022**, *85*, 036601. [CrossRef]
62. Xu, W.; Li, X.; Zheng, Y.; Yuan, W.; Zhou, J.; Yu, C.; Bao, Y.; Shan, G.; Pan, P. Hierarchical Ordering and Multilayer Structure of Poly(ε-Caprolactone) End-Functionalized by a Liquid Crystalline Unit: Role of Polymer Crystallization. *Polym. Chem.* **2021**, *12*, 4175–4183. [CrossRef]
63. Hamley, I.W. Crystallization in Block Copolymers. In *Interfaces Crystallization Viscoelasticity*; Springer: Berlin/Heidelberg, Germany, 1999; pp. 113–137. ISBN 978-3-540-48836-1.
64. Castillo, R.V.; Müller, A.J. Crystallization and Morphology of Biodegradable or Biostable Single and Double Crystalline Block Copolymers. *Prog. Polym. Sci.* **2009**, *34*, 516–560. [CrossRef]

65. Le, T.P.; Smith, B.H.; Lee, Y.; Litofsky, J.H.; Aplan, M.P.; Kuei, B.; Zhu, C.; Wang, C.; Hexemer, A.; Gomez, E.D. Enhancing Optoelectronic Properties of Conjugated Block Copolymers through Crystallization of Both Blocks. *Macromolecules* **2020**, *53*, 1967–1976. [CrossRef]
66. He, Y.; Eloi, J.-C.; Harniman, R.L.; Richardson, R.M.; Whittell, G.R.; Mathers, R.T.; Dove, A.P.; O'Reilly, R.K.; Manners, I. Uniform Biodegradable Fiber-Like Micelles and Block Comicelles via "Living" Crystallization-Driven Self-Assembly of Poly(l-Lactide) Block Copolymers: The Importance of Reducing Unimer Self-Nucleation via Hydrogen Bond Disruption. *J. Am. Chem. Soc.* **2019**, *141*, 19088–19098. [CrossRef] [PubMed]
67. Hamley, I.W. Block Copolymers. In *Encyclopedia of Polymer Science and Technology*; John Wiley & Sons, Ltd.: Hoboken, NJ, USA, 2002; Volume 1, pp. 457–482.
68. Xiong, S.; Li, D.; Hur, S.-M.; Craig, G.S.W.; Arges, C.G.; Qu, X.-P.; Nealey, P.F. The Solvent Distribution Effect on the Self-Assembly of Symmetric Triblock Copolymers during Solvent Vapor Annealing. *Macromolecules* **2018**, *51*, 7145–7151. [CrossRef]
69. Shi, L.-Y.; Yin, C.; Zhou, B.; Xia, W.; Weng, L.; Ross, C.A. Annealing Process Dependence of the Self-Assembly of Rod–Coil Block Copolymer Thin Films. *Macromolecules* **2021**, *54*, 1657–1664. [CrossRef]
70. Ginige, G.; Song, Y.; Olsen, B.C.; Luber, E.J.; Yavuz, C.T.; Buriak, J.M. Solvent Vapor Annealing, Defect Analysis, and Optimization of Self-Assembly of Block Copolymers Using Machine Learning Approaches. *ACS Appl. Mater. Interfaces* **2021**, *13*, 28639–28649. [CrossRef] [PubMed]
71. Atanase, L.I.; Lerch, J.-P.; Caprarescu, S.; Iurciuc, C.E.; Riess, G. Micellization of PH-Sensitive Poly(Butadiene)-Block-Poly(2 Vinylpyridine)-Block-Poly(Ethylene Oxide) Triblock Copolymers: Complex Formation with Anionic Surfactants. *J. Appl. Polym. Sci.* **2017**, *134*, 45313. [CrossRef]
72. Zhao, W.; Su, Y.; Müller, A.J.; Gao, X.; Wang, D. Direct Relationship Between Interfacial Microstructure and Confined Crystallization in Poly(Ethylene Oxide)/Silica Composites: The Study of Polymer Molecular Weight Effects. *J. Polym. Sci. B Polym. Phys.* **2017**, *55*, 1608–1616. [CrossRef]
73. Babutan, I.; Todor-Boer, O.; Atanase, L.I.; Vulpoi, A.; Simon, S.; Botiz, I. Self-Assembly of Block Copolymers on Surfaces Exposed to Space-Confined Solvent Vapor Annealing. *Polymer* **2023**, *273*, 125881. [CrossRef]
74. Babutan, I.; Todor-Boer, O.; Atanase, L.I.; Vulpoi, A.; Botiz, I. Self-Assembly of Block Copolymers in Thin Films Swollen-Rich in Solvent Vapors. *Polymers* **2023**, *15*, 1900. [CrossRef] [PubMed]
75. Reiter, G.; Botiz, I.; Graveleau, L.; Grozev, N.; Albrecht, K.; Mourran, A.; Möller, M. Morphologies of Polymer Crystals in Thin Films. In *Lecture Notes in Physics: Progress in Understanding of Polymer Crystallization*; Reiter, G., Strobl, G.R., Eds.; Springer: Berlin/Heidelberg, Germany, 2007; Volume 714, pp. 179–200. [CrossRef]
76. Soum, A.; Fontanille, M.; Sigwalt, P. Anionic Polymerization of 2-Vinylpyridine Initiated by Symmetrical Organomagnesium Compounds in Tetrahydrofuran. *J. Polym. Sci. Polym. Chem. Ed.* **1977**, *15*, 659–673. [CrossRef]
77. Atanase, L.I.; Riess, G. Stabilization of Non-Aqueous Emulsions by Poly(2-Vinylpyridine)-b-Poly(Butadiene) Block Copolymers. *Colloids Surf. A Physicochem. Eng. Asp.* **2014**, *458*, 19–24. [CrossRef]
78. Botiz, I.; Schlaad, H.; Reiter, G. Processes of Ordered Structure Formation in Polypeptide Thin Film Solutions. In *Self Organized Nanostructures of Amphiphilic Block Copolymers II*; Springer: Berlin/Heidelberg, Germany, 2011; Volume 242, pp. 117–149.
79. Kovacs, A.J.; Straupe, C.; Gonthier, A. Isothermal Growth, Thickening, and Melting of Polyethylene Oxide) Single Crystals in the Bulk. II. *J. Polym. Sci. Polym. Symp.* **1977**, *59*, 31–54. [CrossRef]
80. Lysenko, E.A.; Bronich, T.K.; Slonkina, E.V.; Eisenberg, A.; Kabanov, V.A.; Kabanov, A.V. Block Ionomer Complexes with Polystyrene Core-Forming Block in Selective Solvents of Various Polarities. 2. Solution Behavior and Self-Assembly in Nonpolar Solvents. *Macromolecules* **2002**, *35*, 6344–6350. [CrossRef]
81. Changez, M.; Kang, N.-G.; Koh, H.-D.; Lee, J.-S. Effect of Solvent Composition on Transformation of Micelles to Vesicles of Rod−Coil Poly(n-Hexyl Isocyanate-Block-2-Vinylpyridine) Diblock Copolymers. *Langmuir* **2010**, *26*, 9981–9985. [CrossRef]
82. Ferreiro, V.; Douglas, J.F.; Warren, J.; Karim, A. Growth Pulsations in Symmetric Dendritic Crystallization in Thin Polymer Blend Films. *Phys. Rev. E* **2002**, *65*, 051606. [CrossRef]
83. Ferreiro, V.; Douglas, J.F.; Warren, J.A.; Karim, A. Nonequilibrium Pattern Formation in the Crystallization of Polymer Blend Films. *Phys. Rev. E* **2002**, *65*, 042802. [CrossRef]
84. Riess, G. Micellization of Block Copolymers. *Prog. Polym. Sci.* **2003**, *28*, 1107–1170. [CrossRef]
85. Lerch, J.-P. Synthesis and Characterization of Amphiphilic Block- and Graft Terpolymers. Study of Inorganic Dispersions in Aqueous and Non-Aqueous Media. Ph.D. Thesis, University Haute Alsace, Mulhouse, France, 1996.

Disclaimer/Publisher's Note: The statements, opinions and data contained in all publications are solely those of the individual author(s) and contributor(s) and not of MDPI and/or the editor(s). MDPI and/or the editor(s) disclaim responsibility for any injury to people or property resulting from any ideas, methods, instructions or products referred to in the content.

Article

Microscale Templating of Materials across Electrospray Deposition Regimes

Michael J. Grzenda [1], Maria Atzampou [2], Alfusainey Samateh [3], Andrei Jitianu [3,4], Jeffrey D. Zahn [2] and Jonathan P. Singer [1,5,*]

[1] Department of Materials Science and Engineering, Rutgers University, Piscataway, NJ 08854, USA
[2] Department of Biomedical Engineering, Rutgers University, Piscataway, NJ 08854, USA
[3] Department of Chemistry, Lehman College, of the City University of New York, New York, NY 10468, USA
[4] Ph.D. Program in Chemistry and Biochemistry, The Graduate Center of the City University of New York, 365 Fifth Avenue, New York, NY 10016, USA
[5] Department of Mechanical and Aerospace Engineering, Rutgers University, Piscataway, NJ 08854, USA
* Correspondence: jonathan.singer@rutgers.edu

Abstract: Electrospray deposition (ESD) uses strong electric fields to produce generations of monodisperse droplets from solutions and dispersions that are driven toward grounded substrates. When soft materials are delivered, the behavior of the growing film depends on the film's ability to dissipate charge, which is strongly tied to its mobility for dielectric materials. Accordingly, there exist three regimes of electrospray: electrowetting, charged melt, and self-limiting. In the self-limiting regime, it has been recently shown that the targeted nature of these sprays allows for corona-free 3D coating. While ESD patterning on the micron-scale has been studied for decades, most typically through the use of insulating masks, there has been no comparative study of this phenomenon across spray regimes. Here, we used test-patterns composed of gratings that range in both feature size (30–240 µm) and spacing ($1/3$x–9x) to compare materials across regimes. The sprayed patterns were scanned using a profilometer, and the density, average height, and specificity were extracted. From these results, it was demonstrated that material deposited in the self-limiting regime showed the highest uniformity and specificity on small features as compared to electrowetting and charged melt sprays. Self-limiting electrospray deposition is, therefore, the best suited for modification of prefabricated electrode patterns.

Keywords: electrospray deposition; electronic packaging; fabrication; microtechnology; thin films

Citation: Grzenda, M.J.; Atzampou, M.; Samateh, A.; Jitianu, A.; Zahn, J.D.; Singer, J.P. Microscale Templating of Materials across Electrospray Deposition Regimes. *Coatings* **2023**, *13*, 599. https://doi.org/10.3390/coatings13030599

Academic Editor: Rainer Hippler

Received: 10 February 2023
Revised: 3 March 2023
Accepted: 6 March 2023
Published: 11 March 2023

Copyright: © 2023 by the authors. Licensee MDPI, Basel, Switzerland. This article is an open access article distributed under the terms and conditions of the Creative Commons Attribution (CC BY) license (https://creativecommons.org/licenses/by/4.0/).

1. Introduction

As microscale manufacturing grows in scale, the cost of production can be greatly reduced through manufacturing techniques that operate at ambient pressures and low temperatures. While electrosprays have long been a keystone of certain large-scale manufacturing industries such as paint coating and agriculture, electrospray deposition (ESD) has yet to reach broad industrial-scale application. This distinct regime of electrospray operates at low flowrates and relies on only electrostatic forces to deliver material. ESD has several unique capabilities for thin film [1–4] and nanoparticle [5–8] processing that are more relevant now than ever. However, further understanding and demonstration of the capabilities of ESD are necessary before employing it in, for example, the creation of dielectric coatings, hermetic barriers, and the deposition of active sensor materials. More specifically, this manuscript explores the potential for ESD to be combined with microfabricated templates without the need of a stencil mask through understanding and leveraging of charge dissipation mechanisms.

During ESD, a strong electric field is applied to a capillary containing a solution or dispersion. This causes the exiting fluid to undergo electrostatic breakdown and produce generations of charged monodisperse droplets that deliver their payloads to a grounded target [9]. For ESD of non-conductive soft matter, we have categorized this behavior into three regimes that depend on the ability of the deposited film to dissipate charge [10]. If the deposited film maintains a certain level of mobility and low viscosity, it will thin to increase surface area and conduct charge to uncoated regions, which is called electrowetting. While spraying liquids or solid materials above their glass transition temperature (T_g) is an obvious way to maintain mobility, it has also been observed that higher temperatures can induce greater mobility by increasing the solvent absorption, which, in turn, decreases the T_g. The charged melt regime also depends on mobile material, but in this case, the film's higher viscosity favors instabilities to dissipate charge. Due to the dependence on mobility, these electrospray regimes exist in a balance, and it has been observed that electrowetting materials can transition to charged melts as the material thickness increases. Further, if the surface affinity to the target is poor, either regime can result in dewetted films. On the completely opposite end of the spectrum is the self-limiting (SL) regime where glassy materials below their effective T_g are sprayed in a phenomenon we have named self-limiting electrospray deposition (SLED). In the SL regime, charge is unable to dissipate from the growing film, and this charge build-up eventually results in the repulsion of nearly all the incoming material. This regime also transitions to the charged melt regime as T_g is approached through either increased temperature or blending with solvent vapor or another material, as the charge is better able to dissipate through mass transport [11]. One of the main advantages of SLED is the ability to conformally coat conductive/charge mobile surfaces, and much of our prior work has focused on 3D, corona-free coatings [12]. However, the ability to target micron-scale 2D patterns is also highly valuable, but in need of further characterization.

While several studies have patterned electrosprayed materials simply by controlling local electric fields [13,14], the typical approach has been through some form of masking, with authors depositing nanomaterials (both organic and inorganic), biomaterials, and polymers [15–22]. In one of the earliest examples of patterning with ESD, Buchko et al. used a shadow mask for depositing arrays of polypeptides, with the term shadow mask implying that the mask was conductive, and the substrate and mask were equally coated like a stencil [23]. It was Morozov and Morozova who first described the use of an insulating mask for the deposition of arrays of DNA material with the observation that the electrospray quickly traps charge on the mask, leading to a repulsion that focuses the spray onto conductive regions [24]. Importantly, this focusing increases the deposition thickness relative to an unmasked substrate. Hu et al. also observed that focusing could even be achieved without a mask when spraying block copolymers onto glass using a grounded needle and a high voltage grating beneath the substrate to direct the spray [14]. Interestingly, their (likely charged melt) spray was observed to spread beyond the width of the buried grating.

While material focusing can be useful, especially when attempting to achieve small linewidths, we must also consider spray uniformity. For unpatterned, non-SLED sprays, deposition tends to lead to a peaked distribution. On the other hand, focusing depends greatly on geometry, meaning that the spray of complex 2D patterns will lead to large variations in the amount of material deposited, which can be especially problematic when manufacturing multi-layer materials. This was encountered by Morozov and Morozova, who tried to counteract the effects either by moving the spray platform or by adding a shield to the spray environment [24]. Though successful to a degree, their array features were all of equal size, so it is unclear if this effect would carry over to more complex geometries.

Further, work from our group has shown that even materials sprayed in the SLED regime can be impacted by feature size. In a recent study, focused laser-spike (FlaSk) dewetting was used to pattern circular holes of varying sizes on photoresist-coated silicon wafers as templates [22]. After spraying with polyvinylpyrrolidone in the SL regime, it was observed that larger features tended to have thicker coatings, and that all features had coatings that were thicker than an unmasked substrate. This is curious in light of our initial work that showed that thickness in the SL regime was dependent only on electric field strength, and that below the SL thickness, total sprayed mass was the only determining factor. However, at the time, it was observed that deposited materials were slightly collapsed, indicating that excessive solvent vapor may have been generated due to focusing from the template. As mentioned earlier, increased solvent absorption can lead to charge dissipation and electrowetting/charged melt behavior. These effects can, therefore, be mitigated in three ways: (1) the total patterned area can be increased, (2) the flowrate can be reduced, and (3) the strength of the dielectric layer can be reduced to decrease the effect of focusing. Regarding the third point, Zhu and Chiarot observed that thinner photoresist layers allowed for deposited nanoparticles to strike closer to recessed feature edges, indicating a reduction in the focusing effect [15].

Here, we aim to elucidate the performance of all three spray regimes by spraying substrates composed of silicon chips patterned with Ti/Pt on top of an insulating layer of Parylene. Furthermore, we aim to quantify the interaction of complex 2D geometry with ESD through the use of a metal pattern that includes multiple gratings of varying sizes and spacing. A MATLAB script is used to extract values for average deposition height, volumetric density, along with a simple figure of merit that characterizes the specificity of the spray to the template.

2. Materials and Methods

2.1. Materials

Polystyrene (35 kDa) and 2-Butanone (>99%) were purchased from Sigma Aldrich (St. Louis, MO, USA) and were used as received. Polystyrene (PS) was added to butanone at a mass loading of 1 wt.% and left on a roller overnight. A "melting gel" material was used to demonstrate a fully liquid, electrowetting spray. Melting gels (MGs) are oligomeric silsesquioxanes (synthesized via a sol–gel process described elsewhere [25]) that show thermoplastic-like behavior above their T_g, but cross-link into hybrid organic/inorganic glasses when heated past their consolidation temperature [26]. The melting gel used here was composed of 65% methyltriethoxysilane (MTES) and 35% dimethyldiethoxysilane (DMDES) (in mol%) and had a T_g below 0 °C. The MG was diluted using absolute ethanol to 1 wt.% MG for spray. All sprays were performed on Ti/Pt patterns on 4–5 μm thick Parylene insulation. These substrates were made in-house via photolithography and acetone lift-off using 4-inch silicon wafer substrates.

2.2. Feature Test Patterns

The feature test patterns were designed to assess a series of gratings that range in both grating width and spacing (Figure 1). The gratings were attached to a 1×1 cm^2 grid used as a grounding path connected to a 0.5 cm^2 grounding pad. Every individual grating filled an approximate width of 1.5 mm and the features had a length of 1 mm. On the interior of the grounding grid, the feature width, referred to as the feature size, increased in multiples of 15 (15/30/60/120/240 μm). Around the exterior, the feature sizes increased in multiples of 20 (20/40/80/160 μm). Along each row, the feature size was constant, but the gap between each finger increased as a function of the feature size ($^1/_3$x, 1x, 2x, 3x, and 9x). For example, the 30 μm row had gaps of 10, 30, 60, 90, and 270 μm. For this work, we focused on the 30/60/120/240 μm features. The 20 μm features were included to contrast the 30 μm features due to their similar size but different spatial location on the test pattern. However, for both these smaller feature sizes, the $^1/_3$x gratings were excluded because they were too small for the profilometer to measure.

2.3. Electrospray Setup

Strong electric fields were applied via alligator clips to a vertically suspended 30-gauge needle and a stainless-steel ring (ID 2 cm) using high-voltage power supplies (Acopian, P012HA5M, Easton, PA, USA) to achieve electrospray. The ring was placed 1 cm above the needle tip and acted to stabilize the electric field and focus the spray. The voltage on the needle was held at a constant 6 kV DC, while the voltage on the ring varied from 3 to 5.5 kV

DC to maintain stable cone jet sprays. The solution was pumped to the high-voltage vertical needle using a syringe and Teflon tubing using a syringe pump at a rate of 0.1 mL/hr. During sample fabrication, the spray was first stabilized on a grounded, blank, 4-inch Si wafer (p-type; 0–100 Ω cm), stacked on top of a steel plate and hot plate, and located 6 cm below the needle tip. Once the spray was stabilized, the test pattern was placed in the center of the spray and allowed to coat for 30 min. Separately, the grounding pad of the test pattern was taped with carbon tape to a 2-inch wafer (p-type; 0–100 Ω cm) to create a grounding path. This 2-inch wafer was able to interface directly with the grounded 4-inch wafer. While the PS sprays were sprayed in ambient atmosphere (30% humidity/22 °C), the MG was sprayed in a low-humidity chamber (21% humidity/25 °C) due to its moisture sensitivity. For the PS, two sprays were performed, one at a hotplate temperature of 30 °C, which is in the SL regime, and the other at 100 °C, which is near or above T_g due to solvent swelling. The MG was sprayed at 35 °C to completely ensure it was sprayed as a liquid. After spraying, the MG was consolidated overnight at 150 °C to fully crosslink. Although SLED sprays are highly porous, the PS was densified via solvent swelling using a brief exposure to butanone vapor to allow for profilometry scanning.

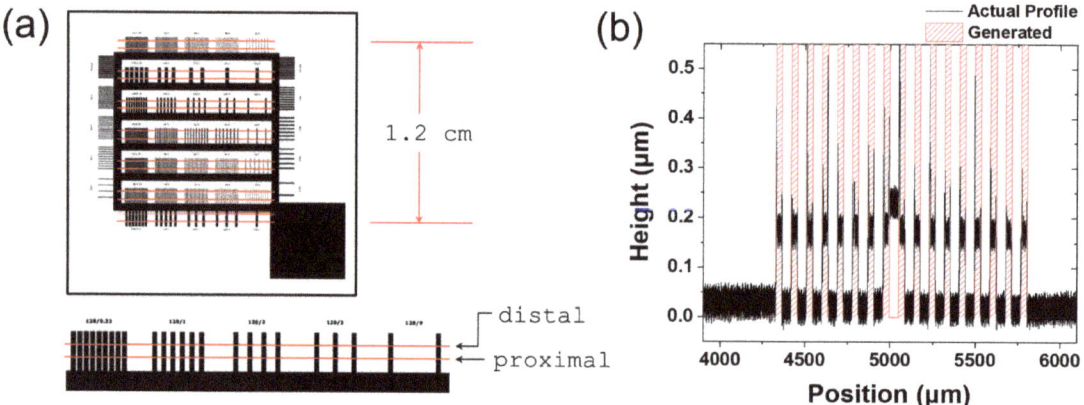

Figure 1. (**a**) (top) Full feature test pattern and (bottom) single row demonstrating the distal/proximal scans. (**b**) Actual uncoated test pattern profile (black) compared to calculated binary profile (shaded red). An error can be observed in the center of the grating that has been corrected in the model.

2.4. Measurement and Analysis

The test patterns were measured using a KLA Tencor P-7 Stylus Profilometer (Milpitas, CA, USA). Each feature was scanned twice, with a 0.5 mm spacing between scans (Figure 1). We differentiate the scans by referring to them as distal or proximal with respect to the grounding grid. Scans were performed with a 100 μm/s scan speed, 200 Hz sampling rate, and a 0.5 μm step size. The stylus had a 2 μm radius and contacted the surface with a pressure of 2 mg per stylus area.

Scans were analyzed using a MATLAB (R2022a, Mathworks®, Natick, MA, USA) script. A blank test pattern was scanned to create a baseline. To remove noise and standardize the analysis, the average feature size (width) was measured from these scans and used to create a binary model of the gratings (Figure 1). The model gaps were calculated for each grating by subtracting the average measured feature size from the total spacing of a gap plus feature size. The measured average width was a few microns larger than the actual feature size due to the stylus diameter. During analysis, the script aligned the collected profile scan with the binary data file (Figures S1–S3, see Supplementary Materials) and calculated 3 parameters: (1) height (H_g) is the average height of the profile directly on top of the features for a given grating; (2) density (ρ_g) is the total amount of material deposited, in area, divided by the length of the features available; (3) specificity (σ_g) is the amount of material on the features divided by the total amount of material deposited, both as areas. The calculation is accomplished through rectangular approximation and array operations:

$$H_g = \frac{\sum (h_p \circ m) \Delta x}{\sum (m) \Delta x} \quad (1)$$

$$\rho_g = \frac{\sum (h_p) \Delta x}{\sum (m) \Delta x} \quad (2)$$

$$\sigma_g = \frac{\sum (h_p \circ m) \Delta x}{\sum (h_p) \Delta x} \quad (3)$$

The height, density, and specificity are given a subscript (g) to denote that each parameter is calculated for every grating. Meanwhile, h_p is the array containing the profile height for the grating, m is the array containing the binary model, Δx is the step size, and ($h_p \circ m$) indicates element-wise multiplication. Although Δx cancels, we leave it here for clarity regarding the rectangular approximation. Despite having the same units, the height is best understood in terms of length (μm) and the density as an area per length (μm^2/μm). On the other hand, the specificity is unitless and ranges from 0 to 1, serving as a representation of how much of the sprayed material landed on the template. A visual representation of these parameters is given in Figure S4. Thanks to the similarity between these equations, the resulting relationship exists that is utilized later:

$$\sigma_g = \frac{H_g}{\rho_g} \quad (4)$$

During analysis, a cutoff value was added so that noise from the profile baseline was not included. These cutoff values were within range of the uncoated feature height (~0.15 μm), so the area of the actual feature was not considered in the density calculation. The analysis was also run with a simulated coating file where the entire profile consisted of $h_p = 1$. This represents a completely indiscriminate coating and is useful for identifying biases in our analyses.

3. Results and Discussion

To determine the importance of the ESD regime in template interactions, we employ several materials, real and simulated, that are expected to deliver highly different behaviors. As mentioned, the simulated data are used to model an indiscriminate coating with a uniform height (labeled $H = 1$ on graphs as we also scale it after analysis), such as would be deposited by, for example, spin coating, highly far-field spray, or vapor deposition. This "material" is useful for identifying biases in our metrics where smaller gaps result in lower densities and higher specificities. Real materials are compared to these biases to ensure observed trends are due to electrostatic phenomena. The melting gel (MG), showing liquid electrowetting behavior, demonstrates the least amount of charge build-up and, therefore, the least amount of repulsion. Without repulsion, a central area receives the most spray. This creates a gradient in spray thickness that, as is shown, almost entirely dominates the behavior observed. The 100 °C polystyrene (PS), demonstrating charged melt behavior, accumulates some charge, largely avoiding the formation of a gradient. However, its ability to spread and form instabilities dissipates some of the charge, leading to higher densities, lower specificities, as well as greater heights on large features. It is also seen that distal areas of the feature are more susceptible to this effect. Finally, 30 °C PS demonstrates a behavior entirely consistent with SLED, with charge accumulation redirecting the spray to uncoated regions. There is no thickness gradient or trend toward metric bias. Its density and height are consistent with the expectation of uniform growth on all features, regardless of size. Meanwhile, its specificity on small features consistently outperforms other materials.

3.1. Density and Height

Before investigating geometry, it is important to demonstrate where macroscopic thickness gradients influence our results. Figure 2 shows a surface map of the test pattern, with the density of each grating plotted based on its relative position on the test pattern while Figure 3 shows the same data plotted by feature size and spacing. MG, unlike the PS sprays, correlates with the position on the test pattern as opposed to feature size and gap. It is apparent that the corner of the test pattern receives a greater amount of spray than the rest of the pattern. As mentioned, the high mobility of the liquid dissipates charge and prevents the repulsion needed to counteract a thickness gradient. From the simulated coating, we can see that there is some bias in the metrics (bottom right) showing a similar trend. It is much easier to have a higher density coating for larger gaps simply because there is less feature surface available. However, this is not the whole story for MG. If we look at the actual profilometry data, we see that the heights of the MG features increase as we move across the test pattern, even within the same grating, which can only be explained by a gradient (Figure S5a).

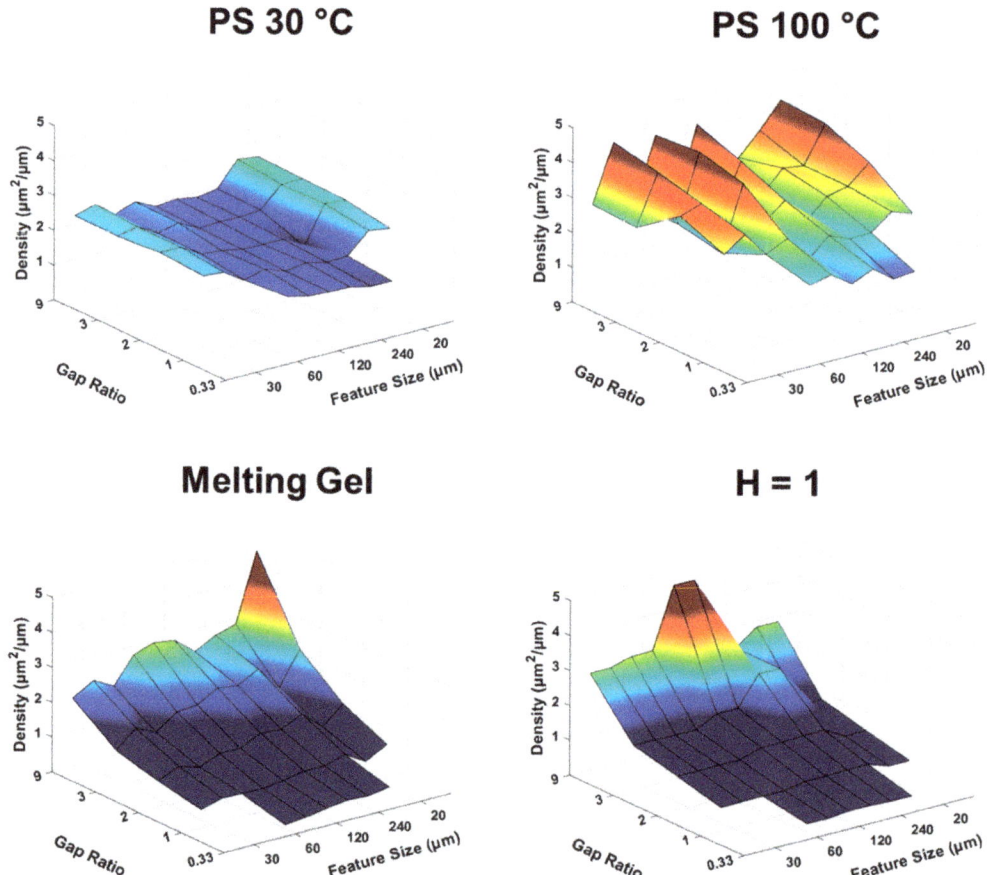

Figure 2. Surface maps of grating density plotted by their relative position on the chip. Note that 20 and 30 μm features are positioned on opposite ends. The simulated indiscriminate coating ($H = 1$) has been scaled to fit on the same axis as the real materials, but the actual values are arbitrary.

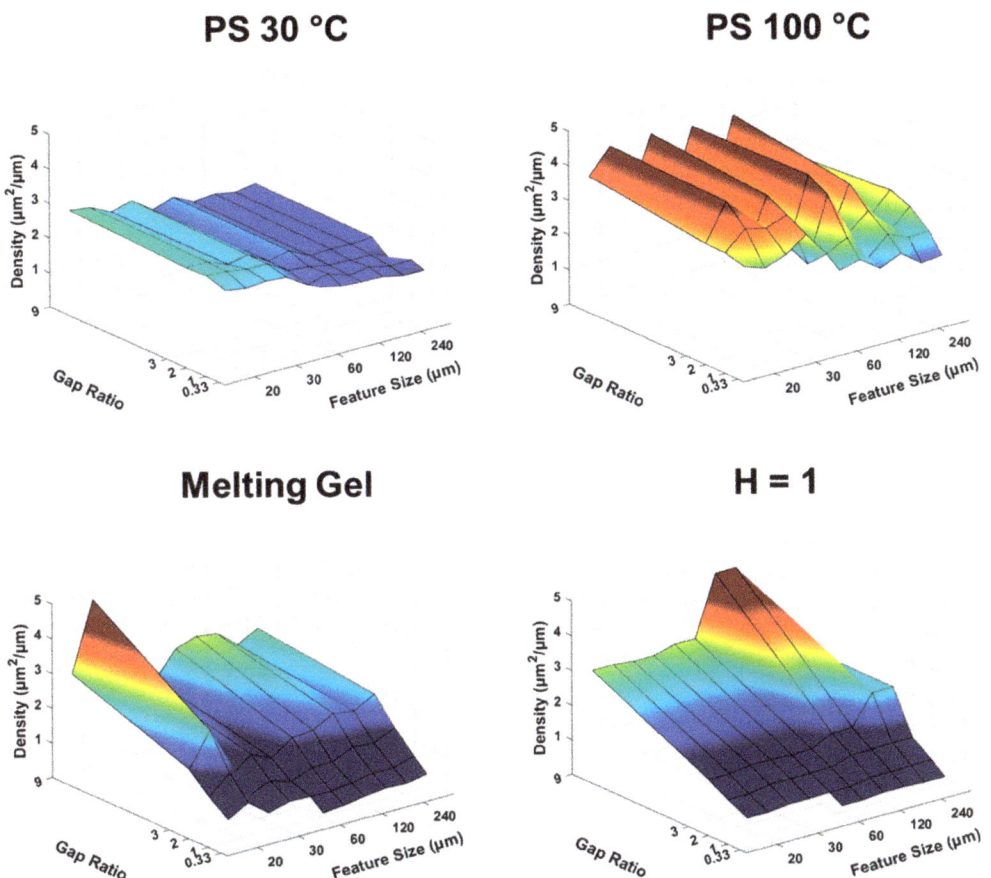

Figure 3. Surface maps of grating density plotted according to their gap ratio and feature size.

Due to this gradient, trends in the MG height data are more anecdotal than quantitative. Comparing the 20 μm large gap features to the 240 μm large gap features, we see an excellent example of the distinction between density and height (Figure S5b). There is clearly much more total material on the 20 μm features, even when there is much less total available surface as seen for the 9× gap. However, only the top left grating (20 micron/3× gap/distal) comes close to the heights seen for the 240 features. Unfortunately, without a single 20-micron feature to compare to, it is difficult to tell if this pattern is due to the feature size, the presence of multiple features, or various focusing or wetting effects.

Moving onto the PS sprays, we do not see evidence for a thickness gradient with relative grating position (Figure 2). While the 100 °C increases in density with increasing gap, this is not seen in the profile data as we saw with the MG, indicating that the trend is due to metric bias. On the other hand, for both sprays, we see a slight trend with respect to feature size with higher densities on smaller features (Figure 3). Importantly, this trend differs from the biases represented by the simulated data, indicating that it is a real effect. We also see that, on average, the 100 °C spray has a higher density than the 30 °C does. However, the 100 °C also has greater variation, appearing to trend with the scan's local position within the grating (distal/proximal) and not with the position on the test pattern.

From the height data, we see further variation between the 30 °C and 100 °C samples (Figure 4a). For 100 °C PS, we see that a definite trend exists for feature size. Demonstrating the lack of thickness gradient, we see that the 30 μm features appear slightly taller than the 20 μm features do, despite the 20 μm feature's close physical proximity to the tallest features, 240 μm (Figure 4b). Looking at all the data points, we see that the 30 °C PS height barely changes with feature size. Averaging these data points, we obtain a height of 1.51 μm ± 0.10 for 30 °C PS features. This differs from our prior results that showed increasing deposition thickness on larger features [22]. This is likely the effect of a reduced spray delivery rate from a combination of (1) distance, (2) flowrate, and (3) overall pattern area, which, for this pattern, is ~0.9 cm^2 and, for the earlier work, was sub-mm^2. For this reason, the current result can be more fully considered SLED. This is reinforced by the templated SL thickness also differing from the SL thickness observed for PS on a bare wafer, which is ~2.5 μm [10,11]. This indicates that the presence of the charged mask makes the exposed template a less favorable target than the supporting ground, reducing the thickness needed to repel incident spray. This is similar to results reported by Kingsley et al., which indicated that small targets receive very little incident spray when depositing in the SL regime [21].

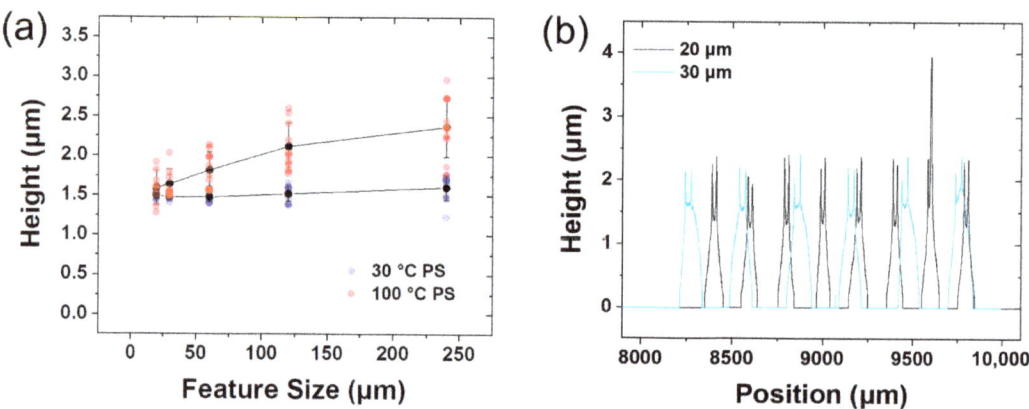

Figure 4. (a) Height data from the PS sprays; (b) Overlay of the 20 μm and 30 μm/9× gap distal features for the 100 °C spray showing a lack of a positional thickness gradient.

To verify and interpret these results, we overlay the profiles for the 30, 60, and 240 μm features for the 30 °C PS proximal as well as the 100 °C PS proximal and distal scans (Figure S6). To compare the profile shape, each feature is centered at zero, and the position data are normalized by the feature size. The profiles show a characteristic shape, with peaks near the feature edge. While profilometry can produce artifacts near sharp edges, as seen in Figure 1b, the raised edges seen here are too broad to fit this description. Instead, this is likely the effect of the surrounding insulating areas focusing material onto the feature edges. The hypothesis is supported through finite element method simulations, which show the electric field lines preferentially directing material toward the edge of the feature when there is a moderate amount of surface charge on the surrounding insulation (Figure S11). It is also important to remember that these coatings are densified, and the actual effect may be even more prominent for undensified material. Comparing between feature sizes, of particular note is the change in shape seen between the 60 and 240 μm features, which is most easily discernable in 30 °C PS. As the edges increase in thickness, they also increase in repulsion. It appears that this repulsion either prevents further deposition on the center of smaller features or acts to focus material onto the centers of larger ones. Comparing across materials, we can gain a much clearer understanding of the height results seen in Figure 4a. There is clearly much greater uniformity in the 30 °C PS profile compared to the 100 °C PS, which explains the relative uniformity of the 30 °C PS height data. While the central regions are increasing in height, this result is tempered due to the small size of the difference and the uniformity of the edge heights. For the 100 °C PS, we can see that the proximal profiles are more uniform than the distal. For both these profiles, the 30 μm features appear to have segregated into two bands, one with a concave central region and another that is relatively flat. This is likely due to the proximity of surrounding features, with smaller gaps allowing for more blending and spreading, creating the flatter band. For the 60 and 240 μm features, there is more room to spread on an individual feature and the variability increases. For the 240 μm features in particular, the curvature becomes distinctly convex, and the central region can grow far beyond the height of the edge peaks.

3.2. Specificity

The specificity of all materials and the simulated data are plotted on surface maps in Figure 5 according to their feature size and gap ratio. For all real materials, the specificity values indicate that it is clearly more difficult to target smaller features, as would be expected. We can also see that our metrics do not bias the results in this direction; rather, the simulated data show us that smaller gap ratios produce a bias toward higher specificity values due to the greater available surface. To probe the specificity further, we turn to a comparison plot that includes both specificity and density, shown in Figure S7 for all materials. Figures 6 and 7 show individual plots of specificity versus the inverse of density with color maps denoting both gap ratio and feature size, respectively. Beginning with the uniform data, we see the justification for the transformation: the relationship between specificity and the inverse of density is now linear, which can be explained by Equation (4) where H_g is equal to 1. For this case, specificity and density are mainly controlled by the amount of the feature available, which is generally controlled by the gap. With the amount of space per grating being relatively constant, larger gaps mean less available surface and vice versa. This explains why the gap-ratio color map is completely ordered.

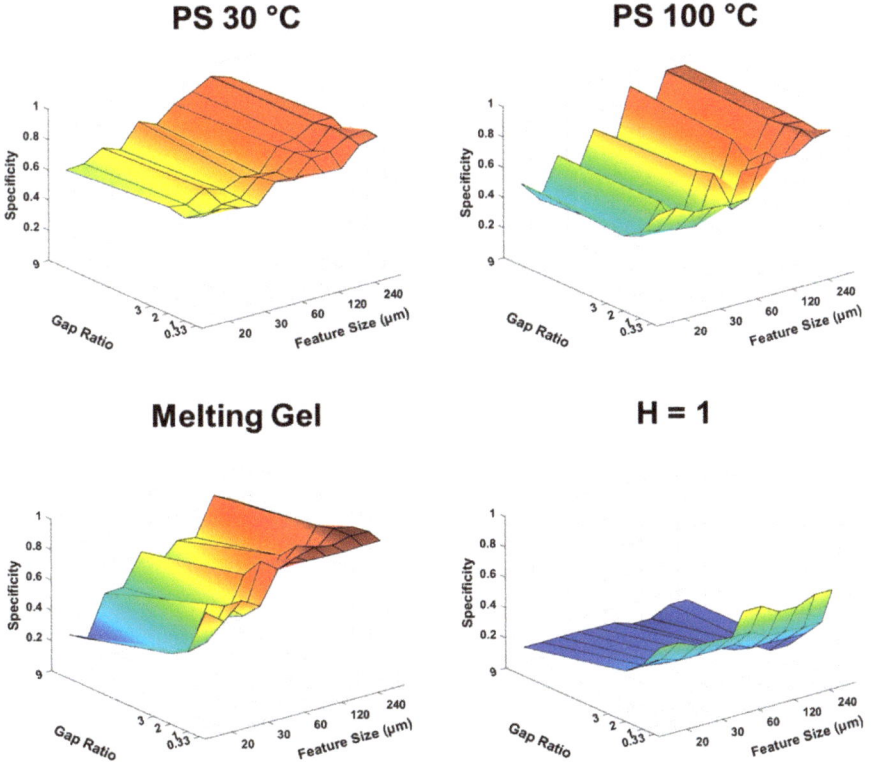

Figure 5. Surface maps of specificity graphed according to the feature size and gap ratio.

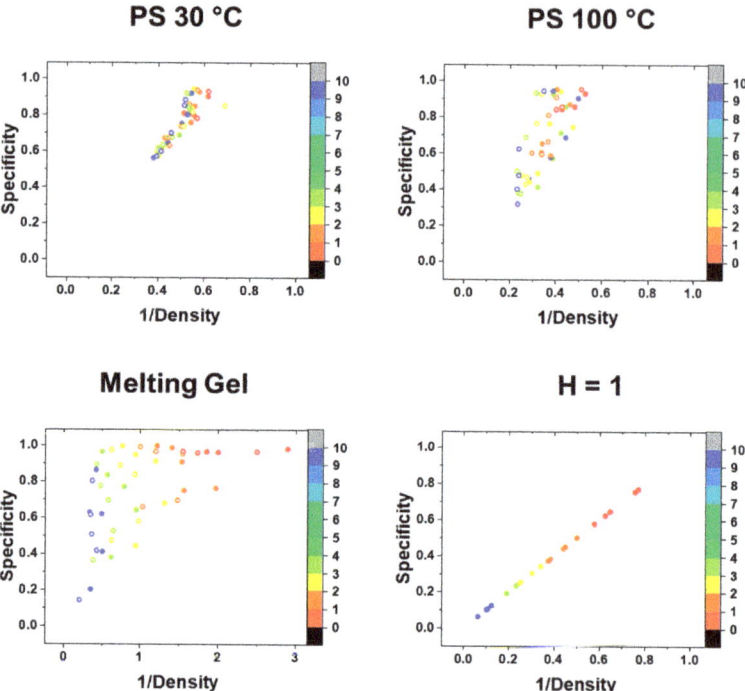

Figure 6. Specificity vs. 1/Density with color maps denoting gap ratio.

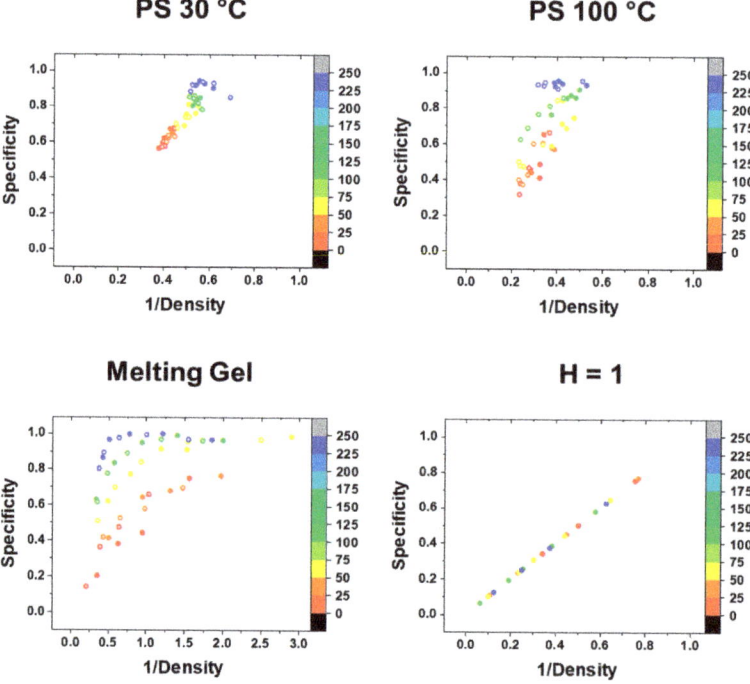

Figure 7. Specificity vs. 1/Density with color maps denoting feature size (μm).

Importantly, we see that the PS sprays differ from these trends. The 30 °C PS shows no trend regarding gap ratio. The 100 °C PS proximal also shows no trend, while 100 °C PS distal does show some relation to the gap ratio (Figure S8). However, both show a strong behavioral correlation to feature size, for the above-stated reason of smaller features being more difficult to target specifically. For the 100 °C PS, we can also see that the high-density, low-specificity data points are mainly derived from smaller features at distal locations. MG, on the other hand, has trends for both the gap ratio and the feature size but is relatively independent of distal/proximal positioning. As shown above, the dependence on the gap is strongly contributed to by a coincidence with the spray thickness gradient. However, we also see in Figure 4b that for the 20 μm/9× gap and 20 μm/3× gap, similar amounts of material are deposited, but there is a large discrepancy in density, demonstrating the metric bias that occurs for coatings trending toward indiscriminate behavior.

One way to think about this analysis is as a spectrum from metric-bias-dominated behavior to phenomenological-dominated behavior. Data from the simulated sample represent what we might expect from a conventional coating method such as spin coating, and it is completely dominated by the inherent biases in our metrics. The MG coating is relatively thin for most of the sample due to the thickness gradient, which is too strong to allow us to hypothesize any other effect for the gap. However, the MG is not completely indiscriminate. Repulsion from the masked region prevents larger features with larger gaps from blending, which is why we see a trend in Figure 7. If we were to increase the MG coating thickness via longer spray times, some of these larger features would likely spread more, and the high specificity values would decrease. However, we would also likely observe a transition to charged melt behavior before material on isolated features spreads far beyond the test pattern. PS sprays, on the other hand, are better able to accumulate charge, even if 100 °C PS can spread and dissipate the charge to some degree. The 100 °C PS distal is less targeted than 100 °C PS proximal and 30 °C PS and, therefore, shows metric bias to a slight degree (Figure S8). The 30 °C PS and 100 °C PS proximal act independent of metric bias, which provides evidence for SL behavior. Further, by considering the 30 °C PS features less than 240 μm, the specificity versus inverse density plot can be fit linearly (R = 0.88) with a slope of 1.49 ± 0.10 and a near-zero intercept of $3.03 \times 10^{-4} \pm 0.05$. As per Equation (4), the slope represents the average height of the features, which is measured directly as 1.51 ± 0.10 μm. This confirms that the SL effect for small features leads to uniform films. For the 240 μm features, the points are more scattered corresponding to the change in profile shape observed for larger features (Figure S6), and the slight increase in height seen in Figure 4a appears to have more of an impact here.

3.3. Feature Crosstalk

A separate figure of merit that is highly relevant to microfabrication is feature crosstalk—i.e., whether a deposition connects two adjacent features. This could be fatal or essential based on the application. For example, interdigitated electrodes have been used for flexible supercapacitors [27,28]. In this case, bridging the gap with a conductive material would short the device. However, if we were to use ESD for encapsulation, we would want to ensure that the entire grating is coated.

We create a new binary model profile with wider features to capture this scenario. The width of a single feature and gap is added, and the feature is expanded to fill 90% of this space (Figure 8a). Gratings that only include a single feature are excluded from this analysis. While specificity increases, only material bridging the space directly in the center of the gap is penalized. Examining the results (Figures 8b and S9) and cross-referencing with the profilometry scans (Figures S1–S3), we see that anything less than 99% has some features bridged by the coating. This gives us more concrete data for the limits of SLED. The 30 °C PS, below T_g, spray allows us to access features separated by gaps of 80–120 μm where 100 °C PS fails. We also see that the limit for 30 °C PS appears to be 60 μm gaps and below. Interestingly, this is irrelevant of feature size, telling us that feature size controls the specificity and the density to some degree, but ultimately, the amount of spreading is relatively similar for all features. Therefore, the size of the gap is the only important factor in this regard.

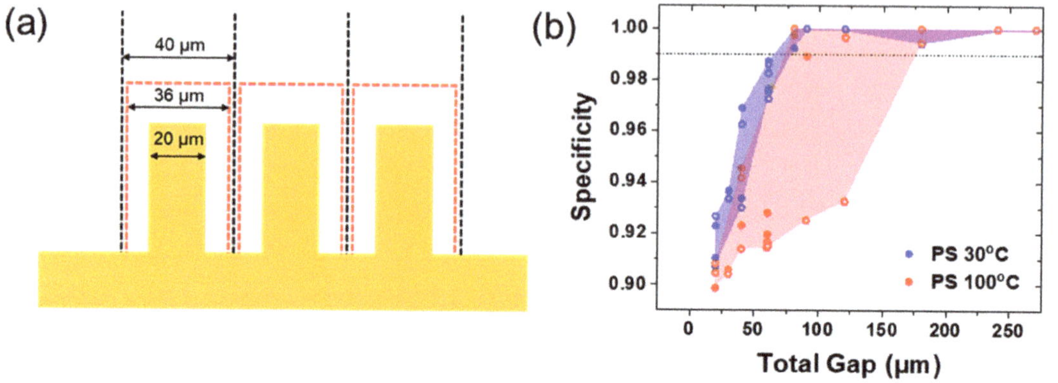

Figure 8. (a) Cross-section drawing of the expanded-feature-generated profile (red) relative to the actual feature (gold) and the total gap (black); (b) Specificity results for the expanded features.

While we would expect the general trends to hold for other sprays and substrate systems, we would expect the actual size limits to vary. As seen in Figure S11, the focusing behavior can vary based on the surrounding surface charge. Increasing the effect of focusing while still remaining in a SL effective flowrate regime could increase specificity. Another option would be to spray for shorter times, though this would decrease the density and risk non-uniformity due to thickness gradients in less developed films. In Figure S10, we show an example of another SL material, methylcellulose. While it appears to have targeted the features with higher fidelity, it is also undensified, so the difference might be due to post-spray flow. Within SL materials, behavior differences can likely occur due to dielectric constant, morphology, and solvent absorption, though we expect general trends compared to electrowetting and charged melt materials to remain the same.

4. Conclusions

In this study, we demonstrate that the mobility of electrosprayed materials strongly impacts the material's interaction with features of varying geometry. For a liquid—electrowetting material—represented by the MG, it is demonstrated that deposition is strongly, though not completely, controlled by the spray thickness gradient. This gradient is observed across the test pattern, even within a single grating, increasing both the measured heights and densities. However, the liquid is still charged and is, therefore, not completely indiscriminate. Accordingly, the effects of charge repulsion still allow for the high-specificity, high-density coatings seen on large features.

The 100 °C PS shows intermediate charged melt behavior between the SL regime and the completely liquid electrowetting regime. The higher temperature allows the material to spread, which alleviates charge buildup, increasing the density and decreasing the specificity, and this effect is more prominent on distal regions of the feature. However, there is still enough charge accumulation to prevent the formation of a strong thickness gradient and provide a more uniform coating, though the feature size appears to control the film thickness.

Finally, 30 °C PS is clearly representative of SL behavior. Its density and specificity do not trend with the model biases whatsoever, and there is relatively little difference between distal and proximal scans. Its specificity, especially for small features, is consistently much higher than any other material shown here, and there is lower variability. Its density is also relatively uniform and, on average, lower than that of 100 °C PS. Most interesting of all, it is found that while smaller features tend to have higher densities and lower specificities, height is relatively constant, indicating that film growth is uniform across features. We also expect that other self-limiting materials would demonstrate this general behavior (Figure S10).

These results differ from our previous findings which showed that for small circular features with diameters ranging from 10 to 40 μm, wider features had thicker coatings, and all these coatings were much thicker than unpatterned electrosprays. This is more similar to what we observed here for 100 °C PS distal. We, therefore, postulate that geometry only impacts the height of non-SL materials. In line with this conclusion, it appears that for our previous results, the deposited material entered the charged-melt regime, likely from an excess accumulation of solvent, allowing for continued material growth that was further compounded by the effects of focusing. The evidence presented in this work suggests that we were able to mitigate this effect, likely through a combination of a lower flow rate, greater spray distance, and larger pattern area.

Finally, we show that the self-limiting material sprayed on these test patterns could target features separated by 80 μm gaps and above without bridging them. The fact that 80 μm is observed as the limit across feature sizes speaks to the material's ability to grow uniformly. However, we do expect this number to be system-dependent, and we predict that it is possible to target features separated by even smaller gaps in other systems. For example, it is likely possible that stronger focusing could be employed at the low flow rates used in this work without great detriment to the uniformity. However, in this case, we can use 80 μm to inform pattern design within this system, and we can use our analysis technique to rapidly assess new systems in the future. Furthermore, we can use the current data to guide the deployment of these materials under different applications. For example, to encapsulate with the melting gel material, we would likely use a moving stage or a focusing collar to ensure more uniform coverage of all features as has been conducted in the past [24]. On the other hand, the uniformity of the SL material on small features is a very promising result that will allow for controlled deposition of multilayered materials.

Supplementary Materials: The following supporting information can be downloaded at: https://www.mdpi.com/article/10.3390/coatings13030599/s1, Figure S1: All aligned profile data for the 30 °C PS; Figure S2: All aligned profile data for the 100 °C PS; Figure S3: All aligned profile data for the MG.; Figure S4: Visual representation of (top) H_g, (middle) ρ_g, and (bottom) σ_g; Figure S5: Profile scan of (a) the 120 μm feature of the MG test pattern and (b) the 20 μm features overlayed with the 240 μm features; Figure S6: Overlays of the 30, 60, and 240 μm feature profiles, standardized by width for (a) PS 30 °C Proximal and (b) PS 100 °C Proximal and Distal; Figure S7: Specificity versus density for all data sets; Figure S8: Specificity vs 1/Density with color maps denoting gap ratio for PS 100 °C distal-only; Figure S9: Expanded gap specificity by feature size. (a) 20 μm, (b) 30 μm, (c) 60 μm and (d) 120 μm; Figure S10: Microscope images of the 60 μm feature/3× gap for (a) blank test pattern, (b) MG, (c) PS 30 °C, and (d) PS 100 °C, and (e) methylcellulose; Figure S11: Pictures of the simulation cell for the electrostatic spray simulation.

Author Contributions: Conceptualization, J.P.S. and M.J.G.; methodology, J.P.S., M.J.G., M.A. and J.D.Z.; software, M.J.G.; investigation M.J.G.; formal analysis, M.J.G.; writing—original draft preparation, J.P.S. and M.J.G.; writing—review and editing, J.P.S., M.A. and J.D.Z.; resources, A.J., A.S. and M.A.; visualization, J.P.S. and M.J.G.; supervision, J.P.S., J.D.Z. and A.J.; project administration, J.P.S.; funding acquisition, J.P.S., J.D.Z. and A.J. All authors have read and agreed to the published version of the manuscript.

Funding: This work was partially funded by NSF Advanced Manufacturing Awards #1911518 and #1911509 and GeneOne Life Science.

Institutional Review Board Statement: Not applicable.

Informed Consent Statement: Not applicable.

Data Availability Statement: Full data are available on reasonable request to the corresponding author.

Acknowledgments: The authors would like to thank Zaynab Hazaveh for contributions to the processing of the profilometry data.

Conflicts of Interest: J.P.S. is an inventor on international patent application PCT/US20/33020 and its derivative regional patent applications that include aspects of self-limiting electrospray technology. The other authors declare no conflict of interest.

References

1. Klein, L.C.; Kallontzi, S.; Fabris, L.; Jitianu, A.; Ryan, C.; Aparicio, M.; Lei, L.; Singer, J.P. Applications of melting gels. *J. Sol.-Gel. Sci. Technol.* **2019**, *89*, 66–77. [CrossRef]
2. Pawliczak, E.E.; Kingsley, B.J.; Chiarot, P.R. Structure and properties of electrospray printed polymeric films. *MRS Adv.* **2022**, *7*, 635–640. [CrossRef]
3. Weiss, F.M.; Töpper, T.; Osmani, B.; Peters, S.; Kovacs, G.; Müller, B. Thin Elastomer Films: Electrospraying Nanometer-Thin Elastomer Films for Low-Voltage Dielectric Actuators (Adv. Electron. Mater. 5/2016). *Adv. Electron. Mater.* **2016**, *2*. [CrossRef]
4. Rietveld, I.B.; Kobayashi, K.; Yamada, H.; Matsushige, K. Electrospray deposition producing ultra-thin polymer films with a regular surface structure. *Soft Matter* **2009**, *5*, 593–598. [CrossRef]
5. Lei, L.; Chen, S.; Nachtigal, C.J.; Moy, T.F.; Yong, X.; Singer, J.P. Homogeneous gelation leads to nanowire forests in the transition between electrospray and electrospinning. *Mater. Horiz.* **2020**, *7*, 2643–2650. [CrossRef]
6. Leeuwenburgh, S.; Wolke, J.; Schoonman, J.; Jansen, J. Electrostatic spray deposition (ESD) of calcium phosphate coatings. *J. Biomed. Mater. Res.* **2003**, *66A*, 330–334. [CrossRef]
7. Kim, J.-Y.; Kim, E.-K.; Kim, S.S. Micro-nano hierarchical superhydrophobic electrospray-synthesized silica layers. *J. Colloid Interface Sci.* **2013**, *392*, 376–381. [CrossRef]
8. Hwang, D.; Lee, H.; Jang, S.-Y.; Jo, S.M.; Kim, D.; Seo, Y.; Kim, D.Y. Electrospray Preparation of Hierarchically-structured Mesoporous TiO_2 Spheres for Use in Highly Efficient Dye-Sensitized Solar Cells. *ACS Appl. Mater. Interfaces* **2011**, *3*, 2719–2725. [CrossRef]
9. Jaworek, A.; Sobczyk, A.T. Electrospraying route to nanotechnology: An overview. *J. Electrost.* **2008**, *66*, 197–219. [CrossRef]
10. Lei, L.; Kovacevich, D.A.; Nitzsche, M.P.; Ryu, J.; Al-Marzoki, K.; Rodriguez, G.; Klein, L.C.; Jitianu, A.; Singer, J.P. Obtaining Thickness-Limited Electrospray Deposition for 3D Coating. *ACS Appl. Mater. Interfaces* **2018**, *10*, 11175–11188. [CrossRef]
11. Green-Warren, R.A.; Bontoux, L.; McAllister, N.M.; Kovacevich, D.A.; Shaikh, A.; Kuznetsova, C.; Tenorio, M.; Lei, L.; Pelegri, A.A.; Singer, J.P. Determining the Self-Limiting Electrospray Deposition Compositional Limits for Mechanically Tunable Polymer Composites. *ACS Appl. Polym. Mater.* **2022**, *4*, 3511–3519. [CrossRef]
12. Kovacevich, D.A.; Lei, L.; Han, D.; Kuznetsova, C.; Kooi, S.E.; Lee, H.; Singer, J.P. Self-Limiting Electrospray Deposition for the Surface Modification of Additively Manufactured Parts. *ACS Appl. Mater. Interfaces* **2020**, *12*, 20901–20911. [CrossRef]
13. Yan, W.-C.; Xie, J.; Wang, C.-H. Electrical Field Guided Electrospray Deposition for Production of Gradient Particle Patterns. *ACS Appl. Mater. Interfaces* **2018**, *10*, 18499–18506. [CrossRef]
14. Hu, H.Q.; Toth, K.; Kim, M.; Gopalan, P.; Osuji, C.O. Continuous and patterned deposition of functional block copolymer thin films using electrospray. *MRS Commun.* **2015**, *5*, 235–242. [CrossRef]
15. Zhu, Y.; Chiarot, P.R. Directed assembly of nanomaterials using electrospray deposition and substrate-level patterning. *Powder Technol.* **2020**, *364*, 845–850. [CrossRef]
16. Khan, S.; Doh, Y.H.; Khan, A.; Rahman, A.; Choi, K.H.; Kim, D.S. Direct patterning and electrospray deposition through EHD for fabrication of printed thin film transistors. *Curr. Appl. Phys.* **2011**, *11*, S271–S279. [CrossRef]
17. Kim, J.W.; Yamagata, Y.; Kim, B.J.; Higuchi, T. Direct and dry micro-patterning of nano-particles by electrospray deposition through a micro-stencil mask. *J. Micromech. Microeng.* **2009**, *19*, 025021. [CrossRef]
18. Al-Milaji, K.N.; Zhao, H. Fabrication of superoleophobic surfaces by mask-assisted electrospray. *Appl. Surf. Sci.* **2017**, *396*, 955–964. [CrossRef]
19. Higashi, K.; Uchida, K.; Hotta, A.; Hishida, K.; Miki, N. Micropatterning of Silica Nanoparticles by Electrospray Deposition through a Stencil Mask. *J. Lab. Autom.* **2014**, *19*, 75–81. [CrossRef]
20. Lee, S.J.; Park, S.M.; Han, S.J.; Kim, D.S. Electrolyte solution-assisted electrospray deposition for direct coating and patterning of polymeric nanoparticles on non-conductive surfaces. *Chem. Eng. J.* **2020**, *379*, 122318. [CrossRef]
21. Kingsley, B.J.; Pawliczak, E.E.; Hurley, T.R.; Chiarot, P.R. Electrospray Printing of Polyimide Films Using Passive Material Focusing. *ACS Appl. Polym. Mater.* **2021**, *3*, 6274–6284. [CrossRef]
22. Lei, L.; Gamboa, A.R.; Kuznetsova, C.; Littlecreek, S.; Wang, J.; Zou, Q.; Zahn, J.D.; Singer, J.P. Self-limiting electrospray deposition on polymer templates. *Sci. Rep.* **2020**, *10*, 17290. [CrossRef]
23. Buchko, C.J.; Kozloff, K.M.; Sioshansi, A.; O'Shea, K.S.; Martin, D.C. Electric Field Mediated Deposition of Bioactive Polypeptides on Neural Prosthetic Devices. *MRS Online Proc. Libr.* **1995**, *414*, 23–28. [CrossRef]
24. Morozov, V.N.; Morozova, T.Y. Electrospray Deposition as a Method for Mass Fabrication of Mono- and Multicomponent Microarrays of Biological and Biologically Active Substances. *Anal. Chem.* **1999**, *71*, 3110–3117. [CrossRef] [PubMed]
25. Jitianu, A.; Doyle, J.; Amatucci, G.; Klein, L.C. Methyl modified siloxane melting gels for hydrophobic films. *J. Sol-Gel Sci. Technol.* **2010**, *53*, 272–279. [CrossRef]
26. Jitianu, A.; Amatucci, G.; Klein, L.C. Organic–inorganic sol-gel thick films for humidity barriers. *J. Mater. Res.* **2008**, *23*, 2084–2090. [CrossRef]

27. Lei, Y.; Zhao, W.; Zhu, Y.; Buttner, U.; Dong, X.; Alshareef, H.N. Three-Dimensional $Ti_3C_2T_x$ MXene-Prussian Blue Hybrid Microsupercapacitors by Water Lift-Off Lithography. *ACS Nano* **2022**, *16*, 1974–1985. [CrossRef]
28. Sollami Delekta, S.; Smith, A.D.; Li, J.; Östling, M. Inkjet printed highly transparent and flexible graphene micro-supercapacitors. *Nanoscale* **2017**, *9*, 6998–7005. [CrossRef]

Disclaimer/Publisher's Note: The statements, opinions and data contained in all publications are solely those of the individual author(s) and contributor(s) and not of MDPI and/or the editor(s). MDPI and/or the editor(s) disclaim responsibility for any injury to people or property resulting from any ideas, methods, instructions or products referred to in the content.

Article

Short-Branched Fluorinated Polyurethane Coating Exhibiting Good Comprehensive Performance and Potential UV Degradation in Leather Waterproofing Modification

Shouhua Su, Juan Wang, Chao Li, Jinfeng Yuan, Zhicheng Pan and Mingwang Pan *

Hebei Key Laboratory of Functional Polymers, School of Chemical Engineering and Technology, Hebei University of Technology, Tianjin 300130, China; shouhuasu@163.com (S.S.); juanwang117@163.com (J.W.); chaoli1129@163.com (C.L.); yuanjf@hebut.edu.cn (J.Y.); panz@hebut.edu.cn (Z.P.)
* Correspondence: mwpan@126.com

Abstract: In the current leather market, waterproof leather occupies a large proportion, where waterproofness has become one of the important standards for leather selection. However, the most advanced fluorine-containing waterproofing agents on the market always have long chains of over eight carbons (C8), whose use has been restricted due to their bioaccumulation and recalcitrance in natural environment. Consequently, creating waterproof materials characterized by their environmentally friendly qualities and high performance is of great significance. Herein, we report a novel strategy for preparation of the fluorinated polyurethanes containing short branched fluorocarbon chains, and apply it in leather waterproofing. Because the fluorine-containing chain segments are enriched on the coating surface, the waterproof agent coating shows good hydrophobicity, low water absorption, high wear resistance and potential photodegradation of performances. Additionally, the water and oil proof performances of the coating are comparable to that of the marketed C8 waterproofing agent. Its solvent-resistant and antifouling performances are also outstanding. Therefore, the coating can meet the property requirements for daily use and has broad application prospects.

Keywords: short branched fluorocarbon chain; fluorinated polyurethane (FPU); waterproof coating; leather surface modification

Citation: Su, S.; Wang, J.; Li, C.; Yuan, J.; Pan, Z.; Pan, M. Short-Branched Fluorinated Polyurethane Coating Exhibiting Good Comprehensive Performance and Potential UV Degradation in Leather Waterproofing Modification. *Coatings* **2021**, *11*, 395. https://doi.org/10.3390/coatings11040395

Academic Editor: Mohor Mihelčič

Received: 11 February 2021
Accepted: 27 March 2021
Published: 30 March 2021

Publisher's Note: MDPI stays neutral with regard to jurisdictional claims in published maps and institutional affiliations.

Copyright: © 2021 by the authors. Licensee MDPI, Basel, Switzerland. This article is an open access article distributed under the terms and conditions of the Creative Commons Attribution (CC BY) license (https://creativecommons.org/licenses/by/4.0/).

1. Introduction

At present, leather has a wide range of applications, but its development is limited by its easy aging and poor properties in terms of water proofing and scratch resistance. The main way to improve the performance of leather materials is to coat its surface with a layer of protective film. The common coatings on the market can roughly be divided into three categories, namely polyurethane [1–6], polyacrylate [7–9], and organic silicon [10–12]. Among them, polyurethane (PU) is the first reported and widely studied due to its strong structural designability and excellent wear resistance [13]. However, the waterproof performance of PU material itself is not enough to meet the requirements of leather materials in some special fields, such as in a long term exposure to a humid environment. To further improve the waterproof, oil-resistant and anti-smudge properties of PU coatings, fluorinated polyurethane (FPU) waterproofing agent is prepared often by introducing fluorine-containing groups into the PU molecular structure. On the one hand, the FPU retains the special structural characteristics of PU itself, which can bring lots of excellent properties to the material, for example good film formation, wear resistance and toughness. On the other hand, the introduction of fluorine-containing segments endows polyurethane-based materials with other special properties (like excellent waterproof, oilproof and anti-smudge properties [14–22]) due to its low surface energy [23], which determines that the FPU will have better application prospects in more fields than the traditional PU. The introduction of long-chain perfluoroalkyl (LF) groups to PU is currently the most used method for FPU preparation through implanting fluorinated polyols as soft chain segments [24,25], perfluorinated alcohols as end sealants [26,27], or fluorinated compounds with multi-functional groups as chain extenders [28–30].

To date, fluorinated chemicals containing LF groups [31], the most advanced waterproofing agent on the market, have been verified, demonstrating that their manufacture originated from straight-chain C8 perfluoroalkyl sulfonic acid or perfluoroalkyl acids, such as C8 fluorocarbons, perfluorooctane sulfonic acid (PFOS) and perfluorooctanoic acid (PFOA) that have potential biotoxicity like bioaccumulation and environmental persistence [32–36]. In view of these worrying dangers, many countries, such as Canada, the United States and China, have adopted certain regulatory measures to prohibit its use in the relevant industrial production. Fortunately, compounds with shorter fluorocarbon (SF) chains are thought to be significantly less toxic, and the subsequent breakdown products are innoxious [37,38]. Although FPU with SF components is more environmentally friendly than FPU with LF components, its hydrophobic, oleophobic and anti-smudge properties are significantly worse than those of long-chain FPU. Thus, it is particularly important to explore a new type of fluorinated polyurethane that is not only environmentally friendly but also has outstanding coating properties in the leather field. As for FPU with short branched fluorocarbon chains, it is unclear whether it can satisfy the above two conditions at the same time, and there have been few reports on this to the best of our knowledge. Therefore, research into the fluorinated polyurethane with short branched chains in the leather field will have strong theoretical and practical significance.

In this work, a novel type of short-branched fluorinated PU was prepared by first synthesizing short-branched chain fluorinated alcohol and then implanting it into the PU structure as an end-capping agent. As a main raw material, hexafluoropropylene trimer is not restricted in use [39]. Thus, we initially synthesized a fluoro-alcohol by reacting hexafluoropropylene trimer with p-hydroxybenzyl alcohol (Figure 1ai) [40]. Next, a new FPU waterproofing agent was prepared by pre-reaction of 1,4-butanediol and toluene diisocyanate, successively end-capping with fluoro-alcohol (R_f-OH) and ethanol respectively (Figure 1 aii). We adjusted the fluorine content in the final mixture by changing the molar ratio of ethanol and fluoro-alcohol, and maintained a molar ratio of approximately 1:1 between isocyanate groups and hydroxy groups. Fluorine-containing short branched chains not only provide FPU with excellent surface performance, low water absorption, and satisfied wear resistance, they also provide it with potential photodegradation. Most notably, the FPU could be compared to the commercial C8 waterproofing agent, and the former also exhibited superior waterproof, oilproof and antifouling properties, prominent solvent resistance, and potential photocatalytic degradation.

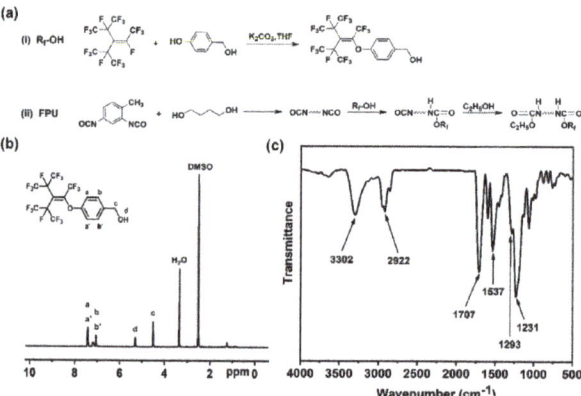

Figure 1. (a) Preparation procedure of (i) R_f-OH and (ii) FPU; (b) 1H-NMR spectrum of R_f-OH in DMSO-d6; (c) FTIR spectrum of FPU.

2. Materials and Methods

2.1. Materials

Hexafluoropropylene trimer (HFPT) was purchased from Quzhou Chemical Industry Co., Ltd. (Quzhou, China). p-Hydroxybenzyl alcohol, potassium carbonate (K_2CO_3), tetrahydrofuran (THF), toluene diisocyanate (TDI), 1,4-butanediol (BDO), N,N-dimethylformamide (DMF) and ethanol (C_2H_5OH) were supplied by Aladdin Chemical Reagent Co., Ltd. (Shanghai, China). C8 waterproofing agent was provided by Taifu Chemical Technology Co., Ltd. (Shanghai, China).

2.2. Synthesis of R_f-OH

p-Hydroxybenzyl alcohol (6.20 g) and K_2CO_3 (6.91 g) were dissolved in 60 mL THF. HFPT (33.75 g) was added dropwise into a four-mouth flask under nitrogen atmosphere, followed by stirring and reflux for 4 h at 60 °C. Afterwards the mixture was slowly cooled to room temperature and THF was removed with rotary evaporation to obtain a yellow liquid (20.78 g, 70% yield).

2.3. Synthesis of FPU

BDO and DMF were mixed in a four-mouth flask at room temperature and then TDI was added dropwise into the reaction flask under a nitrogen atmosphere. The reaction system was kept at 70 °C for 2 h, then polyurethane prepolymer with two NCO groups at the ends was prepared. Next, fluoro-alcohol (R_f-OH) was added to the system as one capping agent, and the reaction was also maintained for 2 h. Finally, ethanol was added as another capping agent under continuous stirring for another 2 h. Thus, the FPU at one end of a molecular chain was sealed with fluoro-alcohol and the other end was successfully sealed with ethanol. In this synthesis stage, we adjusted the fluorine content in the final mixture by changing the molar ratio between ethanol (ET) and fluoro-alcohol (FA) and maintained a molar ratio of approximately 1:1 between isocyanate and hydroxyl groups.

2.4. Leather Waterproof Coating

The leather surface was wiped with an absorbent cotton before coating the FPU waterproofing agent, then a very small amount of FPU solution was dropped and roll-coated on the leather surface with a clean glass rod. Finally, the treated leather was cured at 130 °C for 20 min.

2.5. Characterization and Measurements

2.5.1. Chemical Structure Analysis

For R_f-OH, 1H NMR spectrum was obtained on Bruker 400 nuclear magnetic resonance spectrometer and HRMS-ESI was obtained on Bruker Compact Q-TOF. The composition of FPU coating was analyzed by Fourier transform infrared spectroscopy (FTIR). FTIR spectrum was recorded using a FTIR spectrometer (Tensor-27, Bruker, Karlsruhe, Germany) at room temperature. The mixture of solid sample and desiccative potassium bromide (KBr) was pressed into pellets for FTIR research.

2.5.2. Surface Element Analysis

The contents of chemical elements on the surfaces of PU and FPU coatings were detected by X-ray photoelectron spectroscopy (XPS, ESCALAB 250 Xi, Thermo Fisher Scientific, Waltham, MA, USA). The elements of the coatings and their contents were also studied by energy-dispersive spectrometer (EDS), which was performed on a scanning electron microscope (FEI Nova Nano SEM 450) operating at 10 kV. To avoid charge-loading, the sample surface was sputtered with a thin layer of gold before the EDS experiment.

2.5.3. Wetting Test

Static water drop contact angle (WCA) and surface energy of the FPU membrane were tested by DSA 30 S apparatus (Krüss Co., Hamburg, Germany) with 10 μL of water at room temperature. The WCA was the mean value of five measurements for water droplets at different places on each sample surface.

2.5.4. Water Absorption

The FPU solution was uniformly sprayed on a glass plate with an area of 2.54 × 7.62 cm^2, and then dried and cured at 130 °C to obtain a FPU coating film with a thickness of 80 µm. The sample was placed into distilled water at 25 °C for 120 h. During this soaking process, it was taken out at certain intervals, the water on its surface was quickly wiped with filter paper, and then the weight was measured. The water absorption was calculated by following Formula (1):

$$w = \frac{m_2 - m_1}{m_1} \times 100\% \qquad (1)$$

where m_1 is the mass of the coating before being put into the water, m_2 is the mass of the coating after being put into the water, and w is water absorption.

2.5.5. Abrasion Analysis

Quantitative wear analysis was carried out using a JM-IV wear tester with two abrasive CS-10 wheels. The grinding wheel with each load of 250 g was used to conduct a wear test on the leather treated with the FPU waterproof agent. The contact pressure of the coating film was estimated to be about 12.1 MPa, and the leather was subjected to uniform stress in the slow and uniform rotation of the grinding wheel.

2.5.6. Potential UV Degradation

UV light irradiation was used to investigate the potential degradation performance of the FPU waterproofing agent. The leather treated with the FPU waterproofing agent was placed under a 365 nm UV lamp with a power of 30 W, which was located 0.14 m from the light source.

2.5.7. Waterproof Grade Test

According to BS ISO 23232:2009, the dripping test method was adopted. According to different surface tensions of isopropanol/water solution with different volume fractions, the waterproofing grades are divided into 0~8 grades, which successively increase from 0 to 8, as shown in Table S1 in the Supporting Information. During this test, 5 µL of the liquid was dropped on the leather surface treated with the FPU waterproofing agent at an interval of 5 mm. If the surface of leather sample was not wetted after placement for 10 s, in which the profile morphology and contact angle of the droplet did not significantly change, the waterproof grade was passed. Afterwards, the leather was then tested with a higher grade of water-isopropyl alcohol mixture until the surface was wetted. The final pass grade was the waterproof grade of FPU coated leather.

2.5.8. Oilproof Grade Test

According to BS ISO 14419:2010, oilproof grades can be divided into 1~8 grades on the basis of different types of oil with different surface tensions (Table S2 in the Supporting Information). During this test, 5 µL of oil (analytical reagent) was dropped onto the FPU-coated leather surface in a spacing of 5 mm. If the sample surface was not wetted after 10 s, the profile shape and contact angle of the droplet had no obvious change to suggest the waterproof level pass. Then, the higher grade of oil reagent was used to test until the leather surface was wetted. As a result, the final approved grade was the oil proof grade of the FPU modified leather.

2.5.9. Anti-Smudge Performance

The anti-smudge performance of our designed coating was further investigated. We dropped the ink on the sloping leather to observe the flow mark of the ink print.

2.5.10. Solvent Resistance

According to GB/T 23989-2009, the absorbent cotton was immersed in xylene solvent to obtain a wetted state (no liquid drops should be dropped when it was extruded by hand). Subsequently, using a safe protection, the index finger and thumb clamped the center of the absorbent cotton at a 45° tilt angle against the coating surface. A total of 25 forward and backward wipes were performed at a proper pressure to evaluate the solvent resistance of the coating.

3. Results

3.1. Chemical Structure Analyses

The successful preparation of R_f-OH was confirmed by 1H NMR and HRMS-ESI. Figure 1b shows the 1H NMR spectrum (400 MHz, DMSO-d6) of R_f-OH sample: c, chemical shift 4.50 ppm (s, 2H); d, 5.29 ppm (s, 1H); b and b', 7.02~7.23 ppm (m, 2H); a and a', 7.37~7.47 ppm (m, 2H). HRMS-ESI (m/z): cacld for $C_{16}H_7F_{17}O_2Na$ [M + Na]$^+$ 577.0066, found 577.0061.

To identify the chemical components of FPU coating, its FTIR measurement spectrum was recorded, as is given in Figure 1c, where the characteristic peaks in N-H (3302 cm^{-1}), C=O (1707 cm^{-1}), and saturated C-H (2922 cm^{-1}) stretching vibrations can be observed. The absorption band at 1537 cm^{-1} resulted from N-H bending vibration of secondary amide. The disappearance of absorption bands of TDI's NCO group (2270 cm^{-1}) and alcohol hydroxyl group (3340 cm^{-1}) proved the successful synthesis of FPU oligomer [41]. However, the C-F stretching vibration absorption was not conspicuous as it might be overlapped by the strong absorption of C-O-C group at around 1231 cm^{-1}, and a weak peak at 1293 cm^{-1} may be also due to the C-F stretching vibration. These results confirmed the chemical structure of FPU oligomer.

3.2. Coating Surface Chemistry

XPS is an analysis technique for detecting the surface chemistry of a material in the original state or after some treatment. The surface chemical composition of PU (ET:FA = 1:0 mol/mol) and FPU (ET:FA = 0.5:0.5 mol/mol) coatings were confirmed by XPS, as exhibited in Figure 2i. The surface atomic characteristic signals C 1s (276.8 eV), O 1s (524.4 eV), N 1s (392.3 eV) were observed, while the peak at 682.4 eV assigned to F(1s) was clearly observed only in the FPU sample. This indicated that F atoms were presented smoothly on the surface of the coating.

Generally speaking, fluorinated segments prefer to migrate to the material surface, which contributed to the low surface energy and hydrophobic property [42,43]. When the leather surface was covered with a uniform FPU waterproofing agent, the fluorine-containing chain segments would migrate and enrich onto the surface during the drying process. Here, the XPS detection result of FPU (ET:FA = 0.5:0.5) coating indicated that the fluorine content (31.03 wt%) on its surface was much higher than the theoretically calculated content (20.6 wt%), which clearly confirmed the enrichment effect of fluorine-containing chain segments on the coating surface. The proposed mechanism for dynamic behavior of perfluoroalkyl segments during drying and curing is shown in Figure 2ii. As seen in Figure 2iii, the surface energy of the fluorine-free coating film was 57.61 mN/m, while the surface energy of the coating with the incorporation of 0.25 molar ratio of FA decreased to 30.05 mN/m. As the addition proportion of FA increased, the surface energy of the coating decreased continuously. Intriguingly, after the ET/FA molar ratio reached 0.5/0.5, the surface energy was no longer significantly reduced with the increase in FA proportion. The explanation for this is that at a higher F element content, more fluorine migrated to the surface, meaning that the FPU waterproof coating had a lower surface energy until close to the saturated fluorine content on the surface.

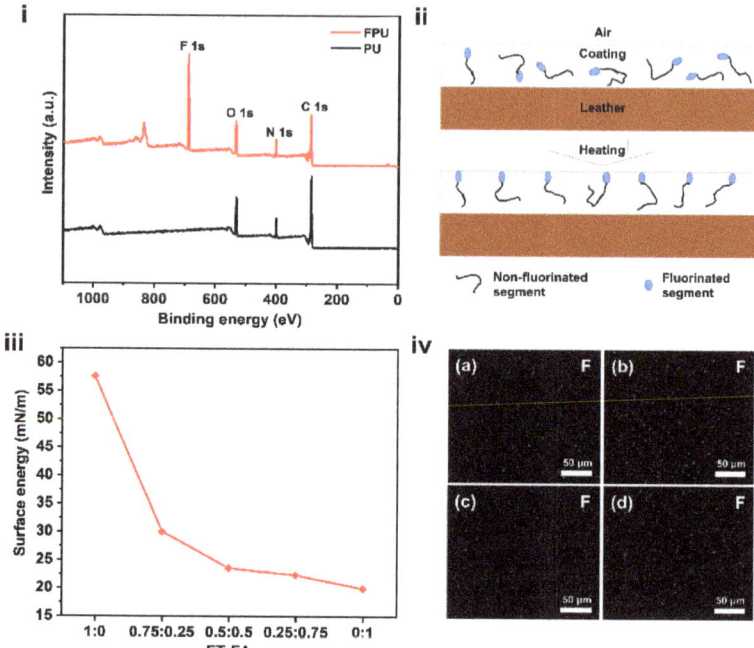

Figure 2. (i) XPS spectra of surface survey of pure PU (ET:FA = 1:0) coating and FPU coating (ET:FA = 0.5:0.5); (ii) Schematic representation of surface enrichment of fluorinated segments on the coating during the drying and curing process; (iii) Surface energy of the coatings against molar feed proportions of ET:FA; (iv) Fluorine atom (F) EDS mappings of the FPU coatings at different molar feed proportions of ET:FA. (a) 0.75:0.25, (b) 0.5:0.5, (c) 0.25:0.75, (d) 0:1.

To further confirm the change in fluorine-containing chain segments on the coating surface, EDS mapping of different ET:FA molar proportions of coatings was performed to further analyze the composition and content of elements on the coating surface. Figure 2iv showed the fluorine atom-based (F) EDS mappings of the FPU samples. It can be seen that the uniform distribution of F element on the surface of the whole coating could clearly be observed, and fluorine content increased significantly with the increase in FA feed proportion. As a result, it is the enrichment of fluorine-containing segments on the surface of materials that made the coating exhibit an outstanding waterproofing performance, which is the main reason it is unique to other materials.

3.3. Surface Wettability

Water droplet contact angle measurement is one of the most common methods used to evaluate surface waterproofing performance. As a direct reflection of the surface energy of the materials, the WCA has great guiding significance for the application of waterproof materials. The smaller WCA is, the better the wettability of liquid on a solid surface is. Consequently, contact angle can be used to measure the wettability. When WCA is less than 90°, the liquid can moisten the solid surface. The change in WCAs of the PU and FPU coatings with increasing FA proportion is given in Figure 3a. It can be seen from Figure 3a that the coating WCA increased with the increase in FA addition ratio. Compared with the pure PU (i.e., ET:FA = 1:0, 75°), the WCA of FPU coating with ET:FA = 0:1 (117°) increased by 42°. In addition, the test result of WCA was similar to that of commercial C8 waterproofing agent. As a result, a hydrophobic surface was already formed due to the enrichment of fluorine-containing chain segments on the surface. Nevertheless, the increase in WCA was not obvious with the ET:FA ratio increasing up to 0.5:0.5, which may be due to the above-mentioned "saturation" of fluorine atoms on the film surface.

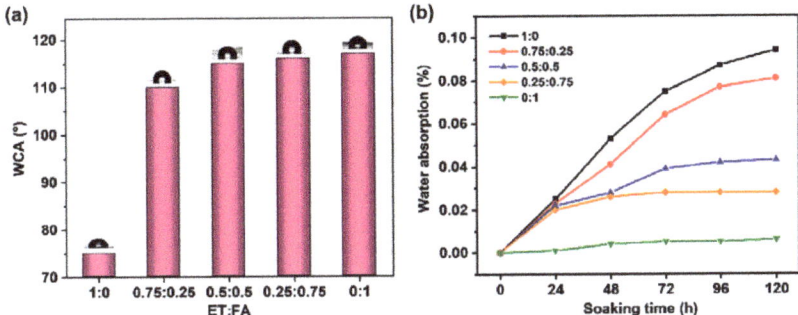

Figure 3. (a) Water drop contact angle (WCA) and (b) water absorption of waterproof coatings with different molar proportions of ET:FA.

Water resistance is also an important investigation index for the practical application of a waterproofing agent. The easy migration of fluorine-containing groups could cover the surface of the coating to reduce the water absorption. As a result, the water absorption for the FPU sample was smaller than the pure PU due to the existence of fluoride chain segments. In addition, the surface of the coating was more compact as the proportion of FA end-sealant increased, which made water molecules' penetration become more difficult. From Figure 3b, after 120 h of continuous immersion, the water absorption rate of PU without FA was 0.094%, whereas the water absorption rates of FPU coatings incorporating FA gradually decreased with the increase in the proportion of FA addition. In detail, the water absorption rates of the FPU coatings at ET:FA of 0.5:0.5 and 0.25:0.75 were 0.048% and 0.028%, respectively, which was lower than half of PU. The FPU coatings showed the outstanding water resistance, and its water absorption rate was at a minimum when ET:FA molar ratio was 0:1, which was almost unaffected by the soaking time. This is also due to the low surface energy of fluorine-containing chain segments migration and enrichment onto the surface. Furthermore, the water absorption did not further change with the extension of soaking time and there is no shedding or foaming phenomenon on the coating surface, which indicated the excellent water resistance of FPU waterproof coating.

3.4. Abrasion Resistance

Abrasion resistance is an important performance index for practical applications of polymer coating. In this work, the waterproofing agent coating showed a satisfactory abrasion resistance. As illustrated in Figure 4a, wearing was tested by a taber abraser at a loading pressure of 12.1 MPa, and the experimental results displayed that the FPU-coated leather still maintained the hydrophobicity even after 600 abrasion cycles (Figure 4b). The coating WCAs increased with the increase in wear time. Among them, the WCA of FPU coating with a large proportion of FA incorporation exceeded 120° after several hundred times of friction. It could be reasonably speculated that this interesting result was ascribed to the joint effects of roughness increase and the enrichment of fluorine-containing segments on the surface of the FPU waterproofing coating. When the number of wear times reached 700, the coating WCA decreased dramatically, no longer exhibiting the hydrophobic performance. This phenomenon can be attributed to the substantive damage of FPU coating on the leather surface.

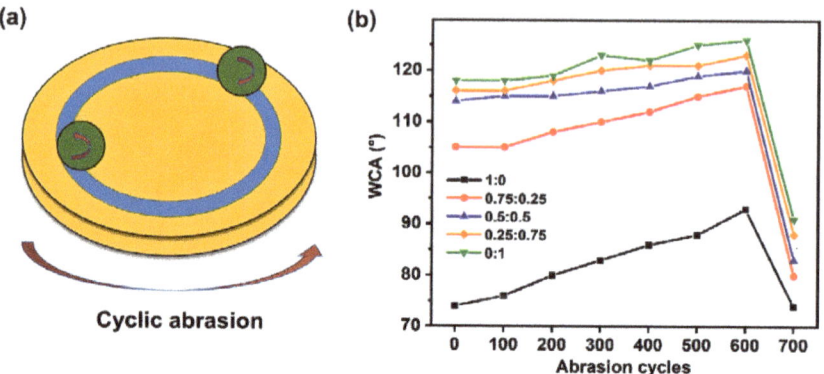

Figure 4. (a) Schematic illustration of abrasion test on a leather and (b) WCA change on the coated leather after repeated abrasion.

3.5. Potential UV Degradation

Long straight-chain fluorocarbon compounds are limited by their non-degradable and bioaccumulative properties, so the materials to be prepared should be environmentally friendly and degradable. In this section, the leather samples treated with our synthesizing FPU waterproofing agent were exposed to ultraviolet light, and their WCAs at different UV irradiation times are compared in Figure 5. During the continuous irradiation of 120 h, the WCAs of FPU coatings decreased by about 15° at the end, but all their WCAs were more than 90°, still showing the hydrophobicity. Different from the PU sample, which was completely terminated by ethanol, the leather samples decorated with the FPU waterproof agent could still achieve the hydrophobic effect, which proved that the FPU waterproof agent had better ultraviolet aging resistance and broad outdoor application prospect. Compared to that the introduction of heptadecafluorodecyl triethoxysilane (FAS), the WCA of the coating barely decreased after 600 min of irradiation, while the WCA of FPU waterproofing agent decreased gradually after continuous UV irradiation. Therefore, this indicated that the FPU waterproofing agent prepared in this research exhibited potential photocatalytic degradation behavior.

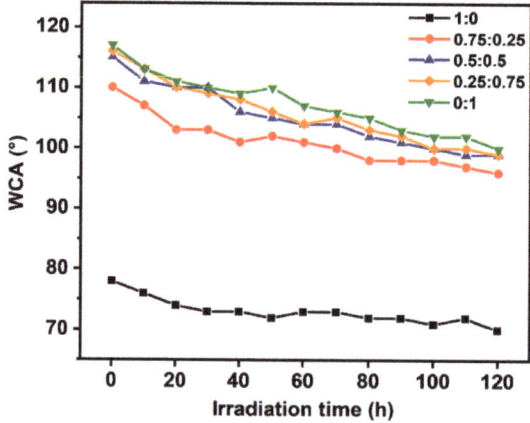

Figure 5. Change of WCA with UV irradiation time for waterproof coatings with different molar ratios of ET:FA.

On the basis of the above test results and cost consideration, when the ET:FA molar ratio was 0.5:0.5, that is, when one end of a molecular chain was sealed with ethanol and the other end was sealed with fluoro-alcohol, we considered that the cost performance was the highest at this moment. Furthermore, we roll-coated this sample on the leather surface and comparatively tested the waterproofing properties of the FPU and the commercial C8 waterproofing agents. The correlative test results are described later.

3.6. Waterproofing Grade

As above discussion, the fluorine-containing chains enriched on the coating surface to make it waterproof. Figure 6(ia–c) are the renderings of droplets of Grade 8 standard test solution on pure leather, the leather coated with the FPU, and the leather coated with the commercial C8 waterproofing agent, respectively. It is obvious that the two coated leather samples were not wetted after 10 s of placement. In consequence, the FPU waterproof agent embellishing leather had the waterproofing effect of Grade 8, almost one grade, as in the commercial C8 waterproofing agent below.

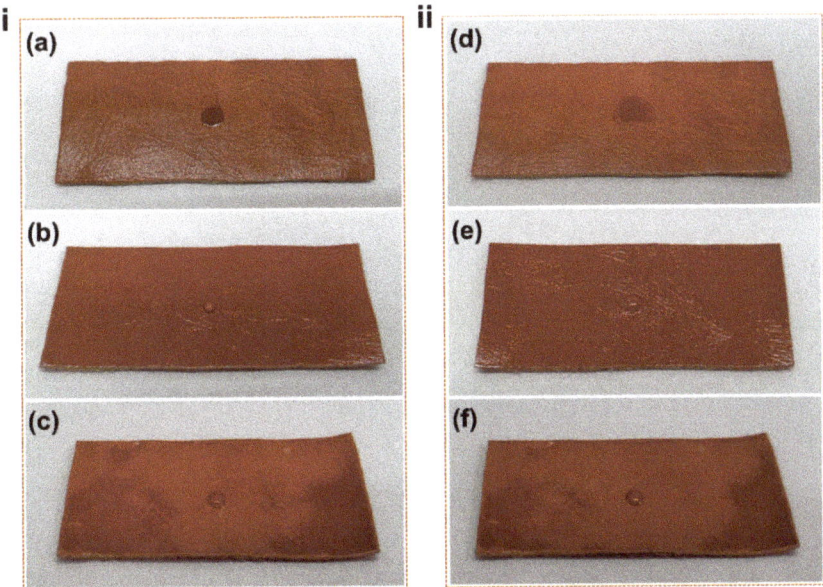

Figure 6. (**i**) Waterproofing grade test with 40/60 (v/v) of water/isopropyl alcohol on uncoated leather (**a**), leather coated with FPU waterproofing agent (**b**), and commercial C8 waterproofing agent (**c**); (**ii**) Oilproofing grade test with n-heptane on uncoated leather (**d**), leather coated with FPU waterproofing agent (**e**), and commercial C8 waterproofing agent (**f**).

3.7. Oilproofing Grade

The FPU coating not only showed good waterproof properties but also special oil-proof properties. Figure 6(iid–f) are the renderings of droplets of Grade 8 standard test solution on pure leather, the leather coated with the FPU, and the leather coated with the commercial C8 waterproofing agent, respectively. It can be clearly seen that the pure leather almost had no oil-proof performance, the test solution immediately wetted the leather, while the FPU modified leather had Grade 8 oil-proof effect, which was the same as that of the commercial C8 waterproofing agent. As a result, it could be used as a substitute for the commercial C8 waterproofing agent to develop its applications, showing a good market prospect.

3.8. Anti-Smudge Performance

In real life, most materials are prone to losing some of their properties after suffering oil pollution, and so do many waterproof coatings. Thus, we investigated the anti-smudge performance of the FPU waterproofing agent modifying leather. Figure S1a,b respectively shows the initial dripping and 2 s sliding traces of ink droplets coated on the unmodified and FPU modified slant leather samples. The dark ink mark was expressly observed on the surface of uncoated leather, but the ink easily slipped off the FPU coated area without leaving a trace. Therefore, the incorporated fluorine-containing chain segments endowed the ink-repellant property so that the as-prepared material showed tremendous potential in anti-graffiti application.

3.9. Solvent Resistance

During testing the waterproofing property of FPU coating by using an isopropanol standard test solution (Figure 7i), we found that the commercial C8 waterproofing agent coating whitened after the test (Figure 7ia), while our prepared FPU waterproofing coating did not show any whitening phenomenon (Figure 7ib). Therefore, it is necessary to carry out solvent-resistant test (Figure 7ii).

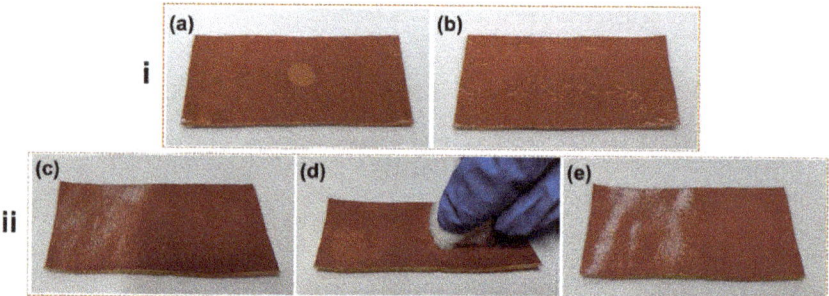

Figure 7. (i) Samples tested with isopropanol/water standard test solution: leathers coated with commercial C8 waterproofing agent (**a**) and FPU waterproofing agent (**b**); (ii) Solvent resistance test of the coating by hand wiping: (**c**) before wiping, (**d**) wiping action illustration, and (**e**) after wiping with the solvent.

It can be observed from Figure 7iie that the FPU coating after the repeated wiping with the standard test solvent showed no damage, the surface gloss was still retained, and the substrate was not exposed to the air at all. This experimental result proved that the FPU waterproofing agent prepared in this work had a superior resistance to organic solvent.

4. Conclusions

In conclusion, we successfully prepared an excellent comprehensive property of a novel leather waterproofing agent by introducing a short-branched fluoro-alcohol into polyurethane oligomer structure as the end-sealant. The investigation results showed that the waterproofing and oilproofing grades of the FPU waterproofing agent coating reached their highest (Grade 8) and the ink easily slipped without leaving a trace, indicating that the coating had excellent water-proof, oil-proof, and anti-smudge properties. Moreover, the coating could sustain the high abrasion resistance and exhibit potential photo-degradation after continuous UV irradiation of 120 h. Compared with the commercial C8 waterproofing agent, the FPU coating also showed significant solvent resistance. The surface chemistry analysis revealed the real reason for this superior performance, which was attributed to the continuous increase in fluorine-containing chain segments with the increasing fluoro-alcohol content and sub-sequent enrichment on the coating surface. This elaborately designed FPU waterproofing agent had many advantages in terms of either structure or property. This may avoid the straight-chain perfluoroalkyl compound (no less than C8) having the defined biotoxicity and environmental persistence. As a result, it is expected to be applied in various base materials. As for an investigation of the bioaccumulation of the FPU waterproofing agent, we will perform further research in a future work.

Supplementary Materials: The following are available online at https://www.mdpi.com/article/10.3390/coatings11040395/s1, Table S1: Standard test liquids for waterproof grade test, Table S2: Standard test liquids for oilproof grade test, Figure S1: Snapshots of sliding ink droplets on uncoated leather (a) and leather coated with FPU waterproofing agent (b).

Author Contributions: Conceptualization, S.S.; methodology, S.S.; software, C.L.; formal analysis, J.Y.; investigation, Z.P.; resources, M.P.; writing—original draft preparation, S.S.; writing—review and editing, M.P.; visualization, S.S.; supervision, J.W. All authors have read and agreed to the published version of the manuscript.

Funding: Please add: This research was funded by the National Natural Science Foundation of China (No. 51973050) and the Hebei Province Natural Science Fund (B2019202114).

Institutional Review Board Statement: Not applicable.

Informed Consent Statement: Not applicable.

Data Availability Statement: The data presented in this study are available on request from the corresponding author.

Acknowledgments: We acknowledge financial support for this work from the National Natural Science Foundation of China (No. 51973050) and the Hebei Province Natural Science Fund (B2019202114). Thanks to Xinwen Jin of Hefei Jinsheng Chemical Co. Ltd., Anhui, China, for his kind help and beneficial discussion of this study.

Conflicts of Interest: The authors declare no conflict of interest.

References

1. Wang, J.; Raza, A.; Si, Y.; Cui, L.; Ge, J.; Ding, B.; Yu, J. Synthesis of superamphiphobic breathable membranes utilizing SiO_2 nanoparticles decorated fluorinated polyurethane nanofibers. *Nanoscale* **2012**, *4*, 7549–7556. [CrossRef]
2. Wu, Z.; Tang, L.; Dai, J.; Qu, J. Synthesis and properties of fluorinated non-isocyanate polyurethanes coatings with good hydrophobic and oleophobic properties. *J. Coat. Technol. Res.* **2019**, *16*, 1233–1241. [CrossRef]
3. Zhuge, W.; Pan, M.; Chang, Y.; Yuan, J.; Wang, X.; Sun, C.; Zhang, G. Monodispersed composite particles prepared via emulsifier-free emulsion polymerization using waterborne polyurethane as microreactors. *J. Appl. Polym. Sci.* **2014**, *131*, 40985. [CrossRef]
4. Khan, F.; Khan, A.; Tuhin, M.; Rabnawaz, M.; Li, Z.; Naveed, M. A novel dual-layer approach towards omniphobic polyurethane coatings. *RSC Adv.* **2019**, *9*, 26703–26711. [CrossRef]
5. Zheng, H.; Pan, M.; Wen, J.; Yuan, J.; Zhu, L.; Yu, H. Robust, transparent, and superhydrophobic coating fabricated with waterborne polyurethane and inorganic nanoparticle composites. *Ind. Eng. Chem. Res.* **2019**, *58*, 8050–8060. [CrossRef]
6. Mao, X.; Chen, Y.; Si, Y.; Li, Y.; Wan, H.; Yu, J.; Sun, G.; Ding, B. Novel fluorinated polyurethane decorated electrospun silica nanofibrous membranes exhibiting robust waterproof and breathable performances. *RSC Adv.* **2013**, *3*, 7562–7569. [CrossRef]
7. Jiang, Y.; Li, L.; Wang, H.; Wang, R.; Tian, Q. Influence of acrylic emulsion on polymer-cement waterproof coating. *Adv. Mat. Res.* **2015**, *1129*, 263–269. [CrossRef]
8. Huang, J.; Meng, W.; Qing, F. Synthesis and repellent properties of vinylidene fluoride-containing polyacrylates. *J. Fluor. Chem.* **2007**, *128*, 1469–1477. [CrossRef]
9. Chen, L.; Shi, H.; Wu, H.; Xiang, J. Preparation and characterization of a novel fluorinated acrylate resin. *J. Fluor. Chem.* **2010**, *131*, 731–737. [CrossRef]
10. Wang, J.; Chen, X.; Kang, Y.; Yang, G.; Yu, L.; Zhang, P. Preparation of superhydrophobic poly (methyl methacrylate)-silicon dioxide nanocomposite films. *Appl. Surf. Sci.* **2010**, *257*, 1473–1477. [CrossRef]
11. Li, Q.; Fan, Z.; Chen, C.; Li, Z. Water and oil repellent properties affected by the crystallinity of fluorocarbon chain in fluorine-silicon containing finishing agent. *J. Text. Inst.* **2020**, *111*, 360–369. [CrossRef]
12. Pan, H.; Li, G. Emulsion waterproof agent and its effects on intrinsic properties of gypsum. *Asian J. Chem.* **2013**, *25*, 5042–5046. [CrossRef]
13. Zhou, Y.; Liu, C.; Gao, J.; Chen, Y.; Yu, F.; Chen, M.; Zhang, H. A novel hydrophobic coating film of water-borne fluoro-silicon polyacrylate polyurethane with properties governed by surface self-segregation. *Prog. Org. Coat.* **2019**, *134*, 134–144. [CrossRef]
14. Wehbi, M.; Banerjee, S.; Mehdi, A.; Alaaeddine, A.; Hachem, A.; Ameduri, B. Vinylidene fluoride-based polymer network via cross-linking of pendant triethoxysilane functionality for potential applications in coatings. *Macromolecules* **2017**, *50*, 9329–9339. [CrossRef]

15. Wang, C.; Li, X.; Du, B.; Li, P.; Lai, X.; Niu, Y. Preparation and properties of a novel waterborne fluorinated polyurethane–acrylate hybrid emulsion. *Colloid Polym. Sci.* **2014**, *292*, 579–587. [CrossRef]
16. Lin, J.; Jiang, F.; Wen, J.; Lv, W.; Porteous, N.; Deng, Y.; Sun, Y. Fluorinated and un-fluorinated N-halamines as antimicrobial and biofilm-controlling additives for polymers. *Polymer* **2015**, *68*, 92–100. [CrossRef] [PubMed]
17. Zhang, L.; Li, Y.; Yu, J.; Ding, B. Fluorinated polyurethane macroporous membranes with waterproof, breathable and mechanical performance improved by lithium chloride. *RSC Adv.* **2015**, *5*, 79807–79814. [CrossRef]
18. Li, N.; Zeng, F.; Wang, Y.; Qu, D.; Zhang, C.; Li, J.; Bai, Y. Synthesis and characterization of fluorinated polyurethane containing carborane in the main chain: Thermal, mechanical and chemical resistance properties. *Chin. J. Polym. Sci.* **2018**, *36*, 85–97. [CrossRef]
19. Zhang, L.; Kong, Q.; Kong, F.; Liu, T.; Qian, H. Synthesis and surface properties of novel fluorinated polyurethane base on F-containing chain extender. *Polym. Adv. Technol.* **2020**, *31*, 616–629. [CrossRef]
20. Yu, F.; Gao, J.; Liu, C.; Chen, Y.; Zhong, G.; Hodges, C.; Chen, M.; Zhang, H. Preparation and UV aging of nano-SiO$_2$/fluorinated polyacrylate polyurethane hydrophobic composite coating. *Prog. Org. Coat.* **2020**, *141*, 105556. [CrossRef]
21. Zhao, D.; Pan, M.; Yuan, J.; Liu, H.; Song, S.; Zhu, L. A waterborne coating for robust superamphiphobic surfaces. *Prog. Org. Coat.* **2020**, *138*, 105368. [CrossRef]
22. Liu, C.; Liao, X.; Shao, W.; Liu, F.; Ding, B.; Ren, G.; Chu, Y.; He, J. Hot-melt Adhesive Bonding of Polyurethane/Fluorinated Polyurethane/Alkylsilane-Functionalized Graphene Nanofibrous Fabrics with Enhanced Waterproofness. *Polymers* **2020**, *12*, 836. [CrossRef]
23. Rabnawaz, M.; Liu, G. Graft-copolymer-based approach to clear, durable, and anti-smudge polyurethane coatings. *Angew. Chem. Int. Edit.* **2015**, *54*, 6516–6520. [CrossRef] [PubMed]
24. Ge, Z.; Zhang, X.; Dai, J.; Li, W.; Luo, Y. Synthesis, characterization and properties of a novel fluorinated polyurethane. *Eur. Polym. J.* **2009**, *45*, 530–536. [CrossRef]
25. Fu, K.; Lu, C.; Liu, Y.; Zhang, H.; Zhang, B.; Zhang, H.; Zhou, F.; Zhu, Q.; Zhu, B. Mechanically robust, self-healing superhydrophobic anti-icing coatings based on a novel fluorinated polyurethane synthesized by a two-step thiol click reaction. *Chem. Eng. J.* **2021**, *404*, 127110. [CrossRef]
26. Zhao, J.; Li, Y.; Sheng, J.; Wang, X.; Liu, L.; Yu, J.; Ding, B. Environmentally friendly and breathable fluorinated polyurethane fibrous membranes exhibiting robust waterproof performance. *ACS Appl. Mater. Interfaces* **2017**, *9*, 29302–29310. [CrossRef]
27. Jeong, H.Y.; Lee, M.H.; Kim, B.K. Surface modification of waterborne polyurethane. Colloids and Surfaces A: Physicochemical and Engineering Aspects. *Colloids Surf. A* **2006**, *290*, 178–185. [CrossRef]
28. Chen, K.; Kuo, J. Synthesis and properties of novel fluorinated aliphatic polyurethanes with fluoro chain extenders. *Macromol. Chem. Phys.* **2000**, *201*, 2676–2686. [CrossRef]
29. Tan, H.; Guo, M.; Du, R.; Xie, X.; Li, J.; Zhong, Y.; Fu, Q. The effect of fluorinated side chain attached on hard segment on the phase separation and surface topography of polyurethanes. *Polymer* **2004**, *45*, 1647–1657. [CrossRef]
30. Wang, L. Experimental and theoretical characterization of the morphologies in fluorinated polyurethanes. *Polymer* **2007**, *48*, 894–900. [CrossRef]
31. Zhao, J.; Wang, X.; Liu, L.; Yu, J.; Ding, B. Human skin-like, robust waterproof, and highly breathable fibrous membranes with short perfluorobutyl chains for eco-friendly protective textiles. *ACS Appl. Mater. Interfaces* **2018**, *10*, 30887–30894. [CrossRef]
32. Li, S.; Huang, J.; Ge, M.; Li, S.; Xing, T.; Chen, G.; Liu, Y.; Zhang, K.; Al-Deyab, S.; Lai, Y. Controlled grafting superhydrophobic cellulose surface with environmentally-friendly short fluoroalkyl chains by ATRP. *Mater. Des.* **2015**, *85*, 815–822. [CrossRef]
33. Soto, D.; Ugur, A.; Farnham, T.A.; Gleason, K.K.; Varanasi, K.K. Short-Fluorinated iCVD Coatings for Nonwetting Fabrics. *Adv. Funct. Mater.* **2018**, *28*, 1707355. [CrossRef]
34. Jiang, J.; Zhang, G.; Wang, Q.; Zhang, Q.; Zhan, X.; Chen, F. Novel fluorinated polymers containing short perfluorobutyl side chains and their super wetting performance on diverse substrates. *ACS Appl. Mater. Interfaces* **2016**, *8*, 10513–10523. [CrossRef] [PubMed]
35. Rodea-Palomares, I.; Leganés, F.; Rosal, R.; Fernández-Pinas, F. Toxicological interactions of perfluorooctane sulfonic acid (PFOS) and perfluorooctanoic acid (PFOA) with selected pollutants. *J. Hazard. Mater.* **2012**, *201*, 209–218. [CrossRef] [PubMed]
36. Gallo, V.; Leonardi, G.; Genser, B.; Lopez-Espinosa, M.J.; Frisbee, S.J.; Karlsson, L.; Ducatman, A.M.; Fletcher, T. Serum perfluorooctanoate (PFOA) and perfluorooctane sulfonate (PFOS) concentrations and liver function biomarkers in a population with elevated PFOA exposure. *Environ. Health Perspect.* **2012**, *120*, 655–660. [CrossRef]
37. Yang, Y.; Shen, J.; Zhang, L.; Li, X. Preparation of a novel water and oil-repellent fabric finishing agent containing a short perfluoroalkyl chain and its application in textiles. *Mater. Res. Innov.* **2015**, *19*, 401–404. [CrossRef]
38. Zhang, Q.; Wang, Q.; Jiang, J.; Zhan, X.; Chen, F. Microphase structure, crystallization behavior, and wettability properties of novel fluorinated copolymers poly (perfluoroalkyl acrylate-co-stearyl acrylate) containing short perfluorohexyl chains. *Langmuir* **2015**, *31*, 4752–4760. [CrossRef]
39. Zhang, D.; Sha, M.; Pan, R.; Lin, X.; Xing, P.; Jiang, B. Synthesis and properties study of novel fluorinated surfactants with perfluorinated branched ether chain. *J. Fluor. Chem.* **2019**, *219*, 62–69. [CrossRef]
40. Jin, H.; Xu, W.; Dai, J. Applications of Hexafluoropropene Trimer. *Zhejiang Chem. Ind.* **2016**, *47*, 1–8.
41. Wen, J.; Jia, Z.; Zhang, X.; Pan, M.; Yuan, J.; Zhu, L. Tough, thermo-Responsive, biodegradable and fast self-healing polyurethane hydrogel based on microdomain-closed dynamic bonds design. *Mater. Today Commun.* **2020**, *25*, 101569. [CrossRef]

42. Zhao, J.; Zhou, T.; Zhang, J.; Chen, H.; Yuan, C.; Zhang, W.; Zhang, A. Synthesis of a waterborne polyurethane-fluorinated emulsion and its hydrophobic properties of coating films. *Ind. Eng. Chem. Res.* **2014**, *53*, 19257–19264. [CrossRef]
43. Xue, D.; Wang, X.; Ni, H.; Zhang, W.; Xue, G. Surface segregation of fluorinated moieties on random copolymer films controlled by random-coil conformation of polymer chains in solution. *Langmuir* **2009**, *25*, 2248–2257. [CrossRef] [PubMed]

MDPI
St. Alban-Anlage 66
4052 Basel
Switzerland
www.mdpi.com

Coatings Editorial Office
E-mail: coatings@mdpi.com
www.mdpi.com/journal/coatings

Disclaimer/Publisher's Note: The statements, opinions and data contained in all publications are solely those of the individual author(s) and contributor(s) and not of MDPI and/or the editor(s). MDPI and/or the editor(s) disclaim responsibility for any injury to people or property resulting from any ideas, methods, instructions or products referred to in the content.

www.ingramcontent.com/pod-product-compliance
Lightning Source LLC
LaVergne TN
LVHW070723100526
838202LV00013B/1159